Colorectal Cancer

This concise text examines the contribution of genetic research to clinical practice in colorectal cancer. It delves into the underlying molecular mechanisms and the genetic and environmental factors contributing to colorectal cancer, as well as into advances in personalized therapies, targeted interventions, and biomarker-based approaches. By providing a comprehensive overview of the subject, the book aims to bridge the gap between scientific research and clinical practice, offering valuable insights to researchers, clinicians, and healthcare professionals dedicated to combating colorectal cancer and improving patient outcomes.

Colorectal Cancer

From Pathogenesis to Precision Medicine

Edited by
Gowhar Rashid
Deena Elsori
Rana Ahmed Youness

CRC Press
Taylor & Francis Group
Boca Raton London New York

CRC Press is an imprint of the
Taylor & Francis Group, an **informa** business

Designed cover image: Shutterstock

First edition published 2025
by CRC Press
2385 NW Executive Center Drive, Suite 320, Boca Raton FL 33431

and by CRC Press
4 Park Square, Milton Park, Abingdon, Oxon, OX14 4RN

CRC Press is an imprint of Taylor & Francis Group, LLC

ISBN: 978-1-032-70447-0 (hbk)
ISBN: 978-1-032-69965-3 (pbk)
ISBN: 978-1-032-70451-7 (ebk)

DOI: 10.1201/9781032704517

Typeset in Palatino
by SPi Technologies India Pvt Ltd (Straive)

Contents

About the Editors

Dr. Gowhar Rashid works as an Assistant Professor in the Amity Medical School at Amity University Haryana, India. Born and raised in the beautiful region of Srinagar, Kashmir, Dr. Rashid has developed a passion for advancing the field of medical science. His educational journey led him to complete his BSc, MSc, and PhD in Biochemistry from the prestigious Sher I Kashmir Institute of Medical Sciences in Kashmir, India.

With a specialization in diagnostics and cancer research, Dr. Rashid has become a recognized expert in his field. His in-depth knowledge and dedication to his work have earned him recognition and respect from both his peers and students. His research endeavors focus on finding innovative solutions and breakthroughs in cancer diagnostics, aiming to improve patient outcomes and contribute to the field's advancements.

Dr. Deena Elsori is an Associate Professor and the Academic Advisor for the Dean of Faculty at Rabdan Academy. Before joining Rabdan Academy, she was the Associate Dean of Student Affairs for the College of Health Sciences at Abu Dhabi University. She achieved several significant milestones and was accredited for being a top-performing professor with excellent team leadership capabilities, constantly instilling a plethora of knowledge in students. Her accolades included achieving the distinguished faculty award several times throughout her career in higher education.

Before moving to the UAE, Dr. Elsori was a postdoctoral fellow at the Cleveland Clinic, Ohio, USA. Her research focused on chronic inflammation and heart disease. Currently her research interest focuses on public health, and she has become an advocate for well-being and health promotion in the UAE.

Dr. Elsori is an academician, researcher, mother, educator, and community builder. She played a key role in building the public health program at Abu Dhabi University. She also shaped the idea of a Happiness course to equip students with essential tools to boost their emotional health. In summary, Dr. Elsori boosts her students' academic success by prioritizing their well-being.

Dr. Rana Ahmed Youness (H index=21) is an Assistant Professor of Molecular Genetics and Biochemistry and Group Leader of Molecular Genetics Research Team-MGRT at the Faculty of Biotechnology at the German International University (GIU). Dr. Youness earned her BSc degree with highest honors from the Faculty of Pharmacy and Biotechnology, German University in Cairo (GUC), in 2013. She was ranked first among her graduating class. In September 2013, she was appointed as a teaching assistant at the pharmaceutical Biology Department, Faculty of Pharmacy and Biotechnology, GUC. In 2015, she earned her MSc degree (Numerical Grade=100%) and then was appointed as an Assistant Lecturer. Her Master thesis work has been acknowledged and she has published a book, *miR-486-5p and Liver Cancer: A Beautiful Mind Act*. Moreover, her work has been ranked in the top 10% of posters at the EASL conference (European Association for the Study of Liver Disease), Vienna, 2015. In 2018, she earned her PhD in Biochemistry and Molecular Biology with Summa cum laude from the GUC. Dr. Youness has worked as a Lecturer of Pharmaceutical Biology at the Faculty of Pharmacy and Biotechnology, GUC, Egypt. In parallel, Dr. Youness had a postdoc position at Dr. Gad's research lab from 2018 to 2020. She was able to publish several publications in high impact factor peer-reviewed journals: based on the work of her MSc project, she published more than two books, three research articles and four abstracts in international conferences such as EASL, AASLD and EACR; while based on the work of her PhD she published three book chapters, four research articles, and ten abstracts in international conferences such as ESMO, EACR, AACR, ENA and EBCC.

Dr. Youness has received several awards and recognitions, most notably, L'Oréal UNESCO for Women in Science 2023, the European Association for the Study of the Liver (EASL) Merit award and Travel Grant for four consecutive years, European Society For Medical Oncology (ESMO) merit award and travel grant for young investigators for eight consecutive years, and several mobility grants funded by the German Academic Exchange Service (DAAD). In 2022, Dr. Youness was awarded the prestigious State of Incentive Award in Medical Sciences from the Egyptian Ministry of Higher Education and the Academy of Scientific Research in Egypt as an acknowledgment for her innovative research projects and her publication record. She founded the Molecular Genetics Research Team- MGRT at the GUC and then at UH-GAF and finally at GIU. The MGRT focuses on (1) unravelling the crosstalk between different classes of non-coding RNAs in different solid malignancies such as breast and liver cancers; (2) deciphering the gasotransmitters associated signaling pathways in breast and liver cancers; and (3) evaluating the immunomodulatory anti-cancer potential of novel natural compounds against breast and liver cancers. The MGRT is now composed of 6+ PhD students, 35+ MSc students, 15+ Pharmaceutical Sciences, Pharmaceutical Engineering and Biotechnology Bachelor students and 10+ interns. Her group is funded by several national and international grants such as STDF, ASRT, SNSF, UKRI, and URB. MGRT is collaborating with several research groups worldwide in Germany, Greece, France, Italy, Poland, Korea, Switzerland, England, KSA, UAE, India and USA.

Dr. Youness is currently an editor and reviewer for several peer-reviewed Q1 journals related to the Biochemistry, Molecular Biology and Biotechnology fields such as *Scientific Reports-Nature*, *BMC Cancer-Nature*, *PlosOne*, *Aging*, *Carcinogenesis*, *Frontiers in Oncology*, and *Frontiers in Immunology*. She has been invited as a guest speaker at more than 30 international conferences in Europe, MENA region, and the USA, published and presented her research work at 60+ international conferences in Europe and the USA, and has published over 60 paper publications in a high impact peer-reviewed international journal. Dr. Youness has been an active member of Empowering Women in Science communities and volunteers in several events, conferences, and media shows to deliver influential talks for the younger generations of undergraduate students and MSc and PhD students, highlighting the importance of empowering women in science.

Contributors

Sonia Tamanna
Department of Biochemistry and Molecular Biology
University of Dhaka
Bangladesh

Taslimul Jannat
Department of Biochemistry and Molecular Biology
University of Dhaka
Bangladesh

Anass M. Abbas
Department of Clinical Laboratory Sciences
College of Applied Medical Sciences
Jouf University, KSA

Saliha Rizvi
Department of Biotechnology
Era University
Lucknow, India

Sanchita Srivastava
Department of Biotechnology
Era University
Lucknow, India

Haider Ali Rizvi
Yognanda School Of Artificial Intelligence
Shoolini University
Solan India

Rizwan Askari
Babu Banarasi Das University Lucknow
India

Syed Tasleem Raza
Department of Biochemistry
Era's Lucknow Medical College & Hospital
Lucknow, India

Farzana Mahdi
Department of Personalized & Molecular Medicine
Era University
Lucknow, India

Rizwan Ahamad
Department of Pharmacology
School of Pharmaceutical Education and Research
Jamia Hamdard
New Delhi, India

Sadma Parveen
Department of Pharmacology
School of Pharmaceutical Education and Research
Jamia Hamdard
New Delhi, India

Abu Safana Biswas
Department of Pharmacology
JSS College of Pharmacy
Mysuru, India

Ayman A. M. Alameen
Department of Clinical Laboratory Sciences
College of Applied Medical Sciences
Jouf University
KSA

Department of Chemical Pathology
Faculty of Medical Laboratory Sciences
University of Khartoum
Sudan

Naseem Fatima
Department of Biochemistry
Era's Lucknow Medical College & Hospital
Lucknow, India

Farheen Khan
Department of Biotechnology
Era University
Lucknow, India

Gowhar Rashid
Amity Medical School
Amity University
Haryana, India

Mohammad Taghi Ahady
Department of Biology
Faculty of Basic Sciences
Ardabil Branch
Azad University
Ardabil, Iran

Marjan Assefi
Marie Curie Science Research Center
Healthcare Department
USA

Ananya Chakraborty
Department of Biotechnology
Adamas University
Kolkata, India

Note: Contributors are listed in chapter order.

Manar G. Shalabi
Department of Clinical Laboratory Sciences
College of Applied Medical Sciences
Jouf University
KSA

Wael Hafez
Internal Medicine Department
Medical Research and Clinical Studies Institute
The National Research Centre
Ad Doqi Cairo, Egypt

Mohd Amir Shafeeque
Department of Biotechnology
Era University
Lucknow, India

Dilshad Ahmad
Era Institute of Allied Health Sciences and Research
Era University
Lucknow, India

Satya N. Sankwar
Director, Institute of Medical Sciences
Banaras Hindu University
Varanasi, India

Noorin Zaidi
Department of Pathology
Era University
Lucknow (Uttar Pradesh)
India

Zainab Siddiqui
Center for Disease Mapping and Therapeutic Research
Era University
Lucknow, India

Subhajit Karmakar
Cancer Biology and Inflammatory Disorder Division
Council of Scientific and Industrial Research-Indian Institute of Chemical Biology (CSIR-IICB)
Kolkata, India

Atharva Anand Mahajan
Academy of Scientific and Innovative Research (AcSIR)
Ghaziabad, India

Subhrojyoti Ghosh
Virology Division, Infection Biology Group
Council of Scientific and Industrial Research-Centre for Cellular and Molecular Biology (CSIR-CCMB)
Hyderabad, India
and
Department of Biotechnology
Indian Institute of Technology, Madras
Chennai, India

Saxena Shivang
Department of Biochemistry
AIIMS Bhopal
Saket Nagar, India

Sreekari Palakodeti
Department of Biochemistry
AIIMS Bhopal
Saket Nagar, India

Thakral Mohit
Department of Biochemistry
AIIMS Bhopal
Saket Nagar, India

Ashwin Kotnis
Department of Biochemistry
AIIMS Bhopal
Saket Nagar, India

Yamna Khurshid
Department of Molecular Medicine
Ziauddin University
Karachi, Pakistan

Kinza Rafi
Department of Biosciences
Shaheed Zulfikar Ali Bhutto Institute of Science and Technology
Karachi, Pakistan

Roy Rillera Marzo
Faculty of Humanities and Health Sciences
Curtin University, Malaysia
Miri, Malaysia

Wasan A.M. Al Taie
Department of Mathematics & Sciences
Rochester Institute of Technology of Dubai
Dubai, United Arab Emirates

Note: Contributors are listed in chapter order.

May K. Ismail
Department of Biomedical Sciences
College of Medicine
Gulf Medical University
Ajman, United Arab Emirates

Arwa Alnahdi
Pharmacy Department
Ibrahim Bin Hamad Obaidallah Hospital
Emirates Health Services
Ras Al Khaimah, United Arab Emirates

Eslam ElNebrisi
Biomedical Sciences Department
Dubai Medical College for Girls
Dubai Medical University
Dubai, United Arab Emirates

Eman Hassan Ibrahim
Department of Biomedical Sciences
College of Medicine
Gulf Medical University
Ajman, United Arab Emirates

Saba Z. Al-Abachi
Department of Chemistry
College of Science
University of Mosul
Mosul, Iraq

Praveen Halagali
Department of Pharmaceutics
Manipal College of Pharmaceutical Sciences
Manipal Academy of Higher Education
Manipal, India

Vamshi Krishna Tippavajhala
Department of Pharmaceutics
Manipal College of Pharmaceutical Sciences
Manipal Academy of Higher Education
Manipal, India

Himanshu Sharma
Teerthanker Mahaveer College of Pharmacy
Teerthanker Mahaveer University
Moradabad (UP), India

Harpreet Singh
School of Pharmaceutical Sciences
IFTM University
Moradabad, India

S. M. Mahfooz Hasan
Research Metabolic Unit
Era's Lucknow Medical College and Hospital
Era University
Lucknow, India

Mohammad Waseem
Department of Zoology
Jagdam College
Chhapra (Bihar), India

Note: Contributors are listed in chapter order.

1 Introduction to Colorectal Cancer

Sonia Tamanna, Taslimul Jannat and Anass M. Abbas

1.1 INTRODUCTION

In the dynamic landscape of medical science, colorectal cancer (CRC) has emerged as a formidable adversary characterized by the unchecked proliferation of aberrant cells within the colon or rectum. The colon and rectum are parts of the digestive system, where the body processes food and eliminates waste. It typically begins as a polyp, a noncancerous growth, which, if left untreated, can develop into cancer over time. The progression of CRC involves the potential spread of cancerous cells to distant organs, leading to metastatic lesions. Among various organs, the liver is commonly affected. This metastatic process is driven by the accumulation of genetic mutations, which can be either acquired (somatic) or inherited (germline) over a span of 10 to 15 years (Lotfollahzadeh et al., 2023; Recio-Boiles and Cagir, 2022). The process of colorectal carcinogenesis involves the gradual build-up of mutations in various genes, leading to the transformation of normal colorectal tissue into adenomas and ultimately progressing to invasive cancer. This multistep process highlights the complexity of the genetic alterations underlying the transformation of healthy colorectal tissue into cancerous tissue (Li et al., 2019). Inheritable susceptibility and environmental variables are essential contributors to this chain of events.

Lower gastrointestinal symptoms may indicate the presence of CRC. The National Institute for Health and Care Excellence (NICE) has released guidelines that medical professionals can use to determine which patients are at high risk for CRC. Suspected CRC is identified and warrants further testing when symptoms such as rectal bleeding, abdominal mass, abdominal pain, changes in bowel habits, unexplained weight loss, and iron deficiency anemia are present (NICE, 2015). However, it is crucial to recognize certain symptoms that are not specific to a particular site, such as deep vein thrombosis and unexplained loss of appetite. A suspected cancer pathway referral or an urgent investigation may be provided for these symptoms, and an assessment of other symptoms, signs, or findings may assist in determining which cancer is most likely to be investigated. However, several guidelines state that in clinical practice, people with gastrointestinal signs and symptoms that raise the possibility of CRC should undergo colonoscopy (NICE, 2015).

A number of variables have been linked to the emergence of CRC. Studies indicate that individuals with a higher risk of CRC often have a history of colon polyps, diabetes mellitus, inflammatory bowel diseases, cancer, or cholecystectomy, either personally or within their family. Additionally, lifestyle factors play a crucial role in the development of CRC. There is substantial evidence indicating that inappropriate dietary habits, characterized by excessive consumption of red and processed meats and inadequate intake of fiber, fruits, vegetables, calcium, and essential nutrients, alongside sedentary lifestyle, smoking, alcohol use, and overweight and obesity, elevate the likelihood of developing CRC. Furthermore, it is known that the gut microbiota, age, sex, race, and socioeconomic status affect the risk of CRC.

Many of the illnesses prevalent in society today are associated with being overweight or obese. It has been noted that visceral adiposity has a negative impact on CRC prognosis in men (Silva et al., 2020). Inflammation, oxidative stress, and metabolic disruption play significant roles in CRC. Although the effects of both short- and long-term food and lifestyle exposures on the gut microbiome and its relationship to colon health have long been hypothesized, research on the subject is just beginning. The development of CRC is a multifaceted, intricate process that most likely involves interactions between genetic and environmental factors. However, there is still much to learn about the role of these interactions in the etiology of CRC. Because CRC typically takes years to develop, it is critical to diagnose the disease at an early stage. Secondary prevention is equally crucial and depends on follow-up exams and nutritional prevention through a balanced diet (Zaytseva, 2021).

1.2 BRIEF HISTORY AND BACKGROUND

The progression of CRC varies among individuals with a genetic predisposition, characterized by a shorter duration of tumor development and frequent manifestations at a younger age. Over a century ago, Dr. Aldred Warthin identified the initial case of hereditary CRC within a family, referred to as Family G, in 1895. His observations, published in 1913, highlighted a recurring pattern of gynecological cancers, particularly endometrial cancer, and gastrointestinal cancers,

DOI: 10.1201/9781032704517-1

including gastric and colon cancers, within this family (Warthin, 1985). In 1971, Lynch and Krush conducted updated studies on Family G, revealing a consistent pattern aligned with what is now recognized as Lynch syndrome (LS) (Lynch and Krush, 1977). Notably, individuals with LS exhibit a 70–80% greater incidence of proximal colon cancers (Lynch et al., 1977).

Further studies on Family G uncovered additional characteristics of LS, including its link to Muir-Torre syndrome, which is marked by skin manifestations like sebaceous adenomas and sebaceous carcinomas. CRC has emerged as the predominant cancer linked to LS, with a notable emphasis on proximal colon cancers (Fusaro et al., 1996). Additionally, endometrial cancer was identified as the second most prevalent cancer associated with LS. This historical account underscores the foundational work in recognizing hereditary CRC syndromes and their diverse clinical manifestations.

In all areas of clinical practice and research, colon cancer (CC) and rectal cancer (RC) are often treated as a single entity, known as CRC. Three distinct factors form the foundation of the term CRC. The first part is the theory that the large intestine, regarded as a single organ, is where CC and RC develop. The second factor is the identical histology and structure of the colonic and rectal walls, which include the mucosa, muscular layer, and, to some extent, the serosa. The comparable roles of the colorectal tract in terms of stool concentration, fluid resorption, transit, and excretion (Paschke et al., 2018) make up the third factor. CC accounts for 72% of all CRCs, and RC accounts for 28%; nonetheless, the incidence of CRC typically reports these values together (Centelles, 2012).

In general, when cancer develops and spreads, patients may start to exhibit symptoms and indicators. Based on the initial site of the tumor and the degree of involvement of other organs, patients with adenocarcinoma present clinically differently (Mullangi and Lekkala, 2020). According to previous studies, blood in the stool (32.8%) was the most common symptom, followed by altered bowel habits (28.9%), other symptoms such as exhaustion, reduced appetite, fever, nausea (22.0%), and stomach discomfort (16.3%) (Jensen et al., 2015). One symptom that was identified as unique to CRC patients was blood in the stool; nonspecific symptoms included fever, nausea, exhaustion, decreased appetite, abdominal discomfort, and bowel obstruction (Mullangi and Lekkala, 2020).

1.3 CLASSIFICATION OF COLORECTAL CANCER

CRC can be classified based on various factors, including the tumor's histopathological characteristics, location within the colorectum, and molecular features.

1.3.1 Histological Classification

The histological subtype of CRC was determined based on the established World Health Organization (WHO) classification for CRC (Nagtegaal et al., 2020) (Figure 1.1).

1.3.1.1 Adenocarcinoma

Most CRCs are adenocarcinomas, constituting more than 90% of CRC cases globally. Adenocarcinomas are cancerous cells that arise from the epithelial cells lining the mucosa of the colorectum and specifically originate from the glands or glandular structures in this region (IARC, 2019; Fleming et al., 2012; Hamilton et al., 2000).

1.3.1.2 Mucinous Carcinoma

Adenocarcinomas with mucinous features are the term used to describe cancers that exhibit a notable mucinous component. The existence of large glandular structures and extracellular mucin pools distinguishes this subtype (Fleming et al., 2012; Verhulst et al., 2012). The second most prevalent subtype of CRC, known as mucinous colorectal adenocarcinoma, is characterized by the existence of extracellular mucinous pools that occupy at least half of the tumor volume. Approximately 5–20% of CRC cases globally are of this subtype (Fleming et al., 2012; Hamilton et al., 2000). Mucinous adenocarcinomas, or high-level microsatellite instability (MSI-H) tumors, are common in people with LS or hereditary nonpolyposis colorectal cancer (HNPCC) (Fleming et al., 2012).

1.3.1.3 Medullary CRC

This particular subtype is identified by the amalgamation of atypical solid proliferation and an accompanying inflammatory response (Nagtegaal et al., 2020).

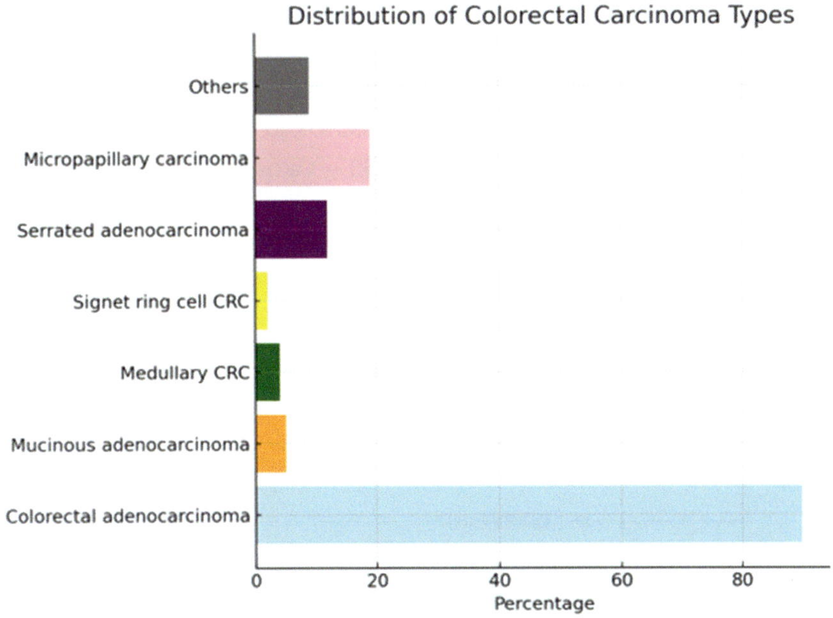

Figure 1.1 A horizontal bar chart displaying the distribution of different colorectal carcinoma types. The chart presents each type's percentage, offering a clear visual comparison of their relative frequencies.

1.3.1.4 Signet Ring Cell CRC

This carcinoma type is distinguished by the presence of more than 50% of cancer cells exhibiting a morphology resembling signet rings (Fleming et al., 2012).

1.3.1.5 Serrated Adenocarcinoma

Serrated adenocarcinoma arises from serrated polyps and is characterized by a serrated or saw-toothed glandular architecture. These tumors often exhibit microsatellite instability and a CpG island methylator phenotype (CIMP) and are associated with the serrated neoplasia pathway of colorectal carcinogenesis (Bettington et al., 2013).

Rare forms of CRC include neuroendocrine, squamous cell, adenosquamous, spindle cell, and undifferentiated carcinomas.

1.3.2 Location-Based Classification

1.3.2.1 Colon Cancer

Tumors arising in the colon are classified into two main types: proximal colon (right-sided) tumors, which occur in the ascending colon, cecum, and transverse colon, and distal colon (left-sided) tumors, which develop in the descending colon, sigmoid colon, and rectum.

1.3.2.2 Rectal Cancer

The tumor specifically arises in the rectum, the final part of the large intestine, just before the anus. It may present differently from colon cancer in terms of symptoms, prognosis, and treatment approach.

1.3.3 Molecular Classification

The classification of colorectal cancer has become more crucial with the increasing importance of molecular pathway categorization, which plays a pivotal role in this classification process (Figure 1.2).

1.3.3.1 Chromosomal Instability (CIN)

Chromosomal instability (CIN) is a molecular anomaly linked to chromosomal count aberrations, resulting in aneuploidy and loss of heterozygosity within tumors. Dysfunctions such as

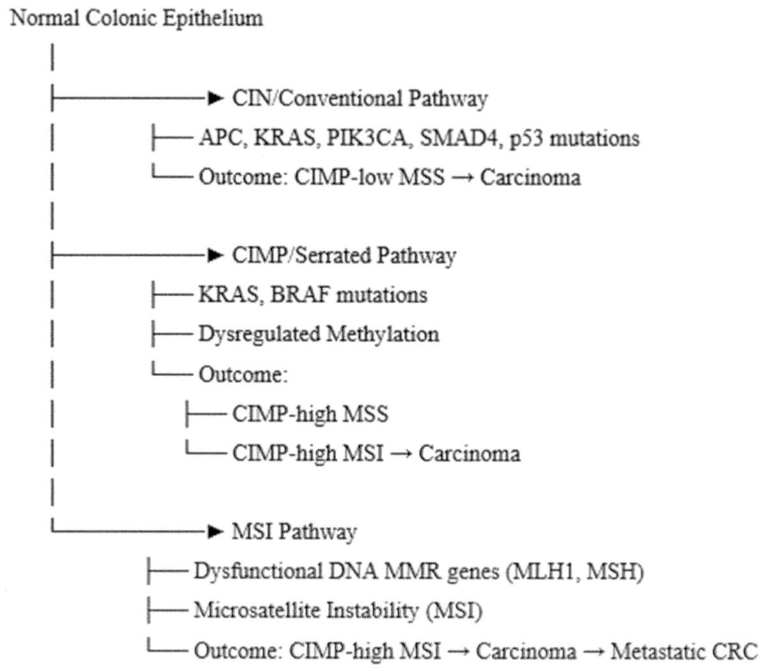

Figure 1.2 Three primary molecular pathways play roles in colorectal cancer (CRC). Normal colonic epithelium: Represents the initial healthy state of colon tissue. CIN/conventional pathway: Involves mutations in genes like APC, KRAS, PIK3CA, SMAD4, p53, leading to chromosomal instability (CIMP-low MSS) and the development of carcinoma. CIMP/Serrated Pathway: Characterized by KRAS, BRAF mutations and dysregulated DNA methylation. Outcomes include CIMP-high MSS or CIMP-high MSI, which can progress to carcinoma. MSI Pathway: Caused by dysfunction in DNA mismatch repair (MMR) genes like MLH1, MSH, leading to microsatellite instability (MSI). This pathway results in CIMP-high MSI, which progresses to carcinoma and may further lead to metastatic colorectal cancer (Huang and Yang, 2022).

chromosome segregation issues, telomere malfunction, and compromised DNA damage response mechanisms contribute to CIN (Mármol et al., 2017; Pino and Chung, 2010). Tumors associated with CIN exhibit specific mutations in genes responsible for suppressing tumors and promoting cancer, such as APC, KRAS, PIK3CA, TGF-β, and TP53, activating pathways critical for the onset and advancement of colorectal cancer (Pino and Chung, 2010) (Figure 1.2). The inactivation of the APC gene precedes the activation of KRAS mutations in CIN CRC. Later, further alterations in the TGF-β, PIK3CA, and TP53 pathways cause malignant transition (Pino and Chung, 2010). Oncogenes such as KRAS and PIK3CA can promote the growth of cancer by expressing oncoproteins, which can lead to an increase in protein levels, a loss of control, or an increase in protein activity. On the other hand, continual MAPK induction results from the mutational activation of KRAS and PIK3CA, which encourages cell growth. Since the primary cell cycle checkpoint protein p53 is encoded by the tumor suppressor gene TP53, its mutational inactivation results in uncontrollable entry into the cell cycle (Mármol et al., 2017; Pino and Chung, 2010). Furthermore, CIN influences the TGF-β pathway in CRC, resulting in abnormalities in two significant tumor suppressor genes, SMAD2 and SMAD4, the loss of which causes escape from apoptosis and dysregulation of the cell cycle (Mármol et al., 2017).

1.3.3.2 Microsatellite Instability (MSI)

Microsatellites (MSs) are small DNA sequences (1–6 base pairs) found in both coding and noncoding regions. They account for approximately 3% of the human genome (Nojadeh et al., 2018). Tumors with MSI have a reduced ability to repair MS due to their repetitive structure. As a result, they are very prone to mutations, which often occur in both coding and noncoding regions of MS (Mármol et al., 2017; Nojadeh et al., 2018). MSI describes the instability of small nucleotide repeat sequences ("microsatellites") in tumor DNA relative to normal DNA. MSI arises from a dysfunctional DNA mismatch repair (MMR) mechanism, resulting in a hypermutable phenotype (Mármol et al., 2017; Nojadeh et al., 2018). Mutations within DNA MMR genes typically cause a DNA MSI phenotype. Mutations in MMR genes, including MutL homolog (MLH) 1, MutS homolog (MSH) 2, MSH6, and the mismatch repair endonucleases PMS1 and PMS2, are observed in tumors expressing MSI (Boland and Goel, 2010) (Figure 1.2). Numerous clinical characteristics, including mucinous differentiation, signet ring cell shape, Crohn's-like lymphoid response, an abundance of tumor-infiltrating lymphocytes, tumor necrosis, and poor differentiation, have previously been associated with a high degree of MSI (Shia et al., 2003).

1.3.3.3 CpG Island Methylator Phenotype (CIMP)

A significant process in the development of human cancer involves the silencing of gene transcription through cytosine methylation at the promoter CpG islands of tumor suppressor genes. It has been demonstrated that promoter methylation silences a number of tumor suppressor genes in CRC (Issa, 2004).

The CpG island methylator phenotype (CIMP) characterizes tumors marked by extensive hypermethylation in CpG island promoters, leading to the inactivation of tumor suppressor genes and subsequent loss of protein expression (Mármol et al., 2017; Boland and Goel, 2010; Morajad et al., 2013). The two types of CIMP-positive tumors are $CIMP^{low}$ tumors, which have KRAS mutations, and $CIMP^{high}$ tumors, which have BRAF mutations, MLH1 methylation, and O-6-methylguanine-DNA methyltransferase (MGMT) or cyclin-dependent kinase inhibitor 2A (CDKN2A) silencing (Figure 1.2). In particular, a serrated pathway linked to both CIMP ($CIMP^{low}$ or $CIMP^{high}$) mutations and either KRAS or BRAF mutations accounts for approximately one-third of CRC cases. Many colorectal tumors classified as CIMP exhibit dual impacts involving genetic and epigenetic modifications, including BRAF mutations and MSI (Weisenberger et al., 2006), which affect the BRAF and MLH1 genes, respectively.

1.3.4 Mutation Origin Subtypes

The mutation origin subtypes are classified into three distinct categories.

1.3.4.1 Sporadic Colorectal Cancer (SCC)

SCC is a type of CRC that does not have a recognized genetic etiology, familial history of CRC, or inflammatory bowel disease (IBD) (Carethers and Jung, 2015). It is a somatic genetic malignant disorder that affects only single cells and their progeny, is not linked to hereditary syndromes, and is triggered by point mutations occurring during a patient's lifetime (Mármol et al., 2017). The genetic basis of SCC involves multiple mutations affecting different genes. Specifically, SCC progresses due to the accumulation of various anomalies in oncogenes and tumor suppressor genes. Inactivation of the APC gene leads to the formation of nonmalignant adenomas (polyps), which can potentially develop into adenocarcinomas. This progression is followed by mutations in the TP53 and KRAS genes, and, ultimately, deletions in genes associated with colorectal cancer (DCC) (Nojadeh et al., 2018). A common somatic mutation in MSI-H SCC (70–95%) is the promotor hypermethylation-induced alteration of the MLH1 gene (Boland and Goel, 2010). Additionally, sporadic somatic mutations in oncogenes such as RAS, SRC, and MYC are associated with CRC.

1.3.4.2 Familial CRC (FCC)

Familial colorectal cancer (FCC) is often considered sporadic in nature, as it typically occurs without a clear association with a well-defined genetic syndrome (Mármol et al., 2017; Boland and Goel, 2010; Nezamalhosseini Mojarad et al., 2013). However, it is important to note that individuals with first-degree relatives who have a history of colorectal cancer are at a significantly increased risk. Specifically, these individuals are two to three times more likely to develop colorectal cancer compared to the general population (Boland and Goel, 2010; Lin, 2012).

1.3.4.3 Inherited CRC (ICC)

This disease results from hereditary mutations that impact one of the gene's mutant alleles (Mármol et al., 2017). Two syndromes, familial adenomatous polyposis (FAP) and hereditary nonpolyposis colorectal cancer (HNPCC) are the most prevalent ICC subtypes (Menon et al., 2024).

The most prevalent hereditary polyposis syndrome is FAP, an inherited CRC syndrome. The development of many possibly malignant polyps in the colon is a defining feature of FAP, and these polyps are primarily located in the proximal colon and infrequently in the rectum in FAP. A mutation (inactivation) in the APC gene, a tumor suppressor gene encoding the APC protein, is the cause of FAP syndrome (Boland and Goel, 2010). A mutation in the APC gene causes the APC protein to function improperly, which causes β-catenin to accumulate inside the cell. Moreover, FAP syndrome has been linked to insertions, deletions, and nonsense mutations in the APC gene (Boland and Goel, 2010).

The most prevalent hereditary CRC syndrome is HNPCC, commonly referred to as LS. Genes encoding DNA repair proteins, such as MLH1, MSH2, MSH6, PMS1, and PSM2, and DNA MMR alleles that have germline changes (inherited mutations) are the cause of LS, an autosomal dominant condition (Fleming et al., 2012; Lin, 2012). A large proportion of LS cases are caused by mutations in the MSH1 and MLH2 genes (Hamilton et al., 2000). Alterations to DNA MMR genes cause MSI, a mutant phenotype associated with LS. Additionally, MSI-H is observed in more than 90% of CRCs linked to LS (Boland and Goel, 2010).

1.3.5 Consensus Molecular Subtypes (CMS)

CRC is classified into four consensus molecular subtypes (CMSs), each characterized by distinct molecular and clinical features (Guinney, 2015) (Table 1.1).

1.3.5.1 CMS1 (MSI Immune)

CMS1 is defined by microsatellite instability (MSI) and is associated with high levels of CpG island methylator phenotype (CIMP) and BRAF mutations, coupled with a low frequency of somatic mutations in SNCA (Włodarczyk et al., 2018). This subtype exhibits pronounced immunological activation, marked by significant lymphocyte infiltration and hypermethylation. The reduced signaling via the WNT pathway further distinguishes CMS1, reflecting its immunogenic profile and association with a more active immune environment (Włodarczyk et al., 2018). This subtype's immunological characteristics often correlate with a better response to immune checkpoint inhibitors (Włodarczyk et al., 2018).

1.3.5.2 CMS2 (Canonical)

CMS2 is characterized by high somatic copy number alterations (SCNA) and chromosomal instability. It features frequent mutations in WNT pathway components and MYC oncogenes, leading to robust activation of these intracellular signaling pathways (Włodarczyk et al., 2018). CMS2 also demonstrates elevated expression of epidermal growth factor receptor (EGFR) and its ligands, such as amphiregulin (AREG) and epiregulin (EREG), alongside mutations in TP53, APC, and RAS genes (Stintzing et al., 2019). This subtype is marked by upregulation of WNT and MYC downstream targets, including ERBB2 (HER2), insulin-like growth factor 2 (IGF2), insulin receptor substrate 2 (IRS2), hepatocyte nuclear factor 4 (HNF4), and various cyclins (Włodarczyk et al., 2018). These molecular features contribute to its aggressive nature and higher likelihood of therapeutic resistance.

1.3.5.3 CMS3 (Metabolic)

CMS3 exhibits a distinctive genomic and epigenomic landscape, enriched in KRAS-activating mutations and metabolic reprogramming (Stintzing et al., 2019). This subtype shows moderate activation of WNT and MYC pathways, along with mutations in PIK3CA and overexpression of IGBP3. It is characterized by low to moderate microsatellite instability (MSI) and intermediate CIMP levels, with a notable absence of BRAF mutations (Włodarczyk et al., 2018). The metabolic alterations in CMS3 include enhanced glutaminolysis and lipidogenesis, which are pivotal in its pathogenesis.

1.3.5.4 CMS4 (Mesenchymal)

CMS4 is distinguished by significant activation of mesenchymal and stromal processes, including neoangiogenesis, complement pathways, and extracellular matrix remodeling (Włodarczyk et al., 2018). This subtype is characterized by high SCNA numbers, overexpression of epithelial-mesenchymal transition (EMT) genes, and pronounced TGF-β activation. Key immune genes, including PROK1, THBS1, FGF11, CRP, S100A14, and CCL19, play a crucial role in its pathogenesis and could potentially serve as biomarkers for risk assessment in colorectal cancer patients (Włodarczyk et al., 2018). CMS4's features suggest a more aggressive disease phenotype with a potential for significant stromal and immune component involvement.

Table 1.1: Consensus molecular subtypes (CMSs) in CRC

CMS Subtype	Alternative name	Incidence	Predominant location	Key molecular features	Dominant cellular processes	Clinical features
CMS1	MSI immune	14%	Proximal colon	High levels of microsatellite instability (MSI), frequent BRAF mutations, CIMP high, hypermutated phenotype	Immune response activation, especially through TH1 and cytotoxic T cells	Generally favorable prognosis but poor survival after relapse
CMS2	Canonical	37%	Distal colon and rectum	Increased chromosomal instability (high SCNA counts), mutations in TP53, APC, and RAS genes	Enhanced epithelial cell differentiation with notable upregulation of WNT and MYC pathway targets	Best survival rates after relapse
CMS3	Metabolic	13%	No specific location	Mixed MSI status, common KRAS mutations, low SCNA, low levels of CIMP	Disruption of metabolic processes	Poor prognosis
CMS4	Mesenchymal	24%	Distal colon and rectum	High SCNA counts, overexpression of epithelial–mesenchymal transition (EMT) genes	Increased stromal infiltration, matrix remodeling through extracellular proteins, TGF-β activation, and angiogenesis	Worst overall survival and relapse-free outcomes

CIMP: CpG island methylator phenotype; **MSI**: Microsatellite instability; **SCNA**: Somatic copy number alteration; **EMT**: Epithelial-mesenchymal transition; **TGF-β**: Transforming growth factor-β

Source: Guinney, 2015.

1.4 PREVALENCE OF COLORECTAL CANCER

CRC stands as the third most widespread cancer globally, representing roughly 10% of all cancer cases (Figure 1.3). It ranks second among the leading causes of cancer-related deaths globally (WHO, 2023). CRC is the third most prevalent cancer among men and the second among women, with 1.9 million new cases recorded in 2020 (Arnold et al., 2017; Bray et al., 2015). According to the GLOBOCAN data for 2022, the maximum number of CRC cases were in Asia, and the lowest were in Africa and Oceania. The highest number of CRC-related deaths occurred in Asia among continents and in China among Asian countries in 2022. The incidence of CRC varies geographically, with discernible patterns among countries (Wong et al., 2021; WHO, 2020; Ferlay et al., 2021). Developed countries, particularly those with higher human development index (HDI) scores, exhibit heightened vulnerability to colon cancer. Noteworthy trends include a rise in incidence in certain countries, particularly in Asia and Europe, linked to lifestyle changes, while others, primarily those with high HDI scores, experience a decline. Fascinating findings emerge from a study covering the periods 2007 to 2016, 2006 to 2015, or 2005 to 2014, contingent on data availability. This research revealed an increase in colon cancer incidence in 10 out of 36 countries under scrutiny, with all these nations located in Asia or Europe. The area of India experienced the most notable increase, closely followed by that of Poland (Arnold et al., 2017; Ferlay et al., 2021). A substantial increase in colon cancer cases is observed among individuals under 50 years old in various nations, including the United Kingdom and India, signifying changing demographics (Bray et al., 2015; Wong et al., 2021; WHO, 2020; Ferlay et al., 2021). For individuals aged 0–74 years, the overall risk of mortality from colon cancer is 0.65% for men and 0.45% for women (Arnold et al., 2017; Bray et al., 2015; Ferlay et al., 2021). At the country level, Denmark, Norway, Hungary, and the Netherlands have the highest age-standardized CRC incidence rates, with rates of 48.1, 45.3, 44.2, and 42.8 cases per 100,000 persons, respectively, in 2022. The lowest age-standardized incidence rates in 2022 were observed in Gambia, Guinea, Bangladesh, and Bhutan, with rates of 2.6, 2.8, 3.6, and 4.2 cases per 100,000 persons, respectively. Projections indicate a worrisome increase in the global burden of CRC, with estimates suggesting over 3.2 million new cases and 1.6 million deaths by 2040 (WHO, 2023). Based on demographic variables, the number of new cases of CRC is predicted to increase constantly over the next 20 years, with China and the United States

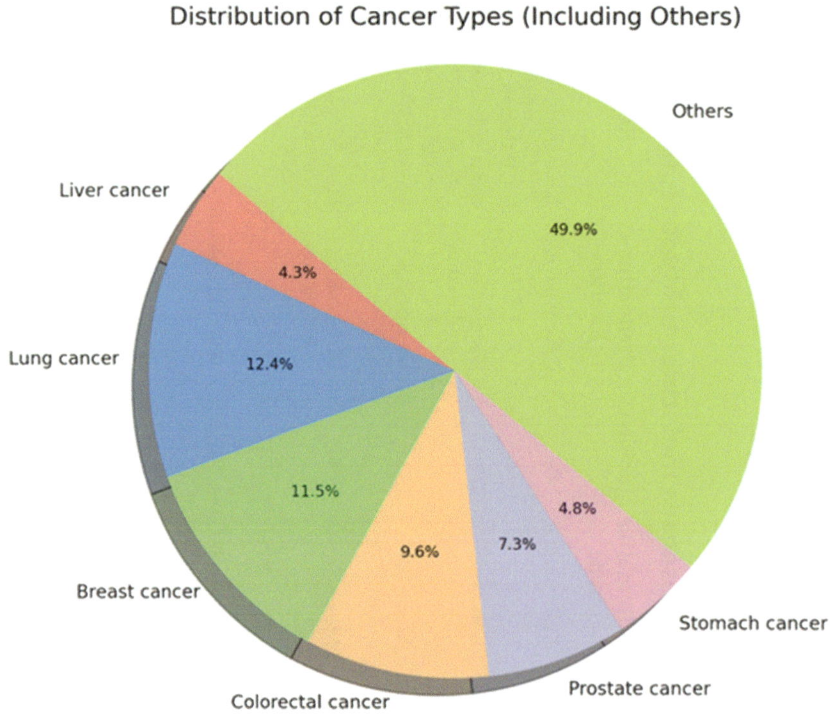

Figure 1.3 Estimated number of new cancer cases in 2022 worldwide (IARC, n.d.).

having the largest estimated number of new cases in 2020 (Xi and Xu, 2021). From 0.56 million in 2020 to 0.91 million in 2040, there will be a 64% increase in CRC cases in China. In the US, 4.4% of males (1 in 23) and 4.1% of females (1 in 25) may eventually develop and become diagnosed with CRC. It is predicted that there will be 0.16 million new instances of CRC in the US in 2020 and that this figure will increase to 0.21 million in 2040. With 60,000 CRC fatalities and a projected 148,500 new CRC cases in 2020, Japan is expected to experience 0.16 million incident cases after 20 years. In 2020, the countries with the highest rates of CRC were France, Italy, Brazil, Germany, Russia, and India (Xi and Xu, 2021). This surge is attributed to economic development, the transition from low- to medium-HDI nations, and generational changes in developed countries.

1.5 PUBLIC HEALTH IMPACT

CRC continues to pose a substantial threat to health globally, causing significant illness and death. Screening programs, diagnostic procedures, and complex treatments contribute to the strain on healthcare resources. The public health impact of CRC varies across diverse racial and ethnic groups (Agrawal et al., 2005; Yan et al., 2009; Du et al., 2007; Doubeni et al., 2007). The multifaceted origins of these racial and ethnic disparities in CRC incidence are traced back to a complex interplay of factors. Variations in environmental exposures, host characteristics, tumor biology, disparities in access to preventive healthcare services, and differential treatment modalities contribute to the intricate disparities observed (Du et al., 2007; Doubeni et al., 2007). Notably, despite concerted efforts, screening rates among the Asian American population lag behind those of other racial groups, with corresponding mortality rates from CRC also remaining lower (Trivers et al., 2008). Beyond mere screening disparities, inequities in accessing specialized CRC care emerge as a crucial factor exacerbating health disparities (Hayanga et al., 2009). Consequently, CRC imposes a substantial burden on healthcare systems, encompassing the considerable costs associated with diagnosis, treatment, and ongoing care.

The increasing "Westernization" of lifestyle—which includes consuming more red and processed meat, being sedentary, and being obese—is reflected in the rising illness load. Rapidly transitioning nations, such as the Baltic countries, have experienced an increase in CRC incidence and CRC mortality. The economic success of these nations has been linked to a shift toward the adoption of unhealthy lifestyle practices, which is accompanied by a rise in the prevalence of overweight and obesity. Some countries (such as Canada and the UK) have experienced a decrease in CRC mortality, which is likely related to best practices in cancer diagnosis and treatment, followed by an increase in CRC incidence, which is likely related to the recent introduction of screening and/or persistently unfavorable lifestyle patterns. The introduction of screening and reductions in the prevalence of risk factors, along with best practices in cancer diagnosis and treatment, are likely responsible for the decrease in CRC incidence and death observed in other countries (such as the United States and Japan). Because of economic transformation and the adoption of Western lifestyle habits, less developed nations are likely to bear the brunt of the growth in CRC burden during the next 25 years. It is necessary to use a combination strategy that includes both CRC screening and primary prevention (Rabeneck et al., 2020).

The financial ramifications of CRC further underscore the enormity of its impact on public health. Projections indicate staggering global economic expenses totaling $25.2 trillion between 2020 and 2050, with colon and rectum cancer ranking among the top five cancers with the highest economic costs. These burdens are disproportionately borne by countries such as China and the United States, highlighting the urgent need for effective interventions to mitigate economic strain. Additionally, the proportion of treatment costs to the overall economic burden of cancers is greater in high-income countries than in low-income countries (Chen et al., 2023). From the standpoint of public health, the most pressing inequity is that underprivileged communities do not have enough access to treatment and prevention services. The greatest improvements in quality of life and quantity and a reduction in the impact of CRC on public health could result from removing socioeconomic obstacles to CRC screening and treatment.

Moreover, CRC treatment, often comprising surgery, chemotherapy, and radiation therapy, has significant potential for adverse side effects that can profoundly impact patients' lives. From body image concerns to social functioning and psychological well-being, the ripple effects of treatment reverberate across various dimensions of patient experience (Krouse et al., 2010). The prospect of a colostomy, affecting up to 35% of patients, introduces additional layers of challenges, including social isolation, intimacy issues, and psychological distress (Krouse et al., 2010). Indeed, psychological distress remains a persistent companion to the CRC journey, with risk factors such as low social support and disease severity amplifying its impact (Chambers et al., 2012). Even years post-diagnosis, anxiety, depression, and traumatic stress symptoms linger, further complicating the management of pain and gastrointestinal distress. Addressing these multifaceted challenges demands a comprehensive approach that encompasses not only medical interventions but also holistic support for patients' psychosocial well-being (Dunn et al., 2013).

1.6 SIGNIFICANCE OF STUDYING COLORECTAL CANCER

CRC research holds significant importance due to its widespread impact on public health, the urgency for improved prevention and treatment strategies, and its complex interplay with various risk factors. Because of its varied nature, CRC involves several carcinogenic mechanisms. Most of the time, early diagnosis (colonoscopy), the use of current information on cancer prevention, and efficient treatment can lower the death rate associated with CRC. While many genetic alterations have been linked to CRC, further investigation is needed to completely understand the underlying causes of CRC. The most promising strategy for preventing the development of CRC in various populations across the globe is avoiding associated risk factors. Ultimately, enhancing the survival rates and well-being of CRC patients in the future necessitates research focused on developing novel medications that are both less invasive and more efficacious than existing therapies. Understanding the significance of studying CRC will help in advancing scientific knowledge and enhancing patient outcomes. CRC ranks among the most prevalent cancers worldwide, affecting millions of individuals and posing a substantial burden on healthcare systems. CRC is largely preventable through effective screening, lifestyle modifications, and early detection. The significance of studying CRC lies in identifying modifiable risk factors, implementing population-wide screening programs, and promoting healthy behaviors to reduce disease incidence and mortality rates. Research into CRC allows for the identification of emerging trends and risk factors, such as sedentary lifestyles, obesity, diet, and genetic predispositions. Understanding these factors will aid in the development of targeted prevention methods and advancements in treatment modalities, including surgery, chemotherapy, immunotherapy, and targeted therapies. By advancing scientific knowledge and implementing evidence-based interventions, researchers and healthcare professionals can mitigate the burden of CRC and improve outcomes for affected individuals. It is evident that different nations, and even areas within nations, need customized strategies for CRC detection and care that consider the political, social, and economic factors that influence the field of gastroenterology.

1.7 CONCLUSION

In summary, CRC is a formidable challenge globally, ranking as the third most prevalent cancer. Its complexity spans histological, location-based, and molecular subtypes, with projections indicating a continued increase in its burden by 2040. The economic impact of CRC underscores the substantial costs associated with diagnosis, treatment, and ongoing care. Moreover, CRC treatment can lead to adverse side effects that impact patients' quality of life, including psychological distress and social functioning. Understanding CRC is paramount for improving public health outcomes and implementing effective prevention strategies. Continued research and innovation are essential for advancing treatment modalities and reducing the global burden of this disease. Through collective efforts, we can strive towards better outcomes and a future where CRC is no longer a pervasive threat to health.

REFERENCES

Agrawal, S., et al. (2005). Colorectal cancer in African Americans. *Am J Gastroenterol*, 100(3):515–523.

Arnold, M., et al. (2017). Global patterns and trends in colorectal cancer incidence and mortality. *Gut*, 66(4):683–691.

Bettington, M., et al. (2013). The serrated pathway to colorectal carcinoma: current concepts and challenges. *Histopathology*, 62(3):367–386.

Boland, C. R., Goel, A. (2010). Microsatellite instability in colorectal cancer. *Gastroenterol*, 138(6):2073–2087.

Bray, F., et al. (2015). Cancer incidence in five continents: inclusion criteria, highlights from Volume X and the global status of cancer registration. *Int J Cancer*, 137(9):2060–2071.

Carethers, J. M., Jung, B. H. (2015). Genetics and genetic biomarkers in sporadic colorectal cancer. *Gastroenterol*, 149(5):1177–1190.

Centelles, J. J. (2012). General aspects of colorectal cancer. *Int Scholarly Res Notices*, 2012(1): 139268.

Chambers, S. K., et al. (2012). A five-year prospective study of quality of life after colorectal cancer. *Quality of Life Research*, 21:1551–1564.

Chen, S., et al. (2023). Estimates and projections of the global economic cost of 29 cancers in 204 countries and territories from 2020 to 2050. *JAMA Oncology*, 9(4):465–472.
Doubeni, C. A., et al. (2007). Racial differences in tumor stage and survival for colorectal cancer in an insured population. *Cancer*, 109(3):612–620.
Du, X. L., et al. (2007). Meta-analysis of racial disparities in survival in association with socioeconomic status among men and women with colon cancer. *Cancer*, 109(11):2161–2170.
Dunn, J., et al. (2013). Trajectories of psychological distress after colorectal cancer. *Psycho-oncology*, 22(8):1759–1765.
Ferlay, J., et al. (2021). Cancer statistics for the year 2020: an overview. *Int J Cancer*, 149(4):778–789.
Fleming, M., et al. (2012). Colorectal carcinoma: Pathologic aspects. *J Gastrointest Oncol*, 3(3):153.
Fusaro, R. M., et al. (1996). Muir-torre syndrome and defective DNA mismatch repair genes [3]. *J Am Acad Dermatol*, 35(3):493–494.
Guinney, J., et al.(2015). The consensus molecular subtypes of colorectal cancer. *Nature Med*, 21/11:1350–1356.
Hamilton, S. R. et al. (2000). Tumours of the colon and rectum: Carcinoma of the colon and rectum. In: Hamilton SR et al., eds., *World Health Organization Classification of Tumours: Pathology and Genetics of Tumours of the Digestive System*. IARC: 105–119.
Hayanga, A. J., et al. (2009). Racial clustering and access to colorectal surgeons, gastroenterologists, and radiation oncologists by African Americans and Asian Americans in the United States: a county-level data analysis. *Arch Surg*, 144(6):532–535.
Huang, Z., Yang, M. (2022). Molecular network of colorectal cancer and current therapeutic options. *Frontiers Oncol*, 12:852927.
International Agency for Research on Cancer. (IARC) (2019). *Colorectal Cancer Screening*. Lyon: International Agency for Research on Cancer
Issa, J. P. (2004). CpG island methylator phenotype in cancer. *Nat. Rev. Cancer*, 4(12):988–993.
Jensen, L. F., et al. (2015). Symptom attributions in patients with colorectal cancer. *BMC Fam. Pract*, 16:1–10.
Krouse, R. S., et al. (eds). (2010). *Gastrointestinal cancer: psycho-oncology*. 2nd edn. Oxford University Press, New York: 140–145.
Li, L., et al. (2019). Activation-induced cytidine deaminase expression in colorectal cancer. *Int J Clin Exper Pathol*, 12(11):4119.
Lin, O. S. (2012). Colorectal cancer screening in patients at moderately increased risk due to family history. *World J Gastrointest Oncol*, 4(6):125.
Lotfollahzadeh, S., et al. (2023). *Rectal cancer [internet]*. Treasure Island, FL: StatPearls Publishing.
Lynch, H. T., Krush, A. J. (1971). Cancer family "G" revisited: 1895–1970. *Cancer*, 27(6):1505–1511.
Lynch, H. T. et al. (1977). Role of heredity in multiple primary cancer. *Cancer*, 40(S4):1849–1854.
Mármol, I., et al. (2017). Colorectal carcinoma: a general overview and future perspectives in colorectal cancer. *Int J Molec Sci*, 18(1):197.
Menon G. et al. Colon Cancer. [Updated 2024 Apr 20]. In: StatPearls [Internet]. Treasure Island, FL: StatPearls Publishing; 2024 Jan. Available from: www.ncbi.nlm.nih.gov/sites/books/NBK470380/
Mojarad, E.N. et al. (2013). The CpG island methylator phenotype (CIMP) in colorectal cancer. *Gastroenterology and Hepatology from Bed to Bench*, 6.3:120.
Mullangi, S., Lekkala, M. R. (2020). *Adenocarcinoma [internet]*. Treasure Island, FL: StatPearls Publishing.
Nagtegaal, I. D., et al. (2020). The 2019 WHO classification of tumours of the digestive system. *Histopathol*, 76(2):182.
National Institute for Health and Care Excellence. (NICE) (2015). *Suspected cancer: recognition and referral [internet]*. London: NICE Pathways.
Nezamalhosseini Mojarad, E., Kuppen, P. J. K., Asadzadeh Aghdaei, H., Zali, M. R. (2013). The CpG island methylator phenotype (CIMP) in colorectal cancer. *Gastroenterol Hepatol Bed Bench*, 6:120–128.
Nojadeh, J. N., et al. (2018). Microsatellite instability in colorectal cancer. *EXCLI Journal*, 17:159.
Paschke, S., et al. (2018). Are colon and rectal cancer two different tumor entities? A proposal to abandon the term colorectal cancer. *Int J Molec Sci*, 19(9):2577.
Pino, M. S., Chung, D. C. (2010). The chromosomal instability pathway in colon cancer. *Gastroenterol*, 138(6):2059–2072.
Rabeneck, L., Chiu, H. M., Senore, C. (2020). International perspective on the burden of colorectal cancer and public health effects. *Gastroenterol*, 158(2):447–452.

Recio-Boiles, A., Cagir, B. (2022). *Colon cancer*. Treasure Island, FL: Stat-Pearls Publishing.
Shia, J., et al. (2003). Value of histopathology in predicting microsatellite instability in hereditary nonpolyposis colorectal cancer and sporadic colorectal cancer. *Am J Surg Pathol*, 27(11):1407–1417.
Silva, A., et al. (2020). Impact of adiposity on staging and prognosis of colorectal cancer. *Crit Rev Oncol Hematol*, 145:102857.
Stintzing, S., et al. (2019). Consensus molecular subgroups (CMS) of colorectal cancer (CRC) and first-line efficacy of FOLFIRI plus cetuximab or bevacizumab in the FIRE3 (AIO KRK-0306) trial. *Ann Oncol*, 30(11):1796–1803.
The International Agency for Research on Cancer (IARC). (n.d.). *Global Cancer Observatory*. Available at: https://gco.iarc.fr/
Trivers, K. F., et al. (2008). Trends in colorectal cancer screening disparities in people aged 50–64 years, 2000–2005. *Am J Prevent Med*, 35(3):185–193.
Verhulst, J., et al. (2012). Mucinous subtype as prognostic factor in colorectal cancer: a systematic review and meta-analysis. *J Clin Pathol*, 65(5):381–388.
Warthin, A. S. (1985). Classics in oncology: heredity with reference to carcinoma as shown by the study of the cases examined in the pathological laboratory of the University of Michigan, 1895–1913. *CA*, 35(6):348–359.
Weisenberger, D. J., et al. (2006). CpG Island methylator phenotype in human colorectal cancer is tightly associated with BRAF mutation and underlies sporadic mismatch repair deficiency. *Nature Genet*, 38:787–793.
Włodarczyk, M., et al. (2018). Genetic molecular subtypes in optimizing personalized therapy for metastatic colorectal Cancer. *Current Drug Targets*, 19(15):1731–1737.
Wong, M. C., et al. (2021). Differences in incidence and mortality trends of colorectal cancer worldwide based on sex, age, and anatomic location. *Clin Gastroenterol Hepatol*, 19(5):955–966.
World Health Organization (WHO). (2020). *World health statistics 2020: monitoring health for the SDGs, sustainable development goals*. Available at: https://iris.who.int/handle/10665/332070
World Health Organization (WHO). (2023). *Colorectal cancer*. Available at: https://www.who.int/news-room/fact-sheets/detail/colorectal-cancer
Xi, Y., Xu, P. (2021). Global colorectal cancer burden in 2020 and projections to 2040. *Translat Oncol*, 14(10):101174.
Yan, B., et al. (2009). Racial differences in colorectal cancer survival in the Detroit Metropolitan Area. *Cancer*, 115(16):3791–3800.
Zaytseva, Y. (2021). Lipid metabolism as a targetable metabolic vulnerability in colorectal cancer. *Cancers*, 13(2):301.

2 Epidemiology and Risk Factors of Colorectal Cancer

Saliha Rizvi, Sanchita Srivastava, Haider Ali Rizvi, Rizwan Askari, Syed Tasleem Raza and Farzana Mahdi

2.1 INTRODUCTION

Colorectal cancer (CRC) poses a significant public health challenge globally, accounting for 10% of cancer cases and 9.4% of cancer-related deaths in 2020, just behind lung cancer (Xi & Xu, 2021). Projections indicate 3.2 million new cases by 2040, driven by aging populations and lifestyle changes. CRC includes various malignant neoplasms such as adenocarcinomas and carcinoid tumors, with symptoms ranging from asymptomatic early stages to advanced disease marked by rectal bleeding and weight loss. The exact cause of CRC remains elusive, leading to low early diagnosis rates, which is crucial for prognosis improvement. Treatment depends on tumor stage and patient characteristics, involving options like immunotherapy, targeted therapy, and chemotherapy.

Epidemiological research is vital for understanding CRC's distribution and risk factors, facilitating early detection and targeted preventive measures. Identifying high-risk populations enhances public health strategies, while insights from epidemiology guide personalized treatment approaches, optimizing outcomes and quality of life for patients. Furthermore, understanding CRC epidemiology aids in effective resource allocation within healthcare systems for screening and treatment, ensuring equitable care access. Ongoing research fueled by these findings drives advancements in CRC etiology and treatment, ultimately improving patient outcomes and global public health.

2.2 EPIDEMIOLOGY OF CRC

2.2.1 Global Incidence and Trends

The global incidence and trends of CRC constitute a multifaceted narrative shaped by a confluence of demographic shifts, socioeconomic disparities, and lifestyle transitions. Over recent decades, CRC has increasingly asserted its prominence as a significant public health challenge, with its prevalence steadily rising across many regions of the world. Epidemiologically, CRC exhibits a multifaceted nature, influenced by a complex interplay of genetic, environmental, and lifestyle factors. It accounts for 10% of all cancer diagnoses globally and holds the position of the second-highest contributor to cancer-related mortality. Familial adenomatous polyposis (FAP) and Lynch syndrome contribute to about 5% of CRC cases, with an additional 10–15% of unselected CRC patients carrying high-risk mutations not associated with these syndromes (Murphy et al., 2019). While the majority of CRC cases (70–80%) are sporadic, 20–30% display hereditary patterns, including susceptibility syndromes such as Lynch syndrome and FAP (Müller et al., 2016). Historically, developed countries have borne a disproportionate share of the CRC burden. This shift is often attributed to the adoption of Westernized lifestyles characterized by sedentary behaviors, high-calorie diets rich in red and processed meats, low intake of fruits and vegetables, and rising rates of obesity – all of which are established risk factors for CRC (World Cancer Research Fund/American Institute for Cancer Research, 2018). In 2020, CRC ranked as the third most common cancer globally, with an estimated 1,931,590 new cases and an age-standardized incidence rate (ASR) of 19.5 per 100,000 person-years. Europe, Oceania, and North America reported the highest ASRs, while Africa and the Eastern Mediterranean Region (EMRO) had the lowest rates. Disparities in CRC incidence were observed between high-income and low- to middle-income countries, as well as among regions with varying human development indices (HDIs) (Global Cancer Observatory, 2020). Early-onset CRC presents a growing concern, with an estimated 188,069 new cases worldwide in individuals under 50 years old. EOCRC rates vary by region, with North America and Oceania reporting the highest incidence. Recent evidence suggests increasing EOCRC rates, particularly in developing countries. Genetic predispositions, lifestyle factors, and alterations in intestinal microbiota composition may contribute to rising EOCRC incidence trends (Global Cancer Observatory, 2020). CRC incidence trends vary globally, with stable or declining rates in high-risk countries and increasing rates in transitioning regions, particularly in parts of Asia, Latin America, and Eastern Europe, where rapid urbanization and

DOI: 10.1201/9781032704517-2

the adoption of Westernized lifestyles—characterized by reduced physical activity, processed food consumption, and rising obesity—are driving this epidemiological shift.

While high-income countries often boast well-established screening programs and advanced treatment options, low- and middle-income countries may face challenges in delivering comprehensive cancer care, resulting in higher mortality rates and poorer survival outcomes (World Health Organization, 2020). Using statistical models for analysis, age-adjusted rates for new CRC cases have been falling an average of 1.2% each year from 2012–2021.

An analysis of SEER data (2000–2014) shows an annual CRC incidence increase of 2.7% among 20–39-year-olds and 1.7% in 40–49-year-olds, while rates decreased in older groups (Ansa et al., 2018). Despite higher percentage increases in younger adults, the absolute number of diagnoses remains lower compared to older populations. From 2012 to 2021, age-adjusted CRC rates declined by 1.2% annually, with death rates dropping by 1.7% from 2013 to 2022, highlighting the need for ongoing surveillance and public health measures (Saad El Din et al., 2020).

Recent trends show stable CRC incidence rates from 2012 to 2021, with average annual percent changes of -0.7% for females and -1% for males. These trends highlight the importance of continuous surveillance and targeted interventions to understand and address the evolving landscape of CRC incidence rates among different genders.

Projections indicate a substantial increase in new CRC cases by 2040, particularly in low-risk regions such as Africa and Latin America. Discrepancies in incidence changes were noted between countries with different HDIs, highlighting the impact of socioeconomic factors on CRC burden (Siegel et al., 2020).

2.2.2 Demographics

In addition to lifestyle changes, demographic factors also play a crucial role in shaping global CRC trends. In developed regions such as North America, Western Europe, and Australia, CRC has long been a significant health concern, with high incidence rates attributed to factors like aging populations, sedentary lifestyles, and diets rich in processed foods and red meat. However, recent decades have witnessed stabilization or even declines in incidence rates in some of these regions, attributed in part to the widespread adoption of screening programs and lifestyle modifications (American Cancer Society, 2022a).

2.2.3 Age

The distribution of CRC cases across age groups reveals a complex interplay influenced by screening interventions. Traditionally, CRC incidence increases with age, with most symptomatic cases occurring in older individuals. Aging populations in developed countries significantly contribute to the rising CRC burden (American Cancer Society, 2022b).

Screening programs have altered the typical age-related trajectory of CRC diagnosis. Although screening was historically recommended starting at age 50, this guideline changed in 2018 (Wolf et al., 2018). Despite a modest 46% screening participation rate among those aged 50 to 54 in 2019, incidence rates in this cohort exceeded projections, reaching 71.2 cases per 100,000 annually, a 4% to 14% increase compared to those aged 51 to 58 (American Cancer Society, 2022b).

These findings highlight the critical role of early detection through screening and age-specific interventions in improving CRC outcomes, underscoring the need for targeted strategies to address the unique challenges faced by different age groups in CRC diagnosis and treatment.

2.2.4 Gender

Gender disparities in CRC reveal notable differences, with men consistently showing higher incidence rates than women. Data from 2015 to 2019 indicate that men had an average yearly incidence rate of 41.5 cases per 100,000, which is 33% higher than women's rate of 31.2 (White et al., 2018). Factors such as age at diagnosis and tumor location contribute to this gender gap.

Similar patterns have been noted in other developed countries, potentially due to differences in carcinogenic exposures, hormonal factors, and unique etiologic pathways among anatomic subsites (Nawa et al., 2008; Lee et al., 2015).

Recent research attributes about one-third of the sex discrepancy in CRC to variations in known risk factors (Jackson et al., 2022). This highlights the need for understanding gender-specific CRC risk factors and developing targeted interventions to address these disparities. The incidence rates of CRC in Asian countries vary significantly, with the highest rates observed in Japan, Singapore, and Korea. Overall, the average age-standardized rate (ASR) for the region is 7.1, indicating a moderate but widespread burden of CRC across different populations.

2.2.5 Regional Variances

Understanding the epidemiology of CRC is essential for effective prevention and targeted interventions. Regional differences in CRC incidence and mortality highlight the complex interactions of genetic, environmental, and lifestyle factors. Epidemiological studies show significant disparities in CRC burden across geographic regions. Higher incidence rates are reported in North America, Australia, and parts of Western Europe, while lower rates are seen in Asia, Africa, and Eastern Europe (World Cancer Research Fund International, 2021).

Variations in CRC risk are linked to dietary habits, healthcare access, screening practices, and genetic predispositions. Diets rich in red and processed meats, common in Western societies, increase CRC risk (World Cancer Research Fund/American Institute for Cancer Research, 2018). Data show wide differences in CRC incidence globally. Afghanistan reports an ASR of 3.6, while Brunei Darussalam has a higher ASR of 17.4, reflecting regional disparities in risk (Ferlay et al., 2020). In the Middle East, the Gaza Strip reports a cumulative risk of 1.2, while Saudi Arabia and Qatar have lower risks of 0.67 and 0.47. Singapore stands out with a high ASR of 16.7, signaling a need for further investigation (Ferlay et al., 2020). Incidence rates of CRC also vary across continents, with North America and Australia showing the highest rates and Africa the lowest.

While taking an overview of the different mortality rates due to CRC globally, Hungary leads with an age-standardized rate (ASR) of 20.23, followed by Croatia (18.49) and Brunei Darussalam (17.39) (Ferlay et al., 2020). These rates indicate the number of CRC deaths per 100,000 population. Notably, countries like Slovakia (17.24), Singapore (16.7), and Moldova (16.27) also demonstrate relatively high mortality rates. Poland (15.64), Romania (15.49), and Barbados (15.17) contribute to the global pattern, emphasizing the impact of this disease. Eastern European nations, including Belarus (14.71) and Ukraine (12.32), exhibit significant mortality rates. New Zealand (13.9), Norway (13.82), and Bosnia Herzegovina (13.79) represent diverse geographical regions, yet share comparable mortality rates (Ferlay et al., 2020). Mortality rates due to CRC vary across continents, with Europe having the highest age-standardized rate (12.1) and Africa the lowest (5.5). Globally, the average mortality rate is 8.1, highlighting the significant burden of CRC worldwide, especially in Europe and Northern America.

2.3 RISK FACTORS

2.3.1 Environmental Factors

Ongoing research explores the impact of environmental toxins on CRC risk. Pesticides like terbufos and chlorpyrifos have been linked to increased risks of colon and rectal cancers (Matich et al., 2021; Kwak et al., 2019). Exposure to industrial pollutants, asbestos, and air pollution, including PM2.5, has also been associated with higher CRC risks (García-Pérez et al., 2020; Pritchett et al., 2022). Endocrine-disrupting chemicals (EDCs), such as bisphenol A and phthalates, have been shown to promote colon cancer cell growth and are found in higher concentrations among CRC patients (Su et al., 2021). Additionally, triclocarban may exacerbate colitis-associated colon cancer through microbiota changes (Yang et al., 2020).

2.3.2 Diet, Physical Activity, and Obesity

High red meat consumption is linked to increased CRC risk, partly due to the low intake of fruits, vegetables, and fiber in high-meat diets. Research shows a significant increase in CRC risk with high intake of red meat, refined starches, eggs, and cheese (Bidoli et al., 1992). Cooking red meat at high temperatures produces heterocyclic amines, which contribute to this risk (Probst-Hensch et al., 1997; Sinha et al., 1999). Meta-analyses indicate a 12% to 33% increase in CRC risk with red and processed meat consumption (Sandhu et al., 2001; Norat et al., 2002). While some studies found no significant link, a modest association cannot be ruled out (Flood et al., 2003).

2.3.3 Smoking

Smoking is a major contributor to lung cancer, with approximately 80% of primary cases linked to cigarette use. It also increases the risk of cancer in non-smoke-exposed organs, including the colon, rectum, pancreas, and kidney (Gandini et al., 2008). Significant associations have been found between smoking and both the incidence and mortality of CRC in humans, as well as increased CRC risk in animal studies (Botteri et al., 2008; Kim et al., 2008).

The mechanisms through which smoking influences CRC initiation and progression are not fully understood. However, studies indicate that bacterial diversity increases after smoking cessation (Biedermann et al., 2013), and smoking is linked to alterations in the microbiome and mucin structure (Biedermann et al., 2014; Allais et al., 2016). There is a well-documented connection between modified gut microbiota and CRC (Yu et al., 2017; Wong et al., 2017), with gut microbes from CRC patients shown to promote colon tumor formation in mice. Despite these findings, the relationship between smoking-induced changes in gut microbiota and CRC remains unclear.

2.3.4 Alcohol

Chronic alcohol consumption is linked to increased CRC risk, with higher intake correlating with various cancers, including CRC (Seitz & Stickel, 2007; Wackernah et al., 2014). No safe threshold for alcohol regarding cancer risk has been established. In 2020, global alcohol-induced colon and rectal cancer incidence rates were 1.0 and 0.7 per 100,000 people, respectively (Rumgay et al., 2021). Ethanol metabolism in the liver and GI tract leads to acetaldehyde production, which promotes cellular proliferation (Lieber et al., 1986; Elamin et al., 2013). Compounds like resveratrol may help reduce CRC risk (Pang et al., 2018).

2.3.5 Geographic Location

From 1982 to 2010, population-based cancer registry (PBCR) data indicated a significant increase in colon and rectum cancer incidence across major registries, with annual percentage changes (APC) of 0.9% to 5.8% for colon cancer and 2.7% to 9.8% for rectal cancer (Ma et al., 2018). This trend has contributed to a substantial burden of these cancers in India, but existing studies are limited and do not provide a comprehensive national overview (Mills et al., 2013; Huang et al., 2011).

Overall, the lifetime risk for females ranges from 1 in 167 to 1 in 500, and for males, it is 1 in 167 to 1 in 250. Incidence is generally higher in males than females, with the Northeast region showing the highest AARs for males (5.36) and females (4.34), followed by the South (4.15 for males, 3.08 for females). The relative proportion of colon cancer among all cancers is 4.07% for males in the East and 2.55% for females in the Northeast, underscoring regional and gender differences in colon cancer incidence across India.

2.3.6 Genetic and Familial Risk

Initial studies on family history and CRC in Utah families found a higher CRC-related death rate (3.9%) among first-degree relatives (FDRs) of deceased patients compared to age- and sex-matched controls (1.2%) (Barone et al., 2008). This trend has been confirmed in diverse populations, showing that FDRs of affected-individuals have a two- to threefold increased risk of developing CRC, regardless of study design or location (Stein et al., 2010; Jeon et al., 2013). A family history of two or more relatives with CRC may indicate a genetic syndrome.

2.3.7 Inheritance of CRC Predisposition

Numerous genes associated with CRC risk have been identified, with most pathogenic variants inherited in an autosomal dominant manner (Dawson, 2004). An exception is MUTYH-associated polyposis (MAP), which follows an autosomal recessive inheritance. Specific family traits indicating autosomal dominant inheritance are critical for assessing heightened risk and potential cancer-causing variants.

Traits include vertical transmission of genetic predisposition, with a 50% inheritance risk for both genders. Other clinical indicators of hereditary CRC syndromes include:

- Earlier cancer onset compared to sporadic cases (Soos et al., 1990).
- Increased risk for other cancers, such as endometrial cancer.
- Development of multiple primary cancers (e.g., CRC and endometrial cancer).
- Nonneoplastic features like congenital hypertrophy of the retinal pigment epithelium and desmoid tumors in FAP.
- Rare tumors (e.g., adrenocortical carcinoma) indicating hereditary syndromes.
- Multiple polyps, with oligopolyposis prompting consideration for multigene testing.

The two primary hereditary causes of CRC are FAP due to pathogenic variants in the APC gene, and Lynch syndrome, linked to pathogenic variants in mismatch repair (MMR) genes (Akagi et al., 1998). Many families display CRC clustering without identifiable syndromes, termed familial CRC.

2.3.8 Diabetes

Patients with T2DM have a 1.18 times higher risk of colorectal cancer (HR: 1.18, 95% CI: 1.04–1.39) according to a large China Kadoorie Biobank study (Pang et al., 2018). The elevated risk of colon and rectal cancers in T2DM patients was slightly higher in males and diminished over time, with a positive correlation observed between random plasma glucose levels and CRC incidence in non-diabetic individuals (Pang et al., 2018).

The Nurses' Health Study and Health Professionals' Follow-up Study found a higher CRC risk in males with T2DM (HR: 1.42) than in females (HR: 1.17) over 26 years (Ma et al., 2018).

In addition to being a risk factor, T2DM also negatively impacts the prognosis of CRC patients. A meta-analysis revealed that CRC patients with preexisting diabetes had higher all-cause and cancer-specific mortality compared to those without diabetes (Mills et al., 2013). Several other studies corroborated these findings, indicating higher overall mortality in CRC patients with preexisting diabetes (Huang et al., 2011; Barone et al., 2008; Stein et al., 2010; Jeon et al., 2013; Dawson, 2004; Soos et al., 1990; Ahmed et al., 2006; Wu et al., 1995; Tran et al., 2006; Kadowaki et al., 1996; Yu & Rohan, 2000; Baxter et al., 1980; Warren et al., 1996; Akagi et al., 1998; André et al., 1999; Jenkins et al., 1999; Jenkins et al., 2000; Orme et al., 1998; Giouleme et al., 2011; Warburg, 1956; Hoxhaj & Manning, 2020; Ancey et al., 2018; Brownlee, 2001; Haak et al., 2020; Rojas et al., 2011). Moreover, T2DM has been linked to worse recurrence-free survival in CRC patients, with many deaths attributed to tumor recurrence. Some studies suggest that T2DM may promote liver and lymphatic metastasis of CRC, significantly affecting patient prognosis and survival (Rojas et al., 2011).

T2DM increases both the risk and worsens the prognosis of CRC, likely due to complex interactions of metabolic, hormonal, and cancer-specific factors.

2.3.9 Polyps

CRC screening programs have significantly improved early detection, leading to a rise in the identification of malignant polyps (MP) (Labianca et al., 2013; Shaukat et al., 2020; Teo et al., 2020). Malignant polyps are defined as those with carcinoma invading through the muscularis mucosae into the submucosa, classified as pT1 according to the TNM system (Teo et al., 2020; National Comprehensive Cancer Network, 2021). Post-endoscopic removal, the prevalence of MP ranges from 0.75% to 5.6%, accounting for 40–60% of stage I CRC cases (Shaukat et al., 2020; Cancer Council Australia, 2021; Kuo et al., 2020).

The risk of lymph node metastasis (LNM) associated with MP varies based on invasion depth. Superficial invasions, such as pedunculated polyps with Haggitt level 1 or sessile polyps with Kikuchi sm1 characteristics, have a negligible risk (0–3%). Advances in imaging techniques and endoscopic treatments have enhanced understanding of MP's significance (Lin et al., 2021; Winawer et al., 1993).

2.4 EARLY DETECTION IMPORTANCE

To significantly increase and boost the chances of successful treatment and survival, the early detection of CRC is pivotal. The role of identifying precancerous lesions at an early stage is crucial.

2.4.1 Significance of Early Detection

2.4.1.1 Improved Treatment Outcomes

Adjuvant treatment combining surgical resection and specific chemotherapy offers a better prognosis for early-stage CRC compared to advanced-stage cases with metastasis, which have a poor outlook. Early detection is crucial, as the American Cancer Society reports a five-year survival rate of approximately 90% for localized CRC, contrasted with around 14% for advanced CRC (Winawer et al., 1993).

2.4.1.2 Prevention of Advanced Disease

Early disease detection underscores the necessity of regular screening, even for asymptomatic individuals (Prorok et al., 2000). Colorectal screening methods like colonoscopy and fecal occult blood tests (FOBT) enhance the identification of precancerous polyps and lesions before symptoms arise. Early screening and removal of these polyps not only impede disease progression but also prevent lesions from developing into cancer.

2.4.1.2.1 *Supporting Evidence*

a. **Screening Trials**
The occurrence and mortality of CRC are directly linked to large-scale randomized controlled trials (RCTs). These trials have significantly aided in preventing CRC through early detection and removal of adenomatous polyps. For instance, the National Polyp Study demonstrated a notable reduction in CRC incidence and mortality via screening colonoscopy (Hewitson et al., 2007). Additionally, participants in the NIH-sponsored Prostate, Lung, Colorectal, and Ovarian (PLCO) Cancer Screening Trial reported decreased CRC mortality following annual FOBT (Lin et al., 2021).

b. **Population-Based Studies**
Numerous studies have demonstrated the effectiveness of CRC screening in reducing mortality. A meta-analysis in the British Medical Journal (BMJ) indicated that CRC screening is linked to a 16% reduction in mortality (Smith et al., 2009). Additionally, the US Preventive Services Task Force (USPSTF) reported that populations undergoing any form of screening experience lower incidence and severity of CRC (Wolf et al., 2018).

2.4.1.2.2 *Guideline Recommendations*

Intending to detect CRC at its inception, a variety of early screening modalities have been recommended by several health agencies some of which are the American Cancer Society (ACS), the United States Preventive Services Task Force (USPSTF), and the European Society of Gastrointestinal Endoscopy (ESGE) (Wolf et al., 2018). With the inculcation of the risk factors and age, these guidelines emphasize screening and early detection of CRCs in individuals.

2.4.2 Screening Guidelines for CRC

The guidelines established by renowned medical associations such as the American Cancer Society (ACS), the United States Preventive Services Task Force (USPSTF), and the European Society of Gastrointestinal Endoscopy (ESGE) outline various modalities for CRC screening at its inception to majorly identify precancerous lesions for resection. With considerations of individual preferences and distinct risk factors, these screening investigations have been declared as the primitive regimen for CRC detection.

2.4.2.1 *Colonoscopy*

Guidelines

Colonoscopy is the ideal screening method for CRC due to its ability to visualize the whole colon and rectum, as well as locate and treat precancerous polyps (Botteri et al., 2008).

Colonoscopy screening for average-risk individuals should begin at 45, repeating every 10 years, with adjustments based on risk factors, as recommended by the American Cancer Society, USPSTF, and ESGE (Wolf et al., 2018).

2.4.2.2 *Fecal Occult Blood Tests (FOBT)*

Guidelines

- The presence of CRC is indicated by the blood cells in the fecal matter for which FOBT is preferred to identify the occult blood cells in the stool (Botteri et al., 2008).
- To initiate the screening, the ACS has recommended the age of 45 to be optimum. Annual FOBT or fecal immunochemical testing (FIT) should be followed by colonoscopy if occult blood is traced (Wolf et al., 2018).
- The USPSTF recommends annual screening with high-sensitivity FOBT or FIT, beginning at age 45 and continuing until 75.
- The ESGE recommends FIT every two years for CRC screening in average-risk patients.

2.4.2.3 *Flexible Sigmoidoscopy*

Guidelines

Flexible sigmoidoscopy visualizes the lower colon and rectum to detect precancerous polyps or CRC. The American College of Surgeons advises flexible sigmoidoscopy every five years, beginning at age 45 or 50, as an alternative to colonoscopy (Botteri et al., 2008). The USPSTF recommends screening with flexible sigmoidoscopy every ten years beginning at age 45, along with annual FOBT testing (Wolf et al., 2018).

2.4.2.4 Computed Tomographic Colonography (CTC)

Guidelines

- CTC, or computer-aided colonoscopy, uses CT scans to create detailed visuals of the colon. The American Cancer Society (ACS) recommends CTC every five years for CRC screening in average-risk adults who decline or have contraindications to colonoscopy (Botteri et al., 2008).
- The USPSTF does not recommend CTC as a main screening technique for CRC due to a lack of adequate evidence (Wolf et al., 2018).

2.4.2.5 Stool DNA Testing

Guidelines

- Stool DNA testing detects DNA changes associated with CRC or precancerous polyps. The American College of Surgeons recommends stool DNA testing every three years as an alternative to colonoscopy for patients who decline or are contraindicated to other screenings (Botteri et al., 2008).
- The USPSTF advises against using stool DNA testing as an elementary screening method for CRC due to insufficient evidence of its usage (Wolf et al., 2018).

2.4.3 Prevention Strategies for CRC

CRC prevention strategies promote alterations to lifestyles and healthy habits that reduce the risk of developing the disease. To prevent CRC, it's crucial to make dietary changes, exercise regularly, maintain a healthy weight, avoid smoking and alcohol, follow screening guidelines, and consider using aspirin or NSAIDs.

a. **Regular Physical Activity and Weight Management**
Maintaining an optimum weight and engaging in physical activities frequently is important in preventing CRC. The regulation of immunological functions and hormonal balance is achieved through exercise promotes the prevention of the disease and reduces inflammation. A weekly engagement of 75 minutes of moderate to vigorous exercise reduces the risks of CRC in adults, according to ACS (Cole et al., 2009).

b. **Smoking Cessation and Alcohol Moderation**
Smoking and heavy alcohol consumption are proven risk factors for CRC. Tobacco smoke contains carcinogens that damage colon and rectal mass, increasing the risk of contracting CRC. Excessive alcohol use can cause inflammation and oxidative stress, which raise the risk of CRC. To avoid CRC, it is vital to quit smoking and consume alcohol in proportion (Cole et al., 2009).

c. **Aspirin and NSAIDs**
Frequent intake of aspirin or non-steroidal anti-inflammatory medicines (NSAIDs) may reduce the likelihood of CRC, especially for those with a genetic inclination of the disease. These drugs are associated with the reduction of inflammation and delaying the formation of precancerous lesions. Aspirin and NSAIDs must be used cautiously as they may cause gastrointestinal bleeding and cardiovascular issues and should be used when medically intended (Cole et al., 2009).

2.5 FUTURE RESEARCH

2.5.1 Emerging Risk Factors

CRC is a major global health issue, with established risk factors such as age, family history, and lifestyle choices. However, new research is identifying additional risk factors crucial for early detection and prevention (Siegel et al., 2020).

2.5.1.1 Microbiome Dysbiosis

Alterations in gut microbiota, or dysbiosis, have been linked to CRC risk, with species like Fusobacterium nucleatum promoting inflammation and carcinogenesis (Yu et al., 2017).

2.5.1.2 Sedentary Lifestyle

Physical inactivity contributes to CRC risk through obesity and chronic inflammation (Boyle et al., 2012).

2.5.1.3 Environmental Exposures

Exposure to air pollution, heavy metals, and other pollutants may promote CRC through oxidative stress and DNA damage (Wu & Lui, 2022).

2.5.1.4 Dietary Factors

Diet plays a critical role in CRC risk; high red and processed meat intake and low fiber consumption are associated with increased risk (Bishop & Brandt, 2017). Certain food additives may also disrupt gut microbiota.

Understanding these emerging factors is essential for effective CRC prevention and intervention strategies.

2.5.2 Advancements in Epidemiological Methods

Advances in epidemiological methods enhance our understanding of CRC etiology and inform prevention and treatment strategies.

2.5.2.1 Large-scale Cohort Studies

Large cohort studies, like the Nurses' Health Study, provide insights into genetic, environmental, and lifestyle risk factors (Garcia-Albeniz et al., 2016).

2.5.2.2 Molecular Epidemiology

Molecular epidemiology explores genetic and epigenetic factors in CRC, identifying biomarkers and pathways involved in tumorigenesis (Huyghe et al., 2019).

2.5.2.3 Geospatial Analysis

Geospatial techniques analyze spatial patterns in CRC incidence, identifying high-risk areas for targeted interventions (Boscoe et al., 2016).

2.5.2.4 Biobanking and Omics Technologies

Biobanks and omics technologies enable detailed molecular analyses, helping to identify biomarkers and personalized treatment strategies (Yin et al., 2019).

2.5.2.5 Machine Learning and Artificial Intelligence

AI and machine learning can analyze complex datasets to predict CRC risk and improve screening accuracy (Liu et al., 2020).

These advancements in epidemiology will lead to a better understanding of CRC and improved prevention and treatment strategies.

2.6 CONCLUSION

In conclusion, CRC presents a growing global health challenge, driven by a complex interplay of genetic, environmental, and lifestyle factors. Key risk contributors include diet, physical inactivity, obesity, smoking, and alcohol, alongside socioeconomic and regional disparities. Early detection and lifestyle interventions remain vital for prevention. Advancements in research, from microbiome studies to machine learning, offer promising avenues for deeper insights. A multidisciplinary approach is essential to improve prevention, intervention, and global outcomes in the fight against CRC.

BIBLIOGRAPHY

Ahmed, R. L., et al. (2006). The metabolic syndrome and risk of incident colorectal cancer. *Cancer, 107*(1):28–36.

Akagi, Y., et al. (1998). Regulation of vascular endothelial growth factor expression in human colon cancer by insulin-like growth factor-I. *Cancer Res, 58*(17):4008–4014.

Allais, L., et al. (2016). Chronic cigarette smoke exposure induces microbial and inflammatory shifts and mucin changes in the murine gut. *Environ Microbiol, 18*(4):1352–1363.

American Cancer Society. (2022a). *Cancer prevention & early detection facts & figures 2022*. American Cancer Society. Available at: www.cancer.org/research/cancer-facts-statistics/all-cancer-facts-figures
American Cancer Society. (2022b). *Colorectal cancer facts & figures*. Available at: www.cancer.org/cancer/colon-rectal-cancer/about/key-statistics.html
Ancey, P.-B., et al. (2018). Glucose transporters in cancer - from tumor cells to the tumor microenvironment. *FEBS Journal, 285*(16):2926–2943.
André, F., et al. (1999). Integrins and E-cadherin cooperate with IGF-I to induce migration of epithelial colonic cells. *Int J Cancer, 83*(4):497–505.
Ansa, B. E., et al. (2018). Evaluation of colorectal cancer incidence trends in the United States (2000–2014). *J Clin Med, 7*(2):22.
Barone, B. B., et al. (2008). Long-term all-cause mortality in cancer patients with preexisting diabetes mellitus: A systematic review and meta-analysis. *J Am Med Assoc, 300*(23), 2754–2764.
Baxter, R. C., et al. (1980). Association between serum insulin, serum somatomedin and liver receptors for human growth hormone in streptozotocin diabetes. *Hormone Metabol Res, 12*(8), 377–381.
Bidoli, E., et al. (1992). Food consumption and cancer of the colon and rectum in northeastern Italy. *Int J Cancer, 50*:223–229.
Biedermann, L., et al. (2014). Smoking cessation alters intestinal microbiota: Insights from quantitative investigations on human fecal samples using fish. *Inflammatory Bowel Dis, 20*(8):1496–1501.
Biedermann, L., et al. (2013). Smoking cessation induces profound changes in the composition of the intestinal microbiota in humans. *PLoS ONE, 8*(11):e59260.
Bishop, K. S., Brandt, H. M. (2017). Colorectal cancer screening and prevention in women. *Women's Health, 13*(5):45–54.
Boscoe, F. P., et al. (2016). The relationship between area poverty rate and site-specific cancer incidence in the United States. *Cancer, 122*(12):3827–3836.
Botteri, E., et al. (2008). Smoking and colorectal cancer: A meta-analysis. *JAMA, 300*:2765–2778.
Boyle, T, et al. (2012). Physical activity and risks of proximal and distal colon cancers: A systematic review and meta-analysis. *J Nat Cancer Instit, 104*(20):1548–1561.
Brownlee, M. (2001). Biochemistry and molecular cell biology of diabetic complications. *Nature, 414*(6865):813–820.
Cancer Council Australia. (2021). *Guidelines: CRC/Colonoscopy surveillance/malignant polyps*. Available at: https://wiki.cancer.org.au/australia/Guidelines:Colorectal_cancer/Colonoscopy_surveillance/Malignant_polyps
Cole, B. F., et al. (2009). Aspirin for the chemoprevention of colorectal adenomas: Meta-analysis of the randomized trials. *J Nat Cancer Instit, 101*(4):256–266.
Dawson, S. I. (2004). Long-term risk of malignant neoplasm associated with gestational glucose intolerance. *Cancer, 100*(1):149–155.
Elamin, E. E., et al. (2013). Ethanol metabolism and its effects on the intestinal epithelial barrier. *Nutrition Rev, 71*(8):483–499.
Ferlay, J., et al. (2020). *Global Cancer Observatory: Cancer today*. Lyon, France: International Agency for Research on Cancer. Available from: https://gco.iarc.who.int/today
Flood, A., et al. (2003). Meat, fat, and their subtypes as risk factors for colorectal cancer in a prospective cohort of women. *Am J Epidemiol, 158*:59–68.
Gandini, S., et al. (2008). Tobacco smoking and cancer: A meta-analysis. *Int J Cancer, 122*:155–164.
Garcia-Albeniz, X., et al. (2016). Cigarette smoking and colorectal cancer risk by molecularly defined subtypes. *JNCI, 108*(4):djv364.
García-Pérez, J., et al. (2020). Residential proximity to industrial pollution sources and colorectal cancer risk: A multicase-control study (MCC-Spain). *Environ Int, 144*:106055.
Giouleme, O. et al. (2011). Is diabetes a causal agent for colorectal cancer? Pathophysiological and molecular mechanisms. *World J Gastroenterol, 17*(4):444–448.
Global Cancer Observatory (GLOBOCAN). (2020). International Agency for Research on Cancer (IARC). Available at: http://gco.iarc.fr/today
Haak, H. E., et al. (2020). Modern MR imaging technology in rectal cancer; there is more than meets the eye. *Frontiers Oncol, 10*:537532.
Hewitson, P., et al. (2007). Screening for colorectal cancer using the faecal occult blood test, Hemoccult. *Cochrane Database Syst Rev, 2007*(1):CD001216.

Hoxhaj, G., Manning, B. D. (2020). The PI3K-AKT network at the interface of oncogenic signalling and cancer metabolism. *Nature Rev Cancer*, *20*(2), 74–88.

Huang, Y.-C., et al. (2011). Diabetes mellitus negatively impacts survival of patients with colon cancer, particularly in stage II disease. *J Cancer Res Clin Oncol*, *137*(2), 211–220.

Huyghe, J. R., et al. (2019). Discovery of common and rare genetic risk variants for colorectal cancer. *Nature Genetics*, *51*(1):76–87.

Jackson, S. S. et al. (2022). Sex disparities in the incidence of 21 cancer types: Quantification of the contribution of risk factors. *Cancer*, *128*(19):3531–3540.

Jenkins, P. J., et al. (1999). Colorectal neoplasia in acromegaly. *Gut*, *44*(5):585–587.

Jenkins, P. J., et al. (2000). Insulin-like growth factor I and the development of colorectal neoplasia in acromegaly. *J Clin Endocrinol Metabol*, *85*(9):3218–3221.

Jeon, J. Y., et al. (2013). Impact of diabetes on oncologic outcome of colorectal cancer patients: Colon vs. rectal cancer. *PLoS ONE*, *8*(2):e55196.

Kadowaki, T, et al. (1996). Signal transduction mechanism of insulin and insulin-like growth factor-1. *Endocrine J*, *43*(Suppl):S33–S41.

Kim, M., et al. (2008). A tobacco-specific carcinogen, NNK, enhances AOM/DSS-induced colon carcinogenesis in male A/J mice. *In Vivo*, *22*:557–563.

Kuo, E., et al. (2020). A focused review on advances in risk stratification of malignant polyps. *Gastroenterol Res*, *13*(3):163–183.

Kwak, K., et al. (2019). Exposure to asbestos and the risk of colorectal cancer mortality: A systematic review and meta-analysis. *Occupat Environ Med*, *76*(11):861–871.

Labianca, R., et al. (2013). Early colon cancer: ESMO Clinical Practice Guidelines for diagnosis, treatment and follow-up. *Ann Oncol*, *24*(Suppl 6):i64–vi72.

Lee, G. H., et al. (2015). Is right-sided colon cancer different to left-sided colorectal cancer?: A systematic review. *Eur J Surg Oncol*, *41*(3):300–308.

Lieber, C. S., et al. (1986). Alcohol and cancer. *Hepatol*, *6*(5):1005–1019.

Lin, J. S., et al. (2021). Screening for colorectal cancer: Updated evidence report and systematic review for the US Preventive Services Task Force. *JAMA*, *325*(19):1978–1998.

Liu, H., et al. (2020). Machine learning in colonoscopy: An emerging frontier. *Eur J Gastroenterol Hepatol*, *32*(6), 632–639.

Ma, Y., et al. (2018). Type 2 diabetes and risk of colorectal cancer in two large U.S. prospective cohorts. *Br J Cancer*, *119*(11):1436–1442.

Matich, E. K., et al. (2021). Association between pesticide exposure and colorectal cancer risk and incidence: A systematic review. *Ecotoxicol Environ Safety*, *219*:112327.

Mills, K. T., et al. (2013). Diabetes mellitus and colorectal cancer prognosis: A meta-analysis. *Dis Colon Rectum*, *56*(11):1304–1319.

Müller, M. F., et al. (2016). Molecular pathological classification of CRC. *Virchows Archiv*, *469*(2), 125–134.

Murphy, N., et al (2019). Lifestyle and dietary environmental factors in CRC susceptibility. *Molec Aspects Med*, *69*:2–9.

National Comprehensive Cancer Network (2021). *Colon cancer (Version 2.2021)*. Available at: www.nccn.org/guidelines/guidelines-detail?category=1&id=1428

Nawa, T., et al. (2008). Differences between right-and left-sided colon cancer in patient characteristics, cancer morphology and histology. *Journal of Gastroenterology and Hepatology*, *23*(3), 418–423.

Norat, T., et al. (2002). Meat consumption and colorectal cancer risk: Dose-response meta-analysis of epidemiological studies. *Int J Cancer*, *98*:241–256.

Orme, S. M., et al. (1998). Mortality and cancer incidence in acromegaly: A retrospective cohort study. *J Clin Endocrinol Metab*, *83*(8), 2730–2734.

Pang, Y., et al. (2018). Diabetes, plasma glucose and incidence of colorectal cancer in Chinese adults: A prospective study of 0.5 million people. *J Epidemiol Community Health*, *72*(10):919–925.

Pritchett, N., et al. (2022). Exposure to outdoor particulate matter air pollution and risk of gastrointestinal cancers in adults: A systematic review and meta-analysis of epidemiologic evidence. *Environ Health Perspectives*, *130*(3):36001.

Probst-Hensch, N. M., et al. (1997). Meat preparation and colorectal adenomas in a large sigmoidoscopy-based case-control study in California (United States). *Cancer Causes Control*, *8*:175–183.

Prorok, P. C., et al. (2000). Design of the Prostate, Lung, Colorectal and Ovarian (PLCO) Cancer Screening Trial. *Controlled Clin Trials*, *21*(6 Suppl):273S–309S.

Rojas, A., et al. (2011). Diabetes and cancer: Looking at the multiligand/RAGE axis. *World J Diabet, 2*(7):108–113.
Rumgay, H., et al. (2021). Global burden of cancer in 2020 attributable to alcohol consumption: A population-based study. *The Lancet Oncol, 22*(7):1071–1080.
Saad El Din, K., et al. (2020). Trends in the epidemiology of young-onset colorectal cancer: a worldwide systematic review. *BMC Cancer, 20*: 1–14
Sandhu, M. S., et al. (2001). Systematic review of the prospective cohort studies on meat consumption and colorectal cancer risk: A meta-analytical approach. *Cancer Epidemiol Biomarkers Prevent, 10*:439–446.
Seitz, H. K., Stickel, F. (2007). Molecular mechanisms of alcohol-mediated carcinogenesis. *Nature Rev Cancer, 7*(7):599–612.
Shaukat, A., et al. (2020). Endoscopic recognition and management strategies for malignant colorectal polyps: Recommendations of the US Multi-Society Task Force on CRC. *Gastroenterol, 159*(6):1916–1934.
Siegel, R. L., et al. (2020). Colorectal cancer statistics, 2020. *CA, 70*(3):145–164.
Sinha, R., et al. (1999). Well-done, grilled red meat increases the risk of colorectal adenomas. *Cancer Res, 59*:4320–4324.
Smith, R. A., et al. (2009). Cancer screening in the United States, 2009: A review of current American Cancer Society guidelines and issues in cancer screening. *CA, 59*(1):27–41.
Soos, M. A., et al. (1990). Receptors for insulin and insulin-like growth factor-I can form hybrid dimers: Characterization of hybrid receptors in transfected cells. *Biochem J, 270*(2):383–390.
Stein, K. B., et al. (2010). Colorectal cancer outcomes, recurrence, and complications in persons with and without diabetes mellitus: A systematic review and meta-analysis. *Digestive Dis Sci, 55*(7):1839–1851.
Su, W. C., et al. (2021). Correlations between urinary monoethylhexyl phthalate concentration in healthy individuals, individuals with colorectal adenomas, and individuals with colorectal cancer. *J Agricult Food Chem, 69*(25):7127–7136.
Teo, N. Z., et al. (2020). Management of malignant colonic polyps. *J Gastrointest Oncol, 11*(3):469–474.
Tran, T. T., et al. (2006). Hyperinsulinemia, but not other factors associated with insulin resistance, acutely enhances colorectal epithelial proliferation in vivo. *Endocrinol, 147*(4):1830–1837.
Wackernah, R. C., et al. (2014). Alcohol use disorder: Pathophysiology, effects, and pharmacologic options for treatment. *Substance Abuse Rehabilitation, 5*:1–12.
Warburg, O. (1956). On the origin of cancer cells. *Science, 123*(3191):309–314.
Warren, R. S., et al. (1996). Induction of vascular endothelial growth factor by insulin-like growth factor 1 in colorectal carcinoma. *J Biolog Chem, 271*(46):29483–29488.
White, A., et al (2018). A review of sex-related differences in colorectal cancer incidence, screening uptake, routes to diagnosis, cancer stage, and survival in the UK. *BMC Cancer, 18*(1):906.
Winawer, S. J., et al. (1993). Prevention of colorectal cancer by colonoscopic polypectomy. *New Engl J Med, 329*(27):1977–1981.
Wolf, A. M. D., et al. (2018). Colorectal cancer screening for average-risk adults: 2018 guideline update from the American Cancer Society. *CA, 68*(4):250–281.
Wong, S. H., et al. (2017). Gavage of fecal samples from patients with colorectal cancer promotes intestinal carcinogenesis in germ-free and conventional mice. *Gastroenterol, 153*(6):1621–1633.
World Cancer Research Fund International. (2021). *Colorectal cancer statistics*. Available at: www.wcrf.org/dietandcancer/cancer-trends/colorectal-cancer-statistics
World Cancer Research Fund/American Institute for Cancer Research. (2018). *Diet, nutrition, physical activity and CRC*. Available at: www.wcrf.org/dietandcancer/colorectal-cancer/
World Health Organization. (2020). *Colorectal cancer*. Available at: www.who.int/news-room/fact-sheets/detail/colorectal-cancer
Wu, C. W., Lui, R. N. (2022). Early-onset colorectal cancer: Current insights and future directions. *World J Gastrointest Oncol, 14*(1):230–241.
Wu, X., et al. (1995). Apoptosis induced by an anti-epidermal growth factor receptor monoclonal antibody in a human colorectal carcinoma cell line and its delay by insulin. *J Clinl Investig, 95*(4):1897–1905.
Xi, Y., Xu, P. (2021). Global CRC burden in 2020 and projections to 2040. *Translational Oncol, 14*(10):101174.
Yang, H., et al. (2020). Triclocarban exposure exaggerates colitis and colon tumorigenesis: Roles of gut microbiota involved. *Gut Microbes, 12*(1):1690364.

Yin, J., et al. (2019). Prospective validation of serum exosomal microRNA-375 as a diagnostic and prognostic biomarker in colorectal cancer. *Cancer Biomarkers, 26*(2):491–504.

Yu, H., Rohan, T. (2000). Role of the insulin-like growth factor family in cancer development and progression. *J Nat Cancer Instit, 92*(18):1472–1489.

Yu, J., et al. (2017). Metagenomic analysis of faecal microbiome as a tool towards targeted non-invasive biomarkers for colorectal cancer. *Gut, 66*(1):70–78.

3 Molecular Pathogenesis of Colorectal Cancer

Rizwan Ahamad, Sadma Parveen, Abu Safana Biswas and Ayman A. M. Alameen

3.1 INTRODUCTION

Colorectal cancer (CRC) is a notable health problem in both industrialized and emerging nations, exhibiting the third highest rate of occurrence among all disorders associated with tumor formation (Issa & Noureddine, 2017). This cancer is a prominent contributor to both illness and death rates in Western countries. A minority of CRC occurrences are attributed to fundamental genetic defects, whereas many instances are associated with particular environmental variables and advancing age (Jeon et al., 2018). If left untreated, the condition has the potential to persist and worsen. The pathogen disseminates to the mucosal lining of the gastrointestinal tract and the subcutaneous muscular layer. The term "environmental" refers to matters related to the natural world and its various components. Furthermore, the combination of hereditary factors might manifest in several ways, hence expediting the progression of cancer. Furthermore, this exacerbates the challenges associated with treatment. Usually, individuals manifest symptoms after the age of 60, accompanied by the presence of precursor lesions known as adenomas.

The condition has the potential to progress into a malignant neoplasm over one to two decades (Carbone et al., 2020). The execution of cancer screening protocols among those 50 years old and above has shown a reduction in mortality rates associated with the illness, as well as an enhanced ability to diagnose cancer. In the first phase, it is important to mitigate the occurrence of the ailment. Four significant ideas have been proposed on the origin of CRC. To start, the identical epigenetic and genetic alterations implicated in the pathogenesis of colon cancer concurrently increase susceptibility to the development of colorectal malignancies. Furthermore, the progression of cancer occurs sequentially, starting with a precancerous stage. The disorder has the potential to evolve into a malignant state. Furthermore, the development of cancer requires a deficiency in the notion of genetic stability relating to the maintenance of the genetic information within an organism or population across time without significant alterations or mutations. Furthermore, within the context of genetic cancer syndromes, it is crucial to consider the germline manifestations that have substantial significance. Genetic aberrations are often linked to the pathogenesis of infrequent forms of colorectal malignancies. The conditions arise as a result of somatic manifestations of genetic disorders (Ishaque et al., 2018).

Epigenetics refers to the phenomenon of heritable alterations in gene regulation that do not entail irreversible changes to the underlying sequence of DNA. Alterations in DNA sequence exert a substantial influence on the genesis of many cancers. This encompasses colorectal cancer as well. Recent studies in the field of epigenetics have uncovered a previously unidentified correlation between certain gene expression levels linked to CRC and the absence of hereditary factors. The discovery of this missing connection has occurred throughout the last two decades (Guo et al., 2019). One such instance is microsatellite instability (MSI), a characteristic feature of a certain molecular subgroup. The occurrence of CRC may be attributed to the failure of DNA mismatch repair (MMR) mechanisms to effectively rectify DNA mismatches (Ganesh et al., 2019). A genetic mutation has the potential to induce a malfunction in any of the MMR genes, leading to the occurrence of hypermethylation, specifically in the MLH1 gene. In general, MSI tumours have a better prognosis than sporadic tumours. An established correlation exists between chromosome instability (CIN) and CRC. The phenomenon of worldwide hypomethylation has been documented in previous studies (Mármol et al., 2017). Additional evidence corroborates the existence of cancer-related pathways at MicroRNAs (miRNAs) that exert an impact on post-transcriptional processes, hence fulfilling a regulatory function. As for the relationships between miRNAs, the two networks did not properly match, indicating that there is room for developing plausible new functional relationships between miRNAs involved in CRC development (Mazza et al., 2016). The term "cell" refers to the basic structural and functional unit of all living organisms. MiR-143 hinders the development process by down-regulating in CRC, as it specifically targets a gene. The KRAS mRNA transcript was mentioned in a previous study (Huang et al., 2017). In addition to enhancing our understanding of the etiology of CRC, these findings have furthermore presented novel prospects for the identification of disease biomarkers and the focus of therapy. There are several CRC biomarkers such as BRAF mutations, NRAS, and KRAS, microsatellite instability (MSI), DNA mismatch repair (MMR) status, and CpG island methylation which have been studied.

DOI: 10.1201/9781032704517-3

The phenomenon of island methylation has been the subject of scientific investigation. AnNRAS or KRAS mutation is related to an unfavorable prognosis, along with the administration of anti-epidermal growth factor receptor (anti-EGFR) antibody treatment. According to Bupathi and Wu (2016), mutation in BRAF V600E is seen in 8–10% of individuals with CRC. These individuals have a more severe pathological condition and a less favorable outlook in the adjuvant and the topic at hand pertains to the metastatic situation. Individuals that possess the BRAF mutation, exhibited microsatellite instability-high (MSI-H). Individuals with tumors had a more favorable overall outcome in comparison to those with a mutation in BRAF and MSI. The identification of individuals who may get advantages from immunotherapy is progressively gaining significance based on the research done on CRC (Mauri et al., 2021).

3.2 EPIDEMIOLOGY

The incidence of CRC has seen a significant increase in recent years on a global scale. In the year 2020, around 1.93 million newly diagnosed cases of CRC were reported worldwide. Additionally, CRC-related causes led to the deaths of approximately 0.94 million individuals. These figures accounted for 10% of the overall global tumor incidence, which amounted to 19.29 million fresh cases, and 9.4% of the total fatalities caused by cancer, which amounted to 9.96 million deaths. According to projections, CRC is anticipated to rank as the third major cause of cancer-related mortality universally in 2020. Australia/New Zealand (38.2 per 100 000), Western Europe (31.4 per 100 000) and Southern Europe (31.1 per 100 000) have the highest incidence rates, whereas South-Central Asia (6.1 per 100 000), Middle Africa (4.8 per 100 000) and Western Africa (4.1 per 100 000) have the lowest incidence rates. In most parts of the world, men have higher incidence rates than women. CRC is the fourth leading cause of cancer death in males and the third in females worldwide (Yu & Hemminki 2020). Globally, the prevalence of CRC is estimated to be around 5.25 million persons, indicating its status as the second-most prevalent illness behind breast tumor, which sees over 7.79 million new instances annually. According to the GLOBOCAN 2020 database, there were an estimated 0.7 million rectal cancer cases, 1.15 million fresh cases of colon cancer, and 50,000 anal cancer cases identified in the year 2020. The aforementioned statistics are projected to increase to 1.92 million, 1.16 million, and 78,000 by the year 2040. CRC rates vary across various countries (Siegel et al., 2020). The nations with the highest age-standardized incidence rates in 2020 were Slovakia, Hungary, Norway, Denmark, and the Netherlands, with rates of 43.9, 45.3, 41.9, 40.9 and 41.0, incidents per 100,000, persons, respectively. In 2020, Gambia, Guinea, Burkina Faso, Bhutan, and Bangladesh are projected to exhibit the minimum age-standardized occurrence rates, with rates of 3.7, 3.3, 3.8, 3.8, and 3.8 cases per 100,000 people, respectively.

3.3 PATHOGENESIS OF COLORECTAL CANCER

The adenoma-carcinoma pathway involves a progressive occurrence of genetic changes in oncogenes and tumor suppressor genes, starting with the normal epithelium and progressing to the dysplastic epithelium. This phenomenon leads to the emergence of CRC. The accumulation of genomic mutations, including both epigenetic and genetic changes, results in the conversion of typical glandular epithelial cells into adenocarcinoma (Müller et al., 2016). The molecular processes behind the persistent growth and dissemination of cancer in the body are still not fully understood. In fact, many genetic paths have been suggested to elucidate the causes of CRC (Figure 3.1). CRC is influenced by various molecular routes, including the three primary routes: CIN, MSI, and CpG island methylator phenotype (CIMP). These routes are pivotal in the progression of CRC since they are linked to genetic and epigenetic alterations (Table 3.1). This study has identified three molecular pathways that are indicative of cancer progression. These pathways include MSI, which is linked with the hypermutated group—chromosomal instability, which is exclusively observed in the non-hypermutated category—and the CpG Island Methylator Phenotype (CIMP), which is linked with both the hypermutated and non-hypermutated groups (Aghagolzadeh & Radpour, 2016). The molecular pathways involved in cancer growth and metastasis may play a crucial role in determining the time it takes for these processes to occur. This is due to variations in epidemiological factors, mutational trials, and immunological responses along different routes. The approaches used for the treatment of a disease may also vary. The biomarker "EMAST" is characterized by a higher frequency of microsatellite mutations at certain tetranucleotide repeats. These modifications have been shown to have a prognostic importance for patient survival by affecting three distinct biological pathways. Individuals diagnosed with sporadic CRC lack the ability to anticipate the development of metachronous adenoma or cancer through previously recognized pathways (Carethers, 2017; Langner, 2014). The importance of clinical

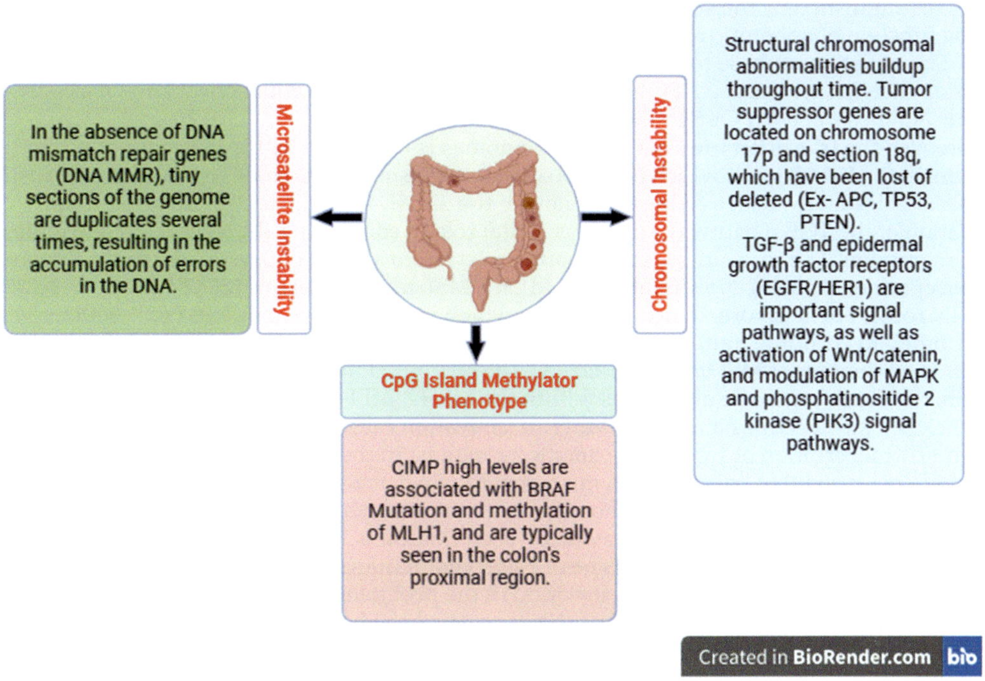

Figure 3.1 The development of CRC involving several genetic mechanisms.

Table 3.1: A comprehensive examination of the CIN, MSI-H, CIMP, and EMAST indices in colorectal cancer

Patho-physiological pathways	Inflammation	Genomic instability	Prognosis	Pathogenesis
CIMP	Without hypermethylation of hMLH1: varying	Primary hypermethylation at certain DNA locations	Low survival in the absence of hMLH1 hypermethylation	Without hypermethylation of hMLH1:unknown
CIN	The degree of differentiation varies at the tumor margin, lamina propria, and intraepithelial sites	Causes of copy number variation, mutation, and aneuploidy (MSS)	Referent	Genetic alterations that result in the loss of heterozygosity
EMAST	This illness has been connected to tumor nests around epithelial components	Most often at MSS and MSI-L, which includes MSI-H, instability was seen.	Later stage; poor survival	Combined with a frameshift mutation in a target gene, chromosomal instability
MSI-H	Similar to Crohn's disease around tumors	Diploid and MSI have been observed	Early stage; Better Survival	BRAFV600E: target gene frameshift mutation

surveillance cannot be overstated when it comes to identifying and eradicating metachronous lesions in these individuals.

The biological characteristics, prognosis, and treatment outcomes of CRC exhibit significant variability. CRC has traditionally been categorized into three distinct categories, including CIN, MSI, and CIMP, which correspond to the three pathophysiological mechanisms of carcinogenesis. This classification is based on a molecular viewpoint (Fischer et al., 2019). The characteristic that sets apart the majority of CRCs is known as CIN, which accounts for around 80–85% of the cases.

CIN is distinguished by the stimulation of growth-promoting routes while concurrently suppressing the function of apoptotic routes, with the second type being higher prevalent (Raskov et al., 2020). The initiation of these tumors occurs when adenomatous polyps develop because of a mutation that deactivates the APC gene, which produces the WNT route effect known as adenomatous polyposis coli (APC). See Table 3.1.

Moreover, the transition towards adenocarcinomas is facilitated by the existence of activating variants in the KRAS gene, which is responsible for transmitting signals from receptor tyrosine kinases. Additionally, deactivating mutations in the SMAD4 gene, which is involved in the regulation of the TGF-β pathway and plays a vital role in controlling the cell cycle, contribute to this progression. The route in question is often referred to as the APC route due to the first event (Koveitypour et al., 2019). No alterations of MLH1, MSI, methylation, or BRAF have been seen in the CIN route. Furthermore, Lynch syndrome, which is the most general kind of hereditary CRC, is attributed to germline mutations in the MMR genes, involving MSH2, MSH6, MLH1, and PMS2. The initiation of MSI is hypothesized to be attributed to a deficiency in the DNA MMR system, resulting in genetic alterations within the MSH2, MLH1, and MSH6 MMR genes. MMR insufficiency may occur as a result of MLH1 deactivation caused by biallelic hyper-methylation of the MLH1 transcription or by double somatic alterations in the MMR genes (Boland et al., 2018). The study observed that patients with mismatch repair-deficient and MSI-high colorectal cancer (dMMR-MSI-H) had a favorable prognosis but did not experience therapeutic advantages from 5-fluorouracil. These patients have previously been classified according to their MSI status (Watson, 2020). CIMPs, a specific subtype of colorectal malignancies, are characterized by a significant level of CpG site hyper-methylation in the promoter region of numerous tumor-inhibiting genes. The identification of CIMPs is mostly based on epigenetic markers, as opposed to the traditional classification methods of CIN or MSI. The presence of CIMP in patients with colorectal carcinogenesis's serrated pathway is closely linked to hypermethylation of the MLH1 promoter. Additionally, CIMP is associated with the female gender, advanced age, and weak histology, and poses a substantial risk in individuals diagnosed with the illness (Mehta et al., 2017). Nevertheless, the absence of a well-said description of CIMP-high poses challenges in applying this pathophysiological characteristic to treatment strategies. The CIMP testing procedure typically focuses on analyzing MLH1, MINT1, CACNA1G, and CDKN2A genes, although as many as 16 distinct genes may be identified. However, there is a lack of consensus regarding the specific cut-off rates that should be used to differentiate between high CIMP and CIMP+ status. Consequently, certain scientific protocols must be used to accurately assess CIMP tests (Mitsuhashi et al., 2015).

Gene changes have a significant influence on the initiation and progression of tumors, even in the early stages of polyp formation in the cancer genesis sequence. The aberrant crypt goal is the primary histological lesion associated with the development of CRC (Szylberg et al., 2015). The APC gene, which functions as a tumor suppressor, is often subject to alterations in colorectal malignancies. Additionally, dysplastic aberrant crypt foci have the potential to disseminate mutations. The activation of the Wingless/Wnt pathway occurs via the inhibition of APC, which serves as a systematic mechanism for initiating the process of polyp cancer growth. The Wnt signaling route facilitates the polyp cells' clonal proliferation, leading to the development of cancer as a result of later mutations in genes such as KRAS or TP53. The cell signaling pathway regulated by transforming growth factor b (TGFB1) may also be influenced by mutations in KRAS and TP53, hence contributing to an increased rate of CRC progression (Novellasdemunt et al., 2015).

Colon cancer progression is characterized by genetic and epigenetic changes that result in the activation of cancer-causing genes and the deactivation of genes that prevent tumor growth. B-Raf transmits signals to EGFR, which in turn activates KRAS, a proto-oncogene responsible for the initiation of the MAPK pathway (Amodio et al., 2020). More than 50% of CRCs have alterations in KRAS or B-Raf, which activate the MAPK signaling pathway, leading to enhanced cell proliferation and inhibition of apoptosis. The PI3K, TGFB1-SMAD, and WNT-APC-CTNNB1, signaling routes are also implicated in CRC. Additionally, the RAS-RAF-MAPK routes are included under this framework (Ahronian et al., 2015).

The examination of epigenetic modifications in cancer has resulted in the discovery of often mutated genes within these pathways, prompting extensive explorations into the biology of CRCs. The results of this study are being used to create novel clinical and prognostic tests for CRC, as well as to explore new therapy options.

3.3.1 Familial Adenomatous Polyposis (FAP)

3.3.1.1 Clinical Features

Familial adenomatous polyposis (FAP) is a hereditary disorder characterized by the presence of many adenomatous polyps. It follows an autosomal dominant pattern of inheritance and has an estimated prevalence of around 1 in 7000 persons. The disease penetrance for polyps is almost 100%, and the majority of individuals affected will progress to develop cancer in the absence of any intervention. In the traditional manifestations of FAP, a substantial number of adenomatous polyps emerge inside the large intestine throughout the 2nd and 3rd centuries of an individual's life. The majority of these people experience the development of benign adenomas, while a subset of these individuals will gradually collect more mutations. The knockout of the p53 gene plays a crucial role in facilitating the transformation of benign lesions into malignant ones (C. R. Boland et al., 1995). Furthermore, individuals diagnosed with FAP may have various extracolonic symptoms as the illness progresses. These manifestations may include ocular lesions, osteomas, polyps in the duodenum or ampulla, desmoid tumors, or brain tumors. While the majority of individuals examined with FAP tend to acquire periampullary adenomas, it is important to note that only a small percentage, around 5–10%, will ultimately advance to malignancy within this specific anatomical area. Desmoid tumors are classified as benign neoplasms; however, they possess the potential to cause mortality. The aforementioned manifestations are seen in around 10–15% of individuals diagnosed with FAP. They exhibit a higher prevalence within certain familial groups and are the primary nonmalignant factors contributing to both morbidity and death in FAP patients.

3.3.2 Lynch Syndrome or Hereditary Nonpolyposis Colorectal Cancer (HNPCC)

3.3.2.1 Clinical Features

Lynch syndrome is an inherited genetic condition that is passed down in an autosomal dominant manner. The condition was defined by the premature development of CRC and the occurrence of other malignant tumors in other relatives, not exclusive to the colon. Notably, this syndrome is not associated with an elevated percentage of colorectal polyps. More than two thirds of colorectal malignancies manifest in the proximal part of colon, with a mean age of diagnosis in the early 40s. This average age aligns with that seen in individuals with FAP. The occurrence of synchronous and metachronous tumor growth, particularly in cases of CRC located in the proximal colon, indicates the potential presence of Lynch syndrome. In addition to their spatial distribution, tumors within the spectrum of Lynch syndrome exhibit certain pathological traits, including lymphocyte infiltration, mucinous properties, and limited differentiation. In Lynch syndrome, the occurrence of colonic neoplasms often arises without prior identification of preceding adenomas. This phenomenon is believed to occur due to an accelerated adenoma-carcinoma sequence within the context of this particular illness.

Hence, there exists a notable distinction between Lynch syndrome and FAP in terms of the quantity and distribution of polyps. In addition to their increased susceptibility to colon and rectal malignancies, persons diagnosed with Lynch syndrome often develop cancers in extracolonic locations, including endometrium, stomach, ovary, pancreas, urinary tract, small intestine, and brain.

3.3.3 Sporadic Colorectal Cancer

Nonhereditary CRC is a prevalent kind of malignancy that is seen globally. The defining feature of this phenomenon is the gradual advancement from healthy colonic epithelium to the development of malignancy, which is accompanied by a series of successive molecular anomalies occurring at each stage (Kinzler & Vogelstein, 1996). During the early 1990s, it was shown in a sophisticated manner that the development of CRCs is a result of the progressive accumulation of oncogenes that have been activated and tumor-suppressor genes that have been inactivated (Fearon & Vogelstein, 1990). Subsequent research on the subject has shown that the production of essential mutations involves several mechanisms, and distinct sets of target genes play a role in mediating various biological activities seen in tumor cells.

3.4 BIOMARKERS BASED ON EPIGENETIC CHANGES FOR CRC

As CRC advances beyond the first stage of adenomas or serrated lesions, it undergoes a series of epigenetic and genetic changes that accumulate gradually within the lesions. These changes result in the development of more dysplastic features, ultimately leading to the formation of an adenocarcinoma (De Palma et al., 2019). In order to decrease the likelihood of developing this ailment, it is essential to identify precancerous conditions. The occurrence of lesions and the development of early-onset CRC in individuals with average risk and no other known predisposing factors. Maintaining good health via regular screening. Regarding the topic of CRC screening, colonoscopy is a commonly used method. The gold standard is very reliable and effective since it can identify and eliminate precursor lesions (US Preventive Services Task Force, 2016). In the context of the present discussion, several factors contribute to individuals opting against colonoscopies. These experiences were marked by feelings of unease, significant costs, and a lack of effectiveness. The rates of compliance may have an impact on the occurrence of complications, such as bleeding and perforation for precursor lesions such as adenomas, non-invasive test approaches such as the fecal immunochemical test (FIT) and fecal occult blood test (FOBT) are extensively used in Europe (Mousavinezhad et al., 2016).

The tumor-node-metastasis (TNM) system is now utilized during the intermediate phases of CRC staging. The current TNM classification technique is deemed insufficient in its ability to accurately predict prognosis and guide clinical decision-making factors should be closely monitored and receive appropriate interventions to mitigate the adverse outcomes(Li et al., 2020). Biomarkers are used to identify individuals who may get therapeutic benefits from immunotherapy, chemotherapy, or targeted interventions. There is a pressing demand for biomarkers in the treatment field to assist in identifying patients who need immediate attention. The inclusion of this element might be seen as superfluous and lacking in need. Epigenetic biomarkers, such as DNA methylation and histone changes, are well-recognized in the field. Long noncoding RNA (lncRNA) and miRNAs have shown potential as therapeutically relevant entities. Biomarkers identification for the prognosis, diagnosis, and prediction of therapeutic response in CRC has been extensively studied, with a particular focus on lncRNAs (Nikolouzakis et al., 2018).

Blood or other bodily fluids and cells may include measurable markers of diseased conditions. Physiological process, illnesses, or disorders, such as molecular markers, are included within the scope of this study. Biomarkers are measurable indicators that can be used to evaluate biological processes or conditions. Molecular markers can facilitate the identification of various disorders or diseases. Biomarkers are measurable indicators that are used to assess biological processes, and disease states, or are crucial screening, prognostic, and therapeutic instruments used in the identification, assessment, and treatment of cancer (Figure 3.2). Furthermore, it can precisely

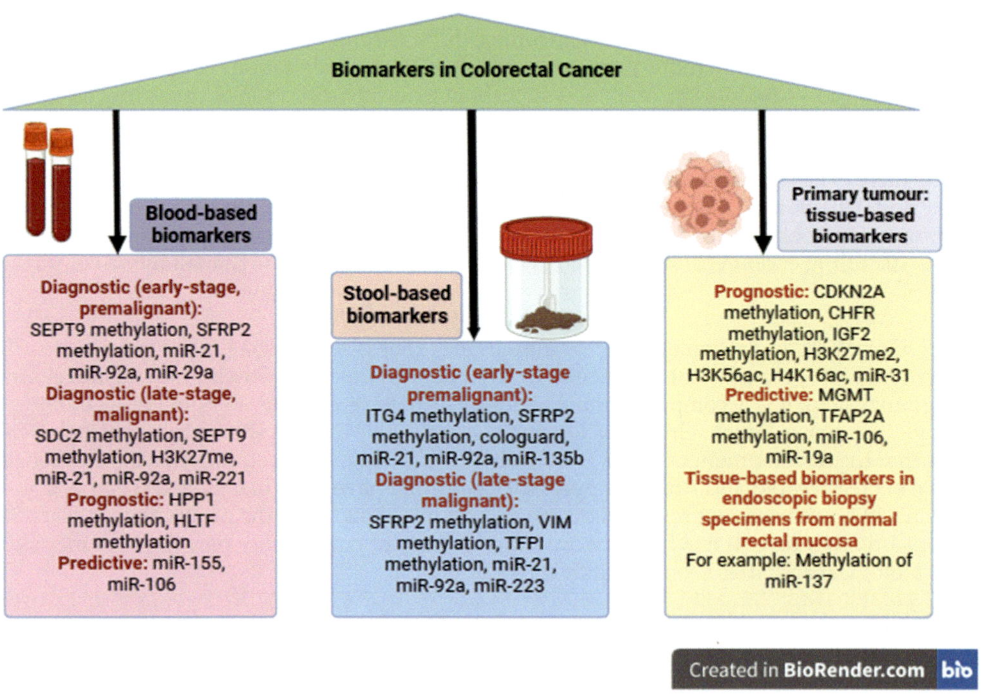

Figure 3.2 Biomarkers based on epigenetic changes for CRC.

determine the specific geographical coordinates of a given site. The aim of this investigation is to detect and characterize tumors, as well as to identify regions that exhibit advanced stages of growth or are susceptible to further development. The treatment in question has been previously mentioned (Ortiz-Quintero, 2016). These molecular markers serve as indicators of the underlying processes involved in the development and progression of neoplastic conditions. The occurrence of cell exfoliation and mucus release in CRC was impacted by the existence of anomalous glycoproteins, which may be observed through reflection. The present inquiry pertains to the elucidation of the underlying mechanisms governing the processes of cell exfoliation and mucus production inside cellular systems. Soon, researchers want to discover innovative, noninvasive molecular markers, such as RNA, DNA, or other protein-based biomarkers, that may be used for the purpose of detection. CRC may be detected in several bodily fluids, including blood, feces, and other fluid specimens. Various prognostic, accurate, and diagnostic indicators have been created with the purpose of determining CRC (Erstad et al., 2015). The term "diagnostic" refers to the process of identifying or determining the nature or cause biomarkers serve as indicators of the probability of disease progression. The implementation of therapeutic interventions for patients. The use of these molecular indicators enables the prediction of the phenomenon. Predictive biomarkers are used to predict the effectiveness of therapy in individuals diagnosed with cancer. Colorectal polyps are capable of early detection, which may be achieved using the following biomarkers.

3.4.1 Molecular Markers for Diagnosis

3.4.1.1 HNPCC (Hereditary Nonpolyposis CRC)

Hereditary Nonpolyposis Colorectal Cancer (HNPCC) is widely recognized as the preeminent autosomal dominant genetic disorder associated with colon cancer. The screening test serves the purpose of early detection of CRC. HNPCC is differentiated by the existence of a hereditary alteration in one of the four DNA mismatch repair genes, specifically PMS2, MLH1, MSH2, and MSH6, which are all naturally found in the human body. A significant proportion of young individuals are diagnosed with a novel genetic disease with a familial history of hereditary colon cancer.

3.4.1.2 Telomerase

Telomerase is an essential enzyme responsible for maintaining telomeres, which are repetitive DNA sequences (TTAGGG/AATCCC in human telomeres). These telomeres play a vital role in safeguarding against chromosomal deterioration, loss of functional genes, and cellular demise. The preservation of telomeres necessitates the presence of telomerase. The stimulation of telomerase serves as a clinical and prognostic marker for malignant tumors, capable of being used as an independent indicator.

3.4.1.3 Insulin-Like Growth Factor Binding Protein 2 (IGFBP2)

Based on empirical evidence, it has been shown that persons diagnosed with CRC have elevated levels of IGFBP2 in their blood and plasma, in comparison to those who are in good condition. The use of this tool has been recognized as a means of diagnosing and assessing CRC since it is associated with the processes of cancer cell growth, movement, and infiltration.

3.4.1.4 Pyruvate Kinase M2 (PKM2)

PKM2 is an enzymatic protein that is found in the cytoplasm of both normal and cancerous cells, and it serves a significant role in the regulation of energy metabolism. PKM2 expression is upregulated in CRC and other gastrointestinal malignancies. The relevance of colorectal cancer inquiry and/or identification using blood or stool specimens as a predictive indication of poor outcomes should also be considered.

3.4.2 4Prognostic Molecular Markers

3.4.2.1 p53

The presence of p53 tumor suppressor protein is also considered a self-sufficient prognostic indicator for colon tumors. The modification of the p53 tumor-inhibiting gene is a crucial and primary event in the development of cancer linked with ulcerative colitis, also known as TP53. The occurrence of a tumor in an adenoma is accompanied by a p53 mutation and an absence of heterozygosity in wild-type variants, providing more evidence of its role in the control of malignancy (Liebl & Hofmann, 2021). Patients diagnosed with CRC who have a deficiency in this particular gene are associated with a worse prognosis.

3.4.2.2 18 q Loss of Heterozygosity (LOH)

Individuals diagnosed with colorectal cancer with a chromosomal 18q loss of heterozygosity experience a diminished likelihood of long-term survival. The prognosis for patients diagnosed with either phase II or phase III colon cancer is more unfavorable if they have somatic heterozygosity loss, compared to those whose tumors maintain both alleles inherited from their genes on chromosome 18 q. Individuals diagnosed with phase II or phase III colon cancer who exhibit an 18qLOH (loss of heterozygosity) genetic alteration tend to have a less favorable prognosis compared to those who possess normal alleles on chromosome 18q (Jernvall et al., 1999).

3.4.2.3 MLH1 Methylation

The detection of DNA MSI or the fall short of MLH1 protein expression by immunohistochemical research serves as confirmation of somatic MLH1 deactivation in core colorectal cancers. This phenomenon is seen more often in initial CRCs compared to advanced stages of the disease. The inactivation of this condition, in lack of adjuvant therapy, may lead to a less severe sickness or a more favorable prognosis.

3.4.2.4 Vascular Endothelial Growth Factor (VEGF)

VEGF is identified as an angiogenic factor in CRC, with its expression seen in around 50% of CRC cases. In contrast, VEGF is shown to be present in far lower quantities in normal colonic mucosa and adenomas. Consequently, the expression of VEGF-1 seems to provide valuable prognostic data in patients with CRC.

3.4.3 Predictive Molecular Markers

3.4.3.1 KRAS

The KRAS mutation identification is used as the primary prospective prognostic indicator in anti-EGFR antibody-based treatment for CRC, including panitumumab and cetuximab.

3.4.3.2 BRaf V600E

The somatic mutation-induced stimulation of this kinase results in an uncontrolled MAPK signaling pathway in the human body. Individuals with stage IV CRC who possess the BRAF V600E operating mutation exhibit resistance to therapy with EGFR inhibitors. The predictive value of BRAF mutation expression as a gene marker is still questionable. However, the role of BRAF mutants as an indicator of resistance to anti-EGFR antibodies has been widely investigated and shows promising predictive potential for BRAF changes.

3.4.3.3 PIK3CA Status

The presence of the mutation of PIK3CA results in an aberrant activation of the AKT process, hence facilitating the proliferation of cancerous cells. The chronic activation of the pathway involving PI3K and AKT is expected to have a significant influence on the course of CRC. In preclinical tests, it was shown that colon cancer cells with a generated PIK3CA mutant exhibited greater sensitivity to cetuximab compared to cells with PIK3CA wild-type.

3.4.3.4 ERCC-1

The repair of proteins involved in DNA excision is a crucial process in maintaining genomic integrity. The ERCC-1 gene is a crucial component of the nucleotide excision repair pathway, playing a significant role in the process of DNA repair. Recent research has investigated the relationship between dissemination and phase II or III colorectal cancer, specifically focusing on the potential of ERCC-1 as a predictive marker in response to oxaliplatin therapy. This study examined the abnormalities of ERCC-1 regulation during oxaliplatin treatment, aiming to determine its suitability as a predictive biomarker (Moosavi et al., 2019; Yao et al., 2018; Masuda et al., 2017; Zarkavelis et al., 2017).

3.5 COLORECTAL CANCER DIAGNOSTIC TOOL

Consuming a diet that is abundant in vegetables, fruits, and whole-grain cereal products, along with consistent physical activity, reduces the likelihood of getting CRC. The risk of environmental variables has been linked to CRC, including obesity, smoking, and excessive alcohol use. Supplementation with calcium and vitamin D3 may potentially exhibit therapeutic benefits in combating diseases. Furthermore, research has shown that the prolonged use of acetylsalicylic acid (NSAIDs) is linked to a decrease in the likelihood of developing cancer (Florensa et al., 2023). CRC is among the several types of malignancies that may be mitigated via smoking cessation or refraining from smoking. Screening procedures (secondary prevention) may be used to identify adenomas for primary prevention and malignancies for early diagnosis.

This is the reason why routine screenings in those at high risk may aid in the identification of CRC, even in the absence of symptoms. Following the identification of colon cancer, testing instruments are further valuable in facilitating the TNM staging process, which is used to ascertain the most suitable course of therapy. Several diagnostic procedures are available for the detection and evaluation of various medical conditions. These procedures include sigmoidoscopy, colonoscopy, magnetic resonance imaging (MRI), stool testing, computed tomography colonography (CTCG), transrectal ultrasonography, double-contrast barium enema, and several other methods. Nevertheless, it is important to acknowledge that many methodologies include inherent limits and so lack accuracy when it comes to effectively diagnosing diseases at their first stages. There exists a significant need to enhance the development of diagnostic techniques that are more sophisticated, with the aim of augmenting survival rates, mitigating complications arising from delayed diagnoses, and accurately determining the stage of the illness to facilitate the prescription of suitable therapeutic interventions. Presented in Figure 3.3 are some frequently used diagnostic methods, accompanied by their respective merits and drawbacks (Vinchhi & Patel, 2021; Ladabaum et al., 2020).

3.5.1 CT Colonography (CTC)

The words "virtual colonoscopy" and "CT colonography" are synonymous when discussing a computerized x-ray imaging process that generates two- and three-dimensional pictures of the colon. CT colonography may potentially serve as a viable substitute for colonoscopy in individuals who present with obstructive or stenotic tumors. Following the full evacuation of the colon during

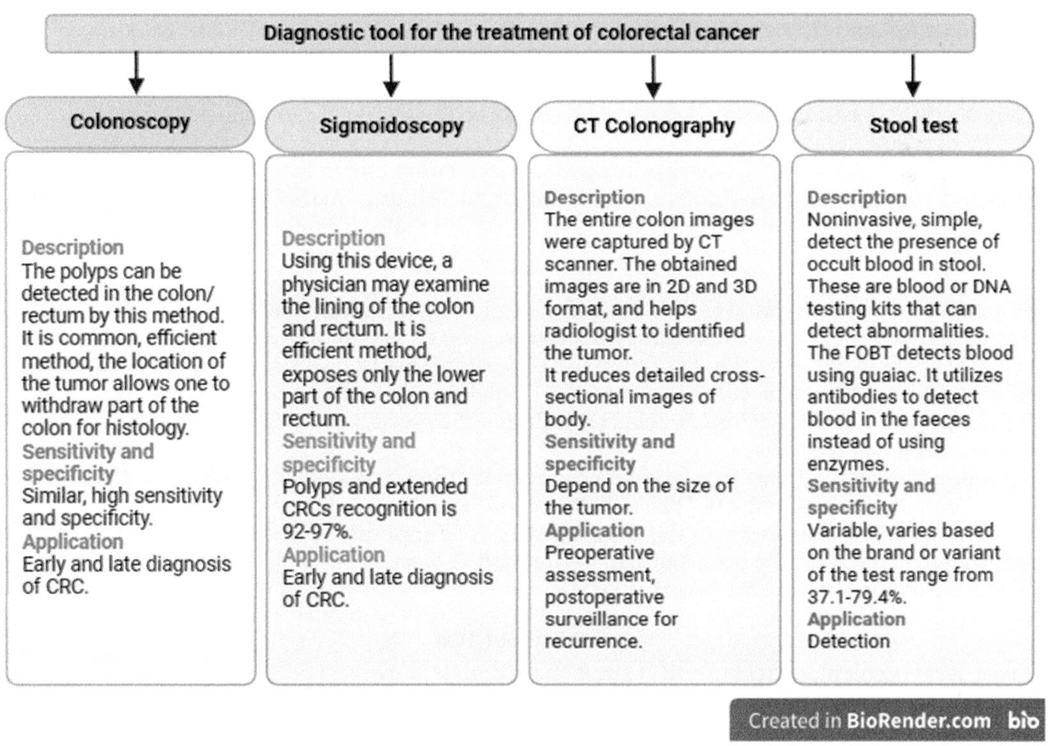

Figure 3.3 Diagnostic tool for the treatment of colorectal cancer.

stool preparation, then it is insufflated with either air or carbon dioxide (CO_2) to enlarge its dimensions, hence enhancing the quality of images during the CT imaging technique. After this insufflation process, the patient is scanned in both the prone and supine positions. CTC and colon endoscopy are similar in case of patient suffering due to the need for colon evacuation and insufflation. However, CTC has the benefit of not necessitating anesthesia. The detection specificity and sensitivity for polyps greater than 10 mm in diameter, often referred to as big polyps, demonstrate good performance at 82%. However, the sensitivity for polyps ranging from 6 to 9 mm in diameter, known as tiny polyps, is comparatively weak at 63%. Moreover, the methodology used lacks the ability to differentiate between flat and serrated tumors. When it comes to determining the local grade of a CRC tumor, MRI is the preferable diagnostic method over CT colonography, since the latter has less effectiveness in this aspect. The use of preoperative MRI is advantageous in surgical planning due to its exceptional precision in assessing the extent of tumor invasion and metastasis. The CTCG technique is associated with many limitations, including potential radiation exposure, limited sensitivity in detecting tiny polyps, the need for intestinal setup, and the necessity for an additional operation to eliminate identified polyps. Consequently, the patient's level of anxiety increases, and there is an associated increase in expenses due to unnecessary surveillance, which is further intensified by the discovery of extracolonic lesions (Vinchhi & Patel, 2021; Santosh et al., 2021).

3.5.2 Colonoscopy and Sigmoidoscopy

Endoscopy is widely recognized as the predominant and very effective method for diagnosing CRC. The technique encompasses both colonoscopy and sigmoidoscopy. These procedures enable the identification of a neoplasm and the extraction of a segment of the colon for further histological examination. The specificity and sensitivity of sigmoidoscopy for the diagnosis of polyps and enlarged CRCs range from 92 to 97%. When conducted with a flexible endoscope, sigmoidoscopy is very limited for seeing the distal region of the colon and the rectum. A colonoscopy can achieve comparable levels of specificity and sensitivity to acquire a comprehensive visualization of the whole intestinal tract. The colonoscopy method offers several benefits compared to other screening tests. One notable advantage is its shorter duration, allowing for a more efficient completion. Additionally, it contributes to the enhancement of patient acceptance and tolerance towards novel sedation techniques. The study, which involved patients with an average risk of CRC after colonoscopy, showed a significant 67% decrease in morbidity and a 65% reduction in death when compared to the control group. A randomized experiment revealed that conducting initial testing with colonoscopy and sigmoidoscopy led to a 33% reduction in colorectal cancer incidence and a 38–59% decrease in mortality (Shaukat et al., 2022).

The popularity of sigmoidoscopy as the primary screening tool in England and other regions has seen a significant increase due to this revelation. Identifying smaller polyps (6–10 mm) poses challenges when using this invasive methodology. Additionally, it imposes limitations on the ability to identify cancer in the transverse colon, ascending colon, and cecum. Endoscopy, specifically sigmoidoscopy, may be utilized as a diagnostic and therapeutic tool for colorectal cancer in individuals who are not suitable candidates for surgical intervention due to factors such as tumor severity or the presence of comorbidities. Procedures aimed at eliminating blockages associated with cancer are included within the repertoire of available methods. Despite the many advantages associated with endoscopy, patients often report experiencing pain during the procedure. Individuals tend to refrain from or delay engaging in the process of assessment due to the possible repercussions that may arise.

The use of virtual colonoscopy has shown a notable surge in popularity during the last several years. Computed tomography (CT) is advantageous due to its ability to provide a three-dimensional (3D) visualization of the large intestine. The application of virtual colonoscopy has been shown to decrease the occurrence of perforation or bleeding in the large intestine (Vinchhi & Patel, 2021; Santosh et al., 2021; Wu et al., 2020).

3.5.3 Stool Test

Guaiac fecal occult blood test (gFOBT) and FIT are both widely used primary screening techniques for the diagnosis of CRC. Stool testing has the capability to diagnose the presence of blood emanating from CRCs in the feces of individuals who do not exhibit any symptoms. FITs are often used in comparison to gFOBTs because of their heightened sensitivity and simplicity of usage. Patients who exhibit a positive result in their stool test are required to have a colonoscopy procedure to obtain a precise diagnosis and effectively eliminate any identified polyps. Identifying

DNA markers that are expelled by malignant cells in an individual's feces holds significant promise as a screening approach for CRC. The sensitivity and specificity of stool DNA testing for colon cancer detection are reported to be high, ranging from 71% to 91% and 93% to 100%, respectively. The prevalence of inaccurate outcomes is often seen due to the limited specificity and sensitivity of the FOBT. Moreover, these tests are ineffective in identifying tumors that do not cause bleeding. Fecal occult blood testing is a diagnostic technique that stands out for its simplicity, cost, and non-invasiveness. The presence of hemoglobin in feces is indicative of gastrointestinal bleeding, as shown by the test. The blood presence in feces is considered an unclear indicator of colon cancer, since it may originate via both malignant changes and polyps bigger than 1 to 2 cm. Repeating the test has the potential to enhance its sensitivity by up to 90%. Human hemoglobin and heme are the fundamental components of hemoglobin, a protein responsible for oxygen transport in the body. The detection of hemoglobin in CRC diagnosis is confirmed using the immunohistochemical FIT. In addition to sDNA, a marker for DNA alterations in colorectal adenocarcinomas, detection techniques such as polymerase chain reaction may be used to differentiate and distinguish bacterial DNA from human DNA in fecal samples. The use of colorectal cancer molecular diagnostics, which rely on genomic and epigenetic testing is limited. The costs associated with this item are rather high due to its limited availability on the market. The molecular diagnoses of CRC possess some limitations, leading researchers to seek alternative markers that may be detected in biological data at a more affordable price (Vinchhi & Patel, 2021; Santosh et al., 2021; Guimarães et al., 2019; Segev et al., 2018).

3.6 FUTURE PROSPECTS AND RECOMMENDATIONS

After conducting research for more than 40 years, a comprehensive comprehension of the genomic and epigenomic fragility in the sequence from colon polyp to CRC has been achieved. The identification of biomarkers that have undergone rigorous validation for regular clinical use in managing cases of CRC is a significant achievement in the field of precision medicine. Notably, the analysis of KRAS mutations to inform anti-EGFR therapy has emerged as one of the first triumphs in this era. The presence of MSI and BRAF mutations now plays a significant role in determining the need for genetic testing at the molecular level for Lynch syndrome. These markers are also expected to play a more prominent role in predicting the prognosis and treatment responses for sporadic colorectal cancers. The recent achievements in using methylation VIM, NDRG4, and BMP3 as tests for early detection of CRC indicate that epigenetic modifications have the potential to serve as valuable molecular markers in a therapeutic context. The potential of using molecular modifications to forecast the likelihood of developing metachronous polyps or CRC is a promising area of research. However, more investigations are required to ascertain the reliability and accuracy of aberrantly methylated CpGs or other genetic changes as indicators of the risk of polyps or CRC.

The absence of agreement in trends in genomic instability or in the methylation status of specific loci within the polyps indicates that relying on markers within the polyps may not yield effective results. However, examining the normal mucosa could be valuable as it has the potential to identify the existence of a field cancerization effect. There is a prevailing expectation that forthcoming investigations during the next 5–10 years will address these unresolved inquiries and uncover more therapeutic implications regarding molecular changes in colon polyps and CRCs.

3.7 CONCLUSIONS

In conclusion, the molecular characterization of CRC has expanded screening and prevention options, yet managing metastatic CRC remains a major challenge. Genetic and epigenetic diversity complicates treatment, particularly in resistant cases. While promising molecular biomarkers are emerging, most lack validation in large patient cohorts. Personalized therapies, integrating molecular insights with lifestyle factors, hold potential for improving outcomes. However, the effectiveness of novel screening and diagnostic techniques must be confirmed through clinical trials to ensure their cost-effectiveness and practical application in patient care.

REFERENCES

Aghagolzadeh, P., Radpour, R. (2016) New trends in molecular and cellular biomarker discovery for colorectal cancer. *World J Gastroenterol* 22; 25: 5678–5693.

Ahronian, L. G. et al. (2015) Clinical acquired resistance to RAF inhibitor combinations in BRAF-mutant colorectal cancer through MAPK pathway alterations. *Cancer Discov* 5; 4: 358–367.

Amodio, V. et al. (2020) EGFR blockade reverts resistance to KRASG12C inhibition in colorectal cancer. *Cancer Discov* 10; 8: 1129–1139.

Boland, C. R. et al. (1995) Microallelotyping defines the sequence and tempo of allelic losses at tumour suppressor gene loci during colorectal cancer progression. *Nat. Med* 1.9: Article 9.
Boland, P. M. et al. (2018) Recent progress in Lynch syndrome and other familial colorectal cancer syndromes. *CA*68; 3: 217–231.
Bupathi, M., Wu, C. (2016) Biomarkers for immune therapy in colorectal cancer: Mismatch-repair deficiency and others. *J Gastrointest Oncol* 7; 5: 713–720.
Carbone, M. et al. (2020) Tumour predisposition and cancer syndromes as models to study gene–environment interactions. *Nat. Rev. Cancer* 20; 9: Article 9.
Carethers, J. M. (2017) Microsatellite instability pathway and EMAST in colorectal cancer. *Curr Colorectal Cancer Rep* 13: 3–80.
De Palma, F. D. E.et al. (2019) The molecular hallmarks of the serrated pathway in colorectal cancer. *Cancers* 11.7:Article 7.
Erstad, D. J. et al. (2015) Prognostic and predictive biomarkers in colorectal cancer: Implications for the clinical surgeon. *Ann Surg Oncol* 22; 11: 3433–3450.
Fearon, E. R., Vogelstein, B. (1990) A genetic model for colorectal tumorigenesis. *Cell* 61; 5: 759–767.
Fischer, J. et al. (2019) Clinical implications of the genetics of sporadic colorectal cancer. *ANZ J Surg* 89; 10: 1224–1229.
Florensa, D. et al. (2023) Acetylsalicylic acid effect in colorectal cancer taking into account the role of tobacco, alcohol and excess weight. *Int J Environ Res Public Health* 20; 5: 4104.
Ganesh, K. et al. (2019) Immunotherapy in colorectal cancer: Rationale, challenges and potential. *Nature Rev Gastroenterol Hepatol* 16; 6: Article 6.
Guimarães, D. P. et al. (2019) Comparison of a new-generation fecal immunochemical test (FIT) with guaiac fecal occult blood test (gFOBT) in detecting colorectal neoplasia among colonoscopy-referral patients. *Anticancer Res* 39; 1: 261–269.
Guo, M. et al. (2019) Epigenetic heterogeneity in cancer. *Biomarker Res* 7; 1: 23.
Huang, F.-T. et al. (2017) The novel long intergenic noncoding RNA UCC promotes colorectal cancer progression by sponging miR-143. *Cell Death Dis* 8; 5: Article 5.
Ishaque, N. et al. (2018) Whole genome sequencing puts forward hypotheses on metastasis evolution and therapy in colorectal cancer. *Nat. Comm* 9; 1: Article 1.
Issa, I. A., Noureddine, M. (2017) Colorectal cancer screening: An updated review of the available options. *World J Gastroenterol* 23; 28: 5086–5096.
Jeon, J.et al. (2018) Determining risk of colorectal cancer and starting age of screening based on lifestyle, environmental, and genetic factors. *Gastroenterol* 154; 8: 2152–2164.
Jernvall, P. et al. (1999) Loss of heterozygosity at 18q21 is indicative of recurrence and therefore poor prognosis in a subset of colorectal cancers. *Br J Cancer* 79; 5–6: 903–908.
Kinzler, K. W., Vogelstein, B. (1996) Lessons from hereditary colorectal cancer. *Cell* 87; 2: 159–170.
Koveitypour, Z. et al. (2019) Signaling pathways involved in colorectal cancer progression. *Cell Biosci* 9; 1: 97.
Ladabaum, U. et al. (2020) Strategies for colorectal cancer screening *Gastroenterol* 158; 2: 418–432.
Langner, C. (2014) Serrated and non-serrated precursor lesions of colorectal cancer. *Digestive Dis* 33; 1: 28–37.
Li, M. et al. (2020) A clinical-radiomics nomogram for the preoperative prediction of lymph node metastasis in colorectal cancer. *J Translat Med* 18; 1: 46.
Liebl, M. C., Hofmann, T. G. (2021) The role of p53 signaling in colorectal cancer. *Cancers* 13; 9: 2125.
Mármol, I.et al. (2017) Colorectal carcinoma: A general overview and future perspectives in colorectal cancer. *Int J Molec Sci* 18; 1: Article 1.
Masuda, T. et al. (2017) MicroRNAs as biomarkers in colorectal cancer. *Cancers* 9; 9: Article 9.
Mauri, G. et al. (2021) The evolutionary landscape of treatment for BRAFV600E mutant metastatic colorectal cancer. *Cancers* 13; 1: Article 1.
Mazza, T. et al. (2016) Multifaceted enrichment analysis of RNA–RNA crosstalk reveals cooperating micro-societies in human colorectal cancer. *Nucleic Acids Res* 44; 9: 4025–4036.
Mehta, R. S. et al. (2017) Dietary patterns and risk of colorectal cancer: Analysis by tumor location and molecular subtypes. *Gastroenterol* 152; 8: 1944–1953. e1.
Mitsuhashi, K. et al. (2015) Analysis of the molecular features of rectal carcinoid tumors to identify new biomarkers that predict biological malignancy. *Oncotarget* 6; 26: 22114–22125.
Moosavi, F. et al. (2019) HGF/MET pathway aberrations as diagnostic, prognostic, and predictive biomarkers in human cancers. *Critical Rev Clin Laborat Sci* 56; 8: 533–566.
Mousavinezhad, M. et al. (2016) The effectiveness of FOBT vs. FIT: A meta-analysis on colorectal cancer screening test. *Med J Islamic Rep Iran* 30: 366.

Müller, M. F. et al. (2016) Molecular pathological classification of colorectal cancer. *Virchows Archiv* 469; 2: 125–134.
Nikolouzakis, T. K. et al. (2018) Improving diagnosis, prognosis and prediction by using biomarkers in CRC patients (Review). *Oncol. Rep* 39; 6: 2455–2472.
Novellasdemunt, L. et al. (2015) Targeting Wnt signaling in colorectal cancer. *Am J Physiology-Cell Physiol* 309; 8: C511–C521.
Ortiz-Quintero, B. (2016) Cell-free microRNAs in blood and other body fluids, as cancer biomarkers. *Cell Prolif* 49; 3: 281–303.
Raskov, H. et al. (2020) Driver Gene Mutations and Epigenetics in Colorectal Cancer. *Ann Surg*. 271; 1: 75.
Santosh, F. et al. (2021) Understanding Diverse signaling pathways involved in colorectal carcinoma and potentials of nanotechnology in their diagnosis and therapeutics. *Pharmacology OnLine* 2: 1226–1243.
Segev, L. et al. (2018) Left-sided dominance of early-onset colorectal cancers: a rationale for screening flexible sigmoidoscopy in the young. *Dis Colon Rectum*. 61; 8: 897.
Shaukat, A. et al. (2022) Current and future colorectal cancer screening strategies. *Nature Reviews: Gastroenterol Hepatol*. 19; 8: 521–531.
Siegel, R. L. et al. (2020) Colorectal cancer in the young: Epidemiology, prevention, management. *American Society of Clinical Oncology Educational Book*. 40: e75–e88.
Szylberg, Ł. et al. (2015) Serrated polyps and their alternative pathway to the colorectal cancer: a systematic review. *Gastroenterol Res. Prac*. 2015(1): 573814.
US Preventive Services Task Force. (2016) Screening for colorectal cancer: US preventive services task force recommendation statement. *JAMA* 315; 23: 2564–2575.
Vinchhi, P., Patel, M. M. (2021) Triumph against cancer: Invading colorectal cancer with nanotechnology. *Expert Opin Drug Deliv*. 18; 9: 1169–1192.
Watson, M. M. (2020). Microsatellite instability at tetranucleotides (EMAST) in colorectal cancer: Clinical relevance, mechanisms and immune markers. [Doctoral thesis, The University of Bergen]; available at: https://bora.uib.no/bora-xmlui/handle/1956/22471
Wu, Z. et al. (2020) Colorectal cancer screening methods and molecular markers for early detection. *Technol Cancer Res Treat*. 19: 1533033820980426.
Yao, Q. et al. (2018) Synergistic role of Caspase-8 and Caspase-3 expressions: Prognostic and predictive biomarkers in colorectal cancer. *Cancer Biomarkers* 21; 4: 899–908.
Yu, H., Hemminki, K. (2020). Genetic epidemiology of colorectal cancer and associated cancers. *Mutagenesis*. 35; 3: 207–219.
Zarkavelis, G. et al. (2017). Current and future biomarkers in colorectal cancer. *Ann Gastroenterol*. 30; 6: 613–621.

4 Genetic Mutations and Inherited Syndromes

Saliha Rizvi, Naseem Fatima, Farheen Khan, Syed Tasleem Raza, Gowhar Rashid and Farzana Mahdi

4.1 INTRODUCTION TO GENETIC MUTATIONS AND INHERITED SYNDROMES

4.1.1 Definition of Genetic Mutations

Genetic mutations are changes to the sequence of DNA that makes up an individual gene. Genes are the instructions for building and maintaining the structures and functions of living organisms. Environmental factors like radiation and certain chemicals, or errors made during DNA replication, can lead to mutations occurring spontaneously. They can also be inherited from one or both parents.

The following are types of mutations:

- ***Point mutations*:** A single nucleotide is altered in this type of mutation (A, T, C, or G). They include substitutions, insertions, and deletions.
- ***Frameshift mutations*:** These types of mutations arise when altered nucleotide (insertion or deletion) shifts the reading frame of the genetic code, affecting the entire sequence downstream.
- ***Chromosomal mutations*:** Mutations occur when there are changes in the number or structure of whole chromosomes; examples are translocations, inversions, duplications, and deletions.
- ***Missense mutations*:** One amino acid is substituted in the protein sequence when a single nucleotide is altered.
- ***Nonsense mutations*:** Add an early stop codon, which causes the protein to be shortened and typically nonfunctional.
- ***Silent mutations*:** Genetic code gets degenerated resulting in no change in the amino acid sequence.
- ***Germline mutations*:** Occur in the eggs or sperm and can be passed from one generation to the next.
- ***Somatic mutations*:** Originate in body cells and are not inherited by progeny. They could play a role in the occurrence of cancer.

Mutations are crucial for genetic diversity and disorders; identifying them aids in diagnosis, disease risk assessment, and personalized medicine.

4.1.2 Overview of Inherited Syndromes

Inherited syndromes are found to be associated with colorectal cancer (CRC) and are characterized by genetic mutations that greatly raise a person's risk of getting CRC. These syndromes are often hereditary. Here is an overview of some of the well-known inherited syndromes associated with CRC:

a. **Familial adenomatous polyposis (FAP):**
   Familial Adenomatous Polyposis (FAP) is an autosomal dominant disorder caused by mutations in the APC gene. It leads to the development of numerous colorectal polyps in adolescence, with a near 100% risk of CRC by age 40 if untreated. Extracolonic features include desmoid tumors and congenital retinal pigment epithelium hypertrophy (CHRPE) (Leoz et al., 2015).

b. **Lynch syndrome (hereditary nonpolyposis colorectal cancer—HNPCC):**
   Lynch syndrome (HNPCC) is an autosomal dominant condition caused by mutations in DNA mismatch repair genes (PMS2, MSH2, MLH1, MSH6). It significantly increases the risk of early-onset CRC, along with elevated risks of endometrial, ovarian, stomach, and urinary tract cancers (Center et al. 2009).

c. **Peutz-Jeghers syndrome:**
   Peutz-Jeghers syndrome is an autosomal dominant disorder caused by STK11 gene mutations, featuring gastrointestinal polyps and pigmented mucosal lesions, with increased risks of colorectal, ovarian, breast, and pancreatic cancers (McGarrity et al. 2000).

 DOI: 10.1201/9781032704517-4

d. **MYH-associated polyposis (MAP):**
MYH-associated polyposis (MAP) is an autosomal recessive condition caused by MYH gene mutations, featuring multiple colon polyps and elevated CRC risk (Jones et al. 2002).

e. **Cowden syndrome:**
Cowden syndrome is an autosomal dominant disorder caused by PTEN gene mutations, associated with increased CRC risk, macrocephaly, skin lesions, and hamartomatous growths (Nelen et al., 1999).

f. **Juvenile polyposis syndrome (JPS):**
Juvenile polyposis syndrome (JPS) is an autosomal dominant disorder caused by mutations in SMAD4 or BMPR1A genes. It features multiple hamartomatous polyps in the gastrointestinal tract, particularly in the colon, and an increased risk of CRC, though lower overall cancer risk compared to FAP (Howe et al., 1998).

4.1.3 Significance of Studying Genetic Mutations and Inherited Syndromes

Genetic predisposition, primarily due to germline mutations in cancer susceptibility genes, accounts for 2–8% of CRC cases (Dominguez-Valentin et al., 2020). This increases to 6–10% when considering high- and moderate-risk genes like APC, TP53, and MLH1 (Pearlman et al., 2017). Individuals with Lynch syndrome or FAP face lifetime CRC risks of up to 80% and nearly 100%, respectively. Precision medicine and genetic testing now enable tailored treatments, screening, and preventive strategies, significantly improving patient outcomes. Genetic research, while promising, raises ethical concerns, particularly around privacy and consent.

4.2 COLORECTAL CANCER: AN OVERVIEW

4.2.1 What Is Colorectal Cancer?

CRC arises from abnormal glandular epithelial cell proliferation in the colon or rectum, driven by genetic, environmental, and behavioral factors. CRC is classified into colitis-associated, sporadic, and inherited forms. Genetic predispositions, including APC, TP53, and KRAS mutations, increase CRC risk (Johnson et al., 2013). CRC development typically involves a gradual accumulation of somatic or germline mutations over 10–15 years. Environmental factors like diet, sedentary lifestyles, and chronic inflammatory conditions also contribute to CRC risk (Johnson et al., 2013).

4.2.2 Epidemiology of Colorectal Cancer

Epidemiologically, CRC accounts for 10% of all cancer diagnoses globally and holds the position of the second-highest contributor to cancer-related mortality. FAP and Lynch syndrome contribute to about 5% of CRC cases, with an additional 10–15% of unselected CRC patients carrying high-risk mutations not associated with these syndromes. While the majority of CRC cases (70–80%) are sporadic, 20–30% display hereditary patterns, including susceptibility syndromes such as Lynch syndrome and FAP (Müller et al., 2016).

Understanding the epidemiology of CRC is crucial for prevention and targeted interventions. The data published by Ferlay et al. in 2020 revealed a diverse landscape of CRC risk, as evidenced by the number of cases, age-standardized rates (ASR), crude rates, and cumulative risks across various countries. Afghanistan reports 739 cases, an ASR of 3.6, and a cumulative risk of 0.41, while Brunei Darussalam shows higher risk with an ASR of 17.4 and a cumulative risk of 1.6. Japan's 1.2 cumulative risk signals significant lifetime CRC risk. High incidence in China and Korea is linked to genetics and lifestyle, whereas lower ASR and risks in Bangladesh and Nepal may reflect healthier lifestyles, better healthcare, or different genetic factors (Ferlay et al., 2020).

Globally, CRC incidence and mortality rates vary significantly. For instance, higher rates are observed in regions like Northern America and Europe, while Asia and Africa show lower cumulative risks. Disparities across countries, such as Japan's substantial lifetime CRC risk and lower rates in countries like Bangladesh, highlight the influence of healthcare access, lifestyle, and genetic factors (Ferlay et al., 2020).

In terms of mortality, countries like Hungary, Croatia, and Brunei report the highest CRC-related deaths. These variations emphasize the need for region-specific public health initiatives, screening programs, and preventive strategies tailored to local epidemiology, genetic factors, and healthcare systems. Global collaboration in CRC research is essential to improve outcomes and reduce the disease's burden (Ferlay et al., 2020).

4.2.3 The Importance of Early Detection

CRC is the third most common solid tumor globally. Precancerous adenomas undergo a prolonged process before turning malignant, making early screening feasible (Dekker et al., 2019). CRC is one of four cancers successfully identified through screening, as recommended by the USPSTF and CDC. Early detection and standard treatments, like surgery and colonoscopy, raise the five-year survival rate to 90%, but this drops significantly in advanced stages, increasing the financial burden (Joranger et al., 2020). Colonoscopy remains the gold standard for CRC diagnosis, allowing for precise lesion removal, though its invasive nature limits adherence to around 50%. Fecal immunochemical testing (FIT), a noninvasive, cost-effective alternative, is widely used but has lower sensitivity for advanced adenomas (Buskermolen et al., 2019). FIT detects CRC with 71% sensitivity at a 20 µg/g threshold but misses about one-third of stage I cases (Niedermaier et al., 2017). Patient preferences lean toward noninvasive tests, with 83% favoring blood-based over stool-based testing.

Circulating tumor DNA (ctDNA) methylation shows promise as a CRC biomarker, with elevated levels of ctDNA linked to poor prognosis, especially in cases with mutations in KRAS, APC, and TP53 (Wang et al., 2004). ctDNA methylation aids early diagnosis and prognosis. Serum carcinoembryonic antigen (CEA) is another biomarker, though its sensitivity for early-stage CRC is limited, ranging from 40% to 60% (Palmqvist et al., 2003). The need for novel noninvasive biomarkers remains critical to improving early detection, compliance, and outcomes in CRC.

4.3 GENETIC MUTATIONS AND COLORECTAL CANCER RISK

4.3.1 The Role of Genetics in Colorectal Cancer

CRC arises from both inherited (germline) and acquired (somatic) mutations. The three major types of genes altered in cancer are oncogenes, tumor suppressor genes, and mismatch repair (MMR) genes. Mutations in oncogenes genes lead to unchecked cell growth and, eventually, cancer. Suppressor genes lose their function when both of their copies, or alleles, are rendered inactive. Before a gene loses function while both copies of the gene are normal, two mutation events, or hits, are needed. This two-hit theory addresses both the idea that suppressor genes cause cancer in a recessive manner at the cellular level and the reason that inherited diseases typically manifest at an earlier age than sporadic diseases. DNA mismatch repair (MMR) systems fix replication faults. Genes that control growth, like the transforming growth factor receptor (*TGFRII*) gene, are impacted by DNA errors that are accumulated throughout the genome in cells that have MMR mutations in both gene copies. The HNPCC has mutations in these genes (Paula & Frucht, 2002).

The multistep model of carcinogenesis explains the molecular basis of selective growth advantage to colon cells in sporadic colon cancer, leading to the formation of clonal tumors. The pathogenesis of CRC involves oncogene activation through mutation and tumor suppressor gene inactivation; somatic mutations in at least four or five genes of a cell; the accumulation of multiple genetic mutations (Fearon & Vogelstein, 1990). Molecular alterations coincide with morphological changes. The "classical" model of carcinogenesis begins with mutations in APC and followed by mutations in SMAD4, KRAS/NRAS genes, and TP53 due to chromosomal instability and microsatellite instability (MSI) in Lynch syndrome. A distinct pathway accounts for the development of some CRCs from serrated polyps, and precursor lesions that result in cancer (CIMP and spontaneous MSI). The first gene to become mutated is *CTNNB1*, which is then followed by PIK3CA, KRAS/BRAF, and TGFBR1 genes (Ciepiela et al., 2024).

CRC is caused by an accumulation of many genetic and epigenetic defects within cells. CRC progression is driven by genomic instability, manifesting in three primary forms: chromosomal instability, MSI, and CpG island methylation. Chromosomal instability results in alterations in chromosome number and structure, leading to loss of key tumor suppressor genes (e.g., *APC, P53, SMAD4*). MSI results from defective MMR genes (MLH1, MSH2, MSH6, PMS2), which cause shifts in repetitive DNA sequences. CIMP, characterized by aberrant DNA methylation, silences genes like MLH1 through epigenetic mechanisms, contributing to CRC development in approximately 15% of cases (Weisenberger et al., 2006).

4.3.2 Common Genetic Mutations Associated with Colorectal Cancer

Environmental and genetic variables interact to create a complex and multifaceted disease like CRC. A number of genetic alterations play a role in the development of CRC (Figure 4.1). Here's a detailed account of some key genetic mutations associated with CRC.

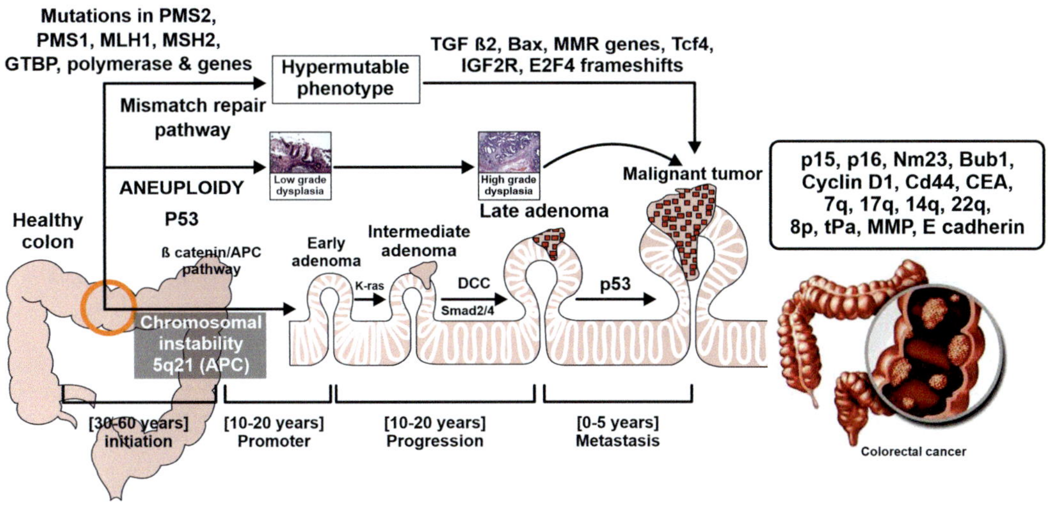

Figure 4.1 Genetic mutations identified as contributors to the development of CRC, which occurs in four stages: initiation, promoter, progression and metastasis. CRC: colorectal cancer.

Adenomatous polyposis coli (APC) gene: The *APC* gene regulates cell division; mutations increase CRC risk due to uncontrolled cell proliferation, linking it to FAP (Sierra et al., 2006).

KRAS gene: *KRAS*, part of the RAS/MAPK pathway, is often mutated in colorectal tumors, promoting continuous cell proliferation and treatment resistance (Malumbres & Barbacid, 2003).

TP53 gene: *TP53*, a tumor suppressor gene, regulates the cell cycle; its mutations lead to loss of tumor-suppressive properties thus uncontrolled cell growth and aggressive CRC (Levine, 1997).

BRAF gene: *BRAF* is implicated in the RAS/RAF/MEK/ERK signaling cascade. Certain *BRAF* mutations (V600E mutant) are linked to a worse prognosis, enhanced cell survival and proliferation (Davies et al., 2002).

Mismatch repair (MMR) gene: *MMR* genes repair DNA replication errors; mutations in which lead to microsatellite instability in some colorectal tumors (Boland & Goel, 2010).

PIK3CA gene: *PIK3CA* gene involved in PI3K/AKT/mTOR signaling pathway controls cell proliferation and survival. *PIK3CA* mutations facilitate tumor development by activating the PI3K pathway (Samuels et al., 2005).

SMAD4 gene: *SMAD4*, a tumor suppressor in TGF-β signaling, is linked to advanced CRC stages and promotes tumor invasion and metastasis when mutated (Hahn et al., 1996).

CHEK2 gene: A gene called *CHEK2* is involved in DNA repair and cell cycle regulation. Studies have looked into the relationship between CRC risk and specific CHEK2 mutations, especially in communities where the disease runs in the family (Bell et al., 1999).

CTNNB1 (beta-catenin) gene: *CTNNB1* encodes beta-catenin, crucial for Wnt signaling; mutations can activate abnormal signaling, promoting CRC development (Morin et al., 1997).

PTEN gene: *PTEN* mutations inactivate its function, promoting PI3K signaling, reducing apoptosis, increasing cell-survival and increasing CRC risk by compromising chromosomal integrity (Munteanu & Mastalier, 2014).

NOD2 gene: NOD2 is involved in inflammation and innate immunity; its variations are linked to colorectal cancer through inflammatory pathways (Hugot et al., 2001).

TGFBR2 gene: TGFBR2 mediates apoptosis and growth arrest; frameshift mutations can inactivate it in mismatch-repair-deficient patients.

(Munteanu & Mastalier, 2014)

Additionally, gene polymorphisms and mutations have been investigated in relation to CRC. A common genetic variant called single nucleotide polymorphism (SNP) can produce distinct

Table 4.1: Gene polymorphisms associated with an increased risk of colorectal cancer

rsID	Gene	Location	Reference
rs6983267	*MYC*	8q24	Houlston et al., 2008
rs10795668	Unknown (LOC105376400)	10p14	Tomlinson et al., 2007
rs4779584	*SMAD7*	15q13.3	Yang et al., 2014
rs16892766	*EIF3H*	8q23.3	Haiman et al., 2007
rs4939827	*SMAD7*	18q21	Broderick et al., 2007
rs3802842	*COLCA1*	11q23.1	Tomlinson et al., 2011
rs10936599	*MYNN*	10p14	Jaeger et al., 2008
rs10795668	Unknown (LOC105376400)	10p14	Tomlinson et al., 2008
rs9929218	*CDH1*	16p22.1	Wang et al., 2021

MYC: MYC Proto-Oncogene; SMAD7: SMAD Family Member 7; EIF3H: Eukaryotic Translation Initiation Factor 3 Subunit H; COLCA1: Colorectal Cancer Associated 1; MYNN: Myoneurin; CDH1: Cadherin1

functional outcomes, influencing an individual's susceptibility to certain diseases. Thus, SNP may be used as a biomarker to estimate the likelihood of sporadic tumors, such as CRC. Over the last 30 years, a large number of SNPs have been shown to be associated with the risk of CRC through both candidate-gene association studies (CGAS) and massive genome-wide association studies (GWAS). Here are a few gene polymorphisms and mutations that have been explored (Table 4.1).

These are just a few examples, and the landscape of CRC genetics is continually evolving. Advances in genomic research and technologies will likely uncover new gene variants, mutations, and polymorphisms that play a role in CRC susceptibility, progression, and response to treatment. Understanding the genetic mutations associated with CRC is crucial for risk assessment, early detection, and the development of targeted therapies. Genetic testing can help identify individuals at higher risk and guide personalized treatment strategies. Ongoing research continues to uncover new insights into the genetic basis of CRC, paving the way for advancements in prevention and treatment.

4.3.3 Genetic Testing and Colorectal Cancer Risk Assessment

Hereditary cancer risk assessment is key to identifying individuals and families at risk for specific cancers, enabling targeted surveillance. High-risk individuals can be identified by family history, age at diagnosis, and tumor phenotype, with genetic mutations confirmed via single or multigene testing. Major inherited CRC syndromes include FAP, MAP, and Lynch syndrome, involving mutations in mismatch repair (MMR) genes (MLH1, MSH2, MSH6, PMS2). Genetic risk factors for CRC have led to advanced diagnostic algorithms using tumor and germline analyses, including techniques such as MLH1 promoter hypermethylation analysis, microsatellite instability (MSI) testing, and MMR protein immunohistochemistry.

4.3.3.1 Lynch Syndrome Testing Criteria

Lynch syndrome risk is determined by Amsterdam II and Bethesda guidelines. The NCCN recommends routine MMR deficiency screening for colorectal and endometrial cancers, regardless of age. MSI testing or immunohistochemistry (IHC) is used to detect MMR protein loss in tumor tissue (Mao et al., 2021). The Amsterdam II criteria require three relatives with CRC or Lynch-associated cancers, spanning two generations, with one diagnosis under age 50. Bethesda criteria focus on CRC diagnosis before age 50, synchronous or metachronous tumors, and family history (Pineda et al., 2010).

4.3.3.2 Molecular Diagnostics

Molecular diagnostics include sequencing of exons, promoters, and 5′- and 3′-UTR regions, along with deletion/duplication assays, to detect germline mutations in MMR genes and EPCAM for Lynch syndrome. Diagnostic findings are considered definitive, though LS-related cancer penetrance is less than 100%, meaning not all carriers develop cancer (Mao et al., 2021).

Hereditary colonic polyposis, including FAP and MAP, is linked to germline mutations in APC and MUTYH genes. Genetic testing is recommended for individuals with 20+ adenomas, congenital hypertrophy of the retinal pigment epithelium (CHRPE), or a history of desmoid tumors or hepatoblastoma. Diagnostic testing involves identifying mutations in APC, MUTYH, POLD1, and POLE (Mao et al., 2021).

4.3.3.3 Multigene Panel Testing

The NCCN advises multigene panel testing (MGPT) for all CRC patients, regardless of age or family history. Panels typically assess 15–100 cancer risk genes, including APC, MLH1, MSH2, MSH6, MUTYH, and PMS2. MGPT may be more efficient and cost-effective than single-gene testing, with clinical sensitivity depending on phenotype and the genes tested (Weiss et al., 2021).

4.4 INHERITED SYNDROMES AND COLORECTAL CANCER

4.4.1 Hereditary Syndromes Predisposing to Colorectal Cancer

CRC is highly prevalent, with family history accounting for about 30% of cases. Research indicates that 5% of CRC cases are caused by highly penetrant hereditary mutations, while 20–30% are attributed to less penetrant, more common genetic variations (Grady, 2003). Inherited CRC may result from cumulative effects of multiple susceptibility loci. Understanding the genetics of hereditary CRC is vital for identifying high-risk individuals, improving surveillance, and developing better diagnostic and treatment strategies. CRC syndromes are classified into polyposis and non-polyposis based on polyp quantity and histology (Figure 4.2).

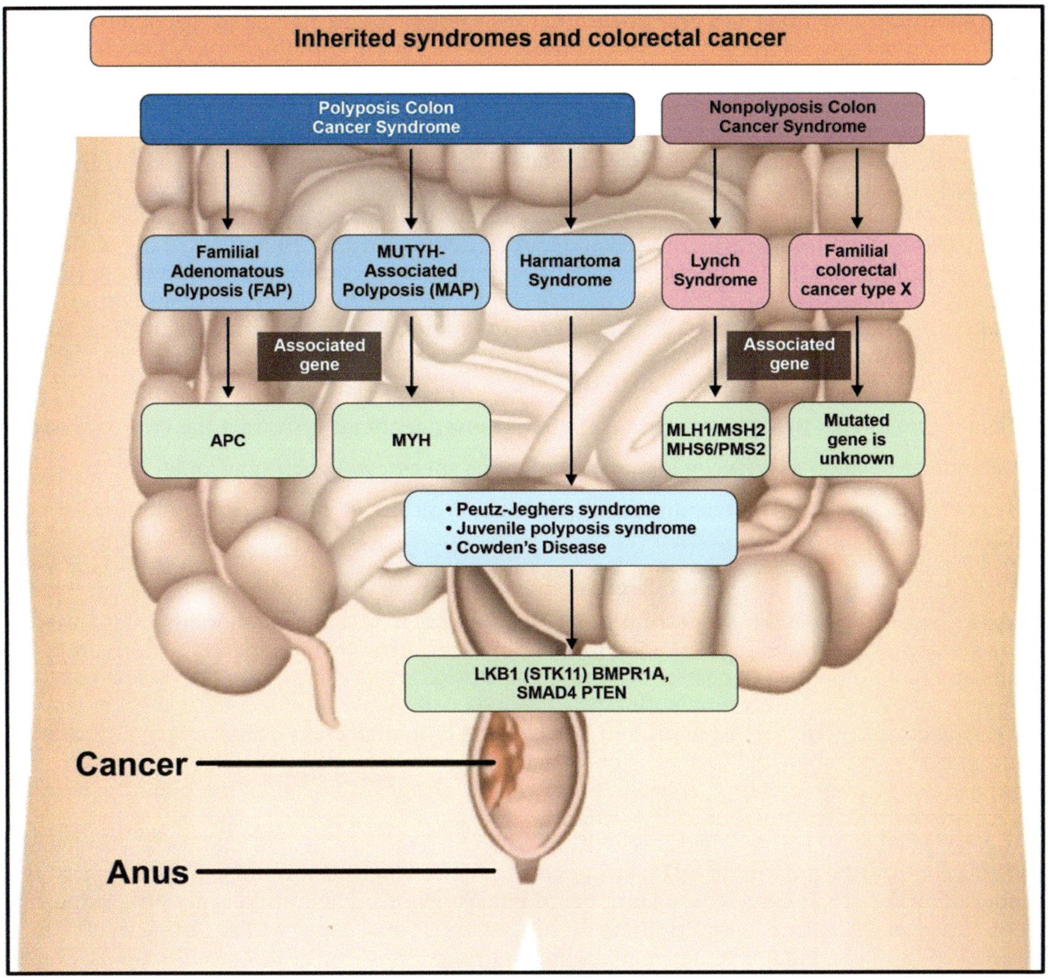

Figure 4.2 Hereditary predisposition to colorectal cancer: Polyposis CRC syndrome and nonpolyposis CRC syndrome.

4.4.2 Familial Adenomatous Polyposis (FAP)

FAP is an autosomal dominant diseased condition that typically manifests as 100–1000 adenomas/benign tumors in the areas of the rectum and colon. The APC gene on chromosome 5q21, is a tumor suppressor involved in 0.5–1% of CRC cases, affecting both males and females equally. Loss of APC function leads to increased β-catenin activity and expression of oncogenes like c-myc and ODC (Sierra et al., 2006) initiating a carcinogenesis cascade with further mutations in KRAS and TP53. In FAP, 80% of tumors show APC mutations, with full-length APC protein absent in 60% of cases. APC mutation location affects polyposis onset and survival rates (Will et al., 2009).

Classic FAP diagnosis is based on clinical evidence and family history, with genetic testing confirming the diagnosis. A complete family history is crucial, as some patients may be asymptomatic. Flexible sigmoidoscopy can detect adenomas early. Attenuated FAP (AFAP) diagnosis is more complex due to phenotypic variation (Half et al., 2009).

Genetic testing screens at-risk family members before symptom onset, with initial testing focused on the index case. APC germline mutations are detected using various methods like full sequencing, CSGE screening, protein truncation tests (70%–90% sensitive), linkage analysis, and direct *APC* gene sequencing (Leoz et al., 2015). Around 20% of classic FAP and 80% of AFAP patients lack APC/MUTYH point mutations, but copy number alterations can be detected through multiplex ligation-dependent probe amplification (MLPA) and real-time fluorescence PCR, which identifies significant APC gene changes during amplification (Rohlin, 2014).

4.4.3 Lynch Syndrome (Hereditary Nonpolyposis CRC)

Lynch syndrome is defined as a genetic disorder responsible for the increased risk of developing cancers, most notably colorectal and endometrial cancers. People with this type of hereditary disorder are also more likely to develop cancers including stomach, gallbladder ducts, small intestine, brain, and skin. It has been observed more in women than men, because most belong to the female reproductive system (Center et al., 2009).

CRC is the most typical histological manifestation of a Lynch syndrome tumor. It often has a combination of signet-ring and/or mucinous cell types, areas of poor differentiation, Crohn-like peritumoral lymphoid aggregates, and significant tumor-infiltrating lymphocytes, occasionally, with a pattern of medullary growth (Alexander et al., 2001). Germline mutations in the mismatch repair genes (MLH1, MSH2, MSH6, and PMS2) result in Lynch syndrome. Errors in duplicated DNA are recognized and fixed by MMR protein complexes during cell division. DNA replication mistakes resulting from mutations causing loss of function in the MMR genes might happen in tumor suppressor genes or proto-oncogenes, which can cause carcinogenesis.

Lynch syndrome leads to CRC through three pathways:

1. Somatic mutations in MMR genes cause spontaneous adenomas activating the Wnt pathway.
2. Adenomas develop from secondary APC mutations specific to Lynch syndrome.
3. Non-dysplastic lesions mutate into flat carcinomas, typically involving TP53 mutations and Wnt activation (Ahadova et al., 2018).

4.4.4 Peutz-Jeghers Syndrome (PJS)

Peutz-Jeghers syndrome (PJS), a cancer syndrome inherited in an autosomal dominant manner, is characterized by abnormal pigmentation (freckling) of certain skin and mucosal regions and results in the development of gastrointestinal polyps (McGarrity et al., 2000). The estimated frequency of PJS, a very uncommon illness, is 1 in 25,000 to 300,000 births (Utsunomiya et al., 1975). Any ethnic group can experience the illness, and both men and women are equally afflicted. Cancer strikes patients with PJS at an average age of 42, and they are more likely to acquire malignancies. Although they can form in other areas of the gastrointestinal tract, small intestine polyps are more frequently seen in PJS patients.

A gene called *STK11* (also called Liver Kinase B1), which is situated at location q13.3 on chromosome 19, controls the polarity of cells. An enzyme known as "serine/threonine kinase 11" that is produced by the *STK11* gene serves a number of vital functions (Hemminki et al., 1998; Jenne et al., 1998). However, in patients with PJS, there is only one functional copy of *STK11* in each cell. The second copy is there, but it is mutated, making it nonfunctional. The afflicted cells may proliferate abnormally as a result, which raises the possibility that they will develop tumors or cancer.

Genetic testing along with genetic counseling, in which blood is tested to check for an *STK11* gene mutation, are available to people who may have PJS. If an *STK11* gene mutation is discovered, testing and the same gene mutation are required for the diagnosis of PJS in other family members. On the other hand, not everyone who receives a clinical diagnosis of PJS will display a recognizable STK11 mutation.

4.4.5 MUTYH-Associated Polyposis (MAP)

Polyposis associated with MUTYH (MAP) is a rare hereditary disorder that results in the occurrence of unusual tissue growth in various a portion of the body. MAP stands for MUTYH-associated adenomatous polyposis, an autosomal recessive condition. It is brought on by a biallelic mutation on chromosome locus 1p34.3-p32.1 in the MUTYH gene (Jones et al., 2002).

The base excision repair (BER) mechanism, which includes the MUTYH gene product, is crucial for cells' defense against oxidative DNA damage (Sampson et al., 2003). The reactive oxygen species formed at the time of aerobic metabolism cause oxidative DNA damage due to which the cellular DNA integrity gets constantly threatened. Oxidative DNA damage leads to the formation of mutagenic products, one of which is the highly stable oxidized base 7, 8-dihydroxy-8-oxoguanine (8-oxo-G) (Cheadle & Sampson 2007).

4.4.5.1 MAP-Associated Colorectal Adenomas and Colorectal Cancer Tumorigenesis

In most cases, when the individual is being observed with adenomatous colonic polyps (tubular adenomas) approximately 20–99 numbers are suspected with MAP. Currently, MAP can also occur unconventionally with hyperplastic polys, sessile serrated adenomas, and mixed polys (Guarinos et al., 2014). According to Boparai et al. 47% of MAP patients also had SSAs or HPs in addition to adenomas. *APC* gene mutations were present in 41% of the adenomas, whereas K-ras mutations were present in 23% of them. However, on further evaluation, they found that none had APC mutations, only 70% had K-ras mutations (Boparai et al., 2008).

4.5 GENETIC COUNSELING AND MANAGEMENT

4.5.1 The Function of Genetic Counseling

Genetic counseling plays a critical role in diagnosing hereditary disorders and assessing disease risk, particularly in CRC, where approximately 30% of cases have a familial component and 5–10% are hereditary (Mao et al., 2021). Specialized counselors provide objective information and support, conducting thorough reviews of personal and family histories to evaluate cancer risks and inheritance patterns.

Genetic counseling involves a thorough case and family history review, education on hereditary disease management, inheritance, and genetic testing, psychosocial support, informed consent, sharing test results with at-risk relatives, and discussions on career, family plans, patient management, and surveillance (Giardiello et al., 2014).

Creating a detailed family tree helps assess cancer risk and examine inheritance patterns, potentially identifying the proband. When direct tumor sequencing is not feasible, computational models like PREMM1 and MMRpro, which evaluate the likelihood of Lynch syndrome, are used (Weissman et al., 2011).

4.5.2 Screening and Surveillance Recommendations Colorectal Cancer Screening

Colonic polyps and a history of Crohn's disease or ulcerative colitis increase CRC risk. The US Preventive Services Task Force recommends starting CRC screening at age 50, with individual decisions for those over 75. For individuals with a family history of CRC, screening should begin at age 40 or 10 years younger than the age of the relative's diagnosis, with colonoscopies every five years for those with two first-degree relatives affected (Rex et al., 2009).

4.5.2.1 Colonoscopy Surveillance after Polyp Excision

Patients with adenomas require surveillance colonoscopies, with intervals based on risk. Low-risk abnormalities should be re-evaluated every five to ten years, while high-risk polyps necessitate surveillance every three years. For serrated polyposis syndrome, a colonoscopy is recommended within a year of diagnosis (Guarinos et al., 2012).

4.5.2.2 Surveillance After Colorectal Cancer

Prior to surgery, remaining colon tissue should be examined, and surveillance colonoscopies should occur three to six months post-resection. Normal findings lead to biennial colonoscopies for five years, with intervals adjusted based on findings (Rex et al., 2006).

4.5.2.3 *Preventive Measures and Treatment Options*

4.5.2.3.1 Strategies for prevention

- **Primary prevention:** Reducing alcohol, smoking, and obesity, along with a healthy diet, can lower CRC risk (Vieira et al., 2017).
- **Secondary prevention:** Early detection through colonoscopy and stool tests is crucial (Reddy, 2007).
- **Tertiary prevention:** Lifestyle changes post-diagnosis and potential chemoprevention may improve survival rates (Li et al., 2015).

4.5.2.3.2 Treatment strategies

- **Medications:** COX-2 inhibitors and NSAIDs may reduce polyp formation but can cause gastrointestinal bleeding.
- **Surgery:** Total colectomy may be indicated for high-risk patients, performed via open or laparoscopic techniques (Safiejko et al., 2021).
- **Gene therapy:** Targeting gene mutations may help manage CRC (Armaghany et al., 2012).
- **Precision therapy:** New treatments like bevacizumab and cetuximab improve survival rates (Xie et al., 2020).
- **Immunotherapy:** Monoclonal antibodies and immune checkpoint inhibitors show promise for treating metastatic CRC (Cunningham et al., 2004).

4.6 CASE STUDIES AND REAL-LIFE EXAMPLES

4.6.1 Case 1: A Family History of Colorectal Cancer

The correlation between CRC susceptibility and family history strengthens with the number of affected relatives, their genetic relatedness, and age at diagnosis. Observational studies show that individuals with a positive family history have approximately double the CRC risk (Fuchs et al., 1994). Key findings include:

- A Danish study reported 1.6 times higher risk in mothers and 1.9 times in fathers of individuals diagnosed with CRC before age 60 (Søndergaard et al., 1991).
- An Australian study indicated an 1.8 times increase in risk with a relative diagnosed with CRC (St John et al., 1993).
- A US study found a 1.72 times higher risk for those with a family history, increasing to 2.75 for individuals with two or more affected relatives, and 5.37 for those under 50 (Fuchs et al., 1994).
- Another US study noted a 2.2 times increased risk among those with any affected ancestors. (Slattery & Kerber, 1994)

Higher risks are noted for individuals with an immediate family member diagnosed under age 55 or multiple relatives with CRC (Slattery & Kerber, 1994; Fuchs et al., 1994). Family predisposition correlates with a 2–4 times increased CRC risk, with about 30% of cases having familial ties and 5% linked to inherited mutations (Siebenhofer et al., 2015). Pathogenic germline mutations account for approximately 5% of all CRC diagnoses and 10–15% of early-onset cases (Stoffel et al., 2015). Emerging genetic findings indicate the need for increased colonoscopy surveillance for high-risk individuals.

4.6.2 Case 2: Identifying a Genetic Mutation

Recent research by Chen et al. (2024) identified TRPSI and METRNL as new genes with carcinogenic roles in CRC, confirming the potential of C14orfl66, further enhancing our understanding of genetic alterations in CRC.

Advancements in genetic testing are transforming the approach to familial cancer syndromes. Current methodologies include comprehensive gene panels for polyposis disorders and Lynch syndrome. Although large deletions may evade detection in exome or panel sequencing, MLPA testing remains essential for reliable identification of specific disorders. Next-generation sequencing is more cost-effective than traditional Sanger sequencing, which often requires sequential testing of multiple genes (Cancer guidelines, 2023). A Moroccan case study highlighted the benefits of oncogenetic consultation and family investigation in managing hereditary CRC (Outtaleb et al., 2021).

4.6.3 Case 3: Management and Outcomes

Management, recurrence, and survival rates in CRC vary by stage (Stintzing, 2014). Personal stories of Lynch syndrome and CRC survivors emphasize the importance of genetic screening:

- A 47-year-old woman with Stage IIIb CRC due to an MSH2 mutation underwent surgery and chemotherapy, opting for reproductive organ removal to mitigate cancer risks (Centers for Disease Control & Prevention, 2023).
- An early-onset CRC survivor diagnosed at 28 discovered a familial predisposition through genetic testing. Early screening later revealed his son carried the same genetic mutation.
- A 40-year-old woman with Lynch syndrome underwent hysterectomy and oophorectomy to reduce her cancer risk (Centers for Disease Control & Prevention, 2023).
- A 45-year-old CRC patient is in good health at 75, thanks to early screening, which also revealed his son carries the mutation allowing for preventive care and regular polyp removal (Centers for Disease Control & Prevention, 2024).

These cases illustrate that successful management and survival depend on early detection and precision therapeutics in CRC.

4.7 FUTURE DIRECTIONS AND RESEARCH

Future directions in CRC research encompass several cutting-edge areas poised to revolutionize diagnosis and treatment. Liquid biopsies offer a non-invasive approach to detect tumor DNA and cells in bodily fluids, enabling early detection, recurrence monitoring, and treatment response assessment. Immunotherapy, while promising, requires further refinement to benefit a broader range of CRC patients. Nanotechnology holds potential for developing novel diagnostics, utilizing nanomaterials to detect CRC biomarkers for early diagnosis and personalized treatment. Larger animal models, such as dogs and pigs, may provide deeper insights into CRC research, surpassing traditional mouse models. In vivo and in vitro models will continue to be crucial for replicating tumor development and testing novel therapies. Research into autophagy, a process involved in cancer progression, will likely focus on its role in the tumor microenvironment and drug resistance. Addressing T-cell exhaustion, a major factor in poor immunotherapy responses, is another key research goal. RNA-based therapies and oncolytic viruses show promise, but challenges like RNA delivery and stability must be addressed through ongoing research and technological advances (Alshaer et al., 2021).

4.8 CONCLUSION

Significant advancements in understanding the genetic and molecular characteristics of CRC have improved its detection, prevention, and treatment. However, the complexity of CRC, particularly in metastatic cases, continues to pose challenges. Genetic testing plays a crucial role in identifying predispositions, guiding early interventions, and personalizing treatments. Next-generation sequencing (NGS) has streamlined the diagnosis of conditions like Lynch syndrome. Looking ahead, novel therapies such as immunotherapy, gene therapy, and targeted treatments show promise, but further validation is needed. Precision medicine, driven by molecular profiling and biomarker development, offers potential for more effective CRC management and patient outcomes.

REFERENCES

Ahadova, A., et al. (2018). Three molecular pathways model colorectal carcinogenesis in Lynch syndrome. *Int J Cancer*, 143(1):139–150.

Alexander, J., et al. (2001). Histopathological identification of colon cancer with microsatellite instability. *Am J Pathol*, 158(2):527–535.

Alshaer, W., et al. (2021). siRNA: Mechanism of action, challenges, and therapeutic approaches. *Eur J Pharmacol.*, 905:174178.

Armaghany, T., et al. (2012). Genetic alterations in colorectal cancer. *Gastrointest Cancer Res*, 5(1):19–27.

Bell, D. W., et al. (1999). Heterozygous germ line hCHK2 mutations in Li-Fraumeni syndrome. *Science*, 286(5449):2528–2531.

Boland, C. R., Goel, A. (2010). Microsatellite instability in colorectal cancer. *Gastroenterol*, 138(6):2073–2087. e3.

Boparai K, et al. (2008). Hyperplastic polyps and sessile serrated adenomas as a phenotypic expression of MYH-associated polyposis. *Gastroenterol*, 135/6:2014–2018.
Broderick, P., et al. (2007). A genome-wide association study shows that common alleles of SMAD7 influence colorectal cancer risk. *Nature Genet*, 39(11):1315–1317
Buskermolen, M., et al. (2019). Colorectal cancer screening with faecal immunochemical testing, sigmoidoscopy or colonoscopy: a microsimulation modelling study. *BMJ Clin Res*, 367:l5383.
Cancer guidelines. (2023). High risk family syndromes. Available at: https://wiki.cancer.org.au/australia/Guidelines:Colorectal_cancer/High-risk_familial_syndromes
Center, M. M., et al. (2009). Worldwide variations in colorectal cancer. *CA*, 59(6):366–378.
Centers for disease control & prevention. (2023). Family stories; available at: https://www.cdc.gov/genomics/disease/colorectal_cancer/family_stories.htm
Centers for disease control & prevention. (2024). Colorectal cancer screening stories. Available at: https://www.cdc.gov/cancer/colorectal/basic_info/stories.htm
Cheadle, J.P., Sampson, J.R. (2007) MUTYH-associated polyposis: From defect in base excision repair to clinical genetic testing. *DNA Repair*, 6:274–279.
Chen, Z., et al. (2024). Novel insights into genetic susceptibility for colorectal cancer from transcriptome-wide association and functional investigation. *J Natl Cancer Inst.*, 116(1):127–137.
Ciepiela, I., et al. (2024). Tumor location matters, next generation sequencing mutation profiling of left-sided, rectal, and right-sided colorectal tumors in 552 patients. *Sci Rep*, 14(1), 4619.
Cunningham, D., et al. (2004). Cetuximab monotherapy and cetuximab plus irinotecan in irinotecan-refractory metastatic colorectal cancer. *New Engl J Med*, 351(4):337–345.
Davies, H., et al. (2002). Mutations of the BRAF gene in human cancer. *Nature*, 417(6892):949–954.
Dekker, E., et al. (2019). Colorectal cancer. *Lancet*, 394(10207):1467–1480.
Dominguez-Valentin, M., et al. (2020). Cancer risks by gene, age, and gender in 6350 carriers of pathogenic mismatch repair variants: findings from the Prospective Lynch Syndrome Database. *Genet Med*, 22(1), 15–25.
Fearon, E. R., Vogelstein, B. (1990). A genetic model for colorectal tumorigenesis. *Cell*, 61(5):759–767.
Ferlay, J. et al. (2020). *Global Cancer Observatory: cancer today*. Lyon, France: International Agency for Research on Cancer. Available from: https://gco.iarc.who.int/today (accessed October 2024).
Fuchs, C. S., et al. (1994). A prospective study of family history and the risk of colorectal cancer. *New Engl J Med*, 331:1669–1674.
Giardiello, F.M., et al. (2014). Guidelines on genetic evaluation and management of Lynch syndrome: a consensus statement by the US multi-society task force on colorectal cancer. *Am J Gastroenterol*, 109(8):159–179.
Grady W. M. (2003). Genetic testing for high-risk colon cancer patients. *Gastroenterol*, 124(6):1574–1594.
Guarinos C, et al., (2014). Prevalence and characteristics of MUTYH-associated polyposis in patients with multiple adenomatous and serrated polyps. *Clin Cancer Res*, 20/5:1158–1168.
Guarinos, C., et al. (2012). Serrated polyposis syndrome: molecular, pathological and clinical aspects. *World J Gastroenterol*, 18(20):2452–2461.
Hahn, S. A., et al. (1996). DPC4, a candidate tumor suppressor gene at human chromosome 18q21.1. *Science*, 271(5247):350–353.
Haiman, C. A., et al. (2007). A common genetic risk factor for colorectal and prostate cancer. *Nat Genet*, 39(8):954–956.
Half, E., et al. (2009) Familial adenomatous polyposis. *Orphanet J Rare Dis*, 4:22.
Hemminki, A., et al. (1998). A serine/threonine kinase gene defective in Peutz-Jeghers syndrome. *Nature*, 391(6663):184–187.
Houlston, R. S., et al. (2008). Meta-analysis of genome-wide association data identifies four new susceptibility loci for colorectal cancer. *Nature Genet*, 40(12):1426–1435.
Howe, J. R., et al. (1998). Mutations in the SMAD4/DPC4 gene in juvenile polyposis. *Science*, 280(5366):1086–1088.
Hugot, J. P., et al. (2001). Association of NOD2 leucine-rich repeat variants with susceptibility to Crohn's disease. *Nature*, 411(6837):599–603.
Jaeger, E., et al. (2008). Common genetic variants at the CRAC1 (HMPS) locus on chromosome 15q13.3 influence colorectal cancer risk. *Nature Genet*, 40(1):26–28.
Jenne DE et al. (1998). Peutz-Jeghers syndrome is caused by mutations in a novel serine threonine kinase. *Nat Genet*, 18(1):38–43.
Johnson, C. M., et al. (2013). Meta-analyses of colorectal cancer risk factors. *Cancer Causes Control*, 24(6):1207–1222.

Jones, S, et al. (2002). Biallelic germline mutations in MYH predispose to multiple colorectal adenoma and somatic G:C–>T:A mutations. *Hum Mol Genet*, 11(23):2961–2967.
Joranger, P., et al. (2020). Survival and costs of colorectal cancer treatment and effects of changing treatment strategies: A model approach. *Eur J Health Econom*, 21(3):321–334.
Leoz, M. L., et al. (2015). The genetic basis of familial adenomatous polyposis and its implications for clinical practice and risk management. *Application Clin Genet*, 8:95–107.
Levine A. J. (1997). p53, the cellular gatekeeper for growth and division. *Cell*, 88(3):323–331.
Li, P., et al. (2015). Aspirin use after diagnosis but not prediagnosis improves established colorectal cancer survival: A meta-analysis. *Gut*, 64(9):1419–1425.
Malumbres, M., Barbacid, M. (2003). RAS oncogenes: the first 30 years. Nature reviews. *Cancer*, 3(6):459–465.
Mao, R., et al. (2021). Genetic testing for inherited colorectal cancer and polyposis, 2021 revision: A technical standard of the American College of Medical Genetics and Genomics (ACMG). *Genet Med*, 23(10):1807–1817.
McGarrity, T. J., et al. (2000). Peutz-Jeghers syndrome. *Am J Gastroenterol*, 95(3), 596–604.
Morin, P.J., et al. (1997). Activation of beta-catenin-Tcf signaling in colon cancer by mutations in beta-catenin or APC. *Science*, 275(5307):1787–1790.
Müller, M.F. et al. (2016). Molecular pathological classification of colorectal cancer. *Virchows Archiv*, 469(2):125–134.
Munteanu, I., Mastalier, B. (2014). Genetics of colorectal cancer. *J Med Life*, 7(4):507–511.
Nelen, M. R., et al. (1999). Novel PTEN mutations in patients with Cowden disease: absence of clear genotype-phenotype correlations. *Eur J Human Genet*, 7(3):267–273.
Niedermaier, T., et al. (2017). Diagnostic performance of flexible sigmoidoscopy combined with fecal immunochemical test in colorectal cancer screening: meta-analysis and modeling. *Eur J Epidemiol*, 32(6):481–493.
Outtaleb, F. Z. et al. (2021). Lynch syndrome or hereditary non polyposis colorectal cancer (HNPCC) in a Moroccan family: Case report. *Annals of Medicine and Surgery*, 62:123–126.
Palmqvist, R., et al. (2003). Prediagnostic levels of carcinoembryonic antigen and CA 242 in colorectal cancer: a matched case-control study. *Dis colon Rectum*, 46(11): 1538–1544.
Paula, M. C., Frucht, H. (2002). The genetics of colorectal cancer. *Ann Intern Med*, 137:603–612.
Pearlman, R., et al. Ohio Colorectal Cancer Prevention Initiative Study Group (2017). Prevalence and Spectrum of Germline Cancer Susceptibility Gene Mutations Among Patients With Early-Onset Colorectal Cancer. *JAMA Oncol*, 3(4):464–471.
Pineda, M., et al. (2010). Detection of genetic alterations in hereditary colorectal cancer screening. *Mutation Res*, 693(1–2):19–31.
Reddy B. S. (2007). Strategies for colon cancer prevention: combination of chemopreventive agents. *Sub-cellular Biochem*, 42:213–225.
Rex, D. K., et al. (2009). American College of Gastroenterology guidelines for colorectal cancer screening 2009 [corrected]. *Am J Gastroenterol*, 104(3):739–750.
Rex, D. K., et al. (2006). Guidelines for colonoscopy surveillance after cancer resection: a consensus update by the American Cancer Society and the US Multi-Society Task Force on Colorectal Cancer. *Gastroenterol*, 130(6):1865–1871.
Rohlin, A. (2014). *Hereditary colorectal cancer*. Gothenburg, Sweden: Ineko.
Safiejko, K., et al. (2021). Robotic-assisted vs. standard laparoscopic surgery for rectal cancer resection: a systematic review and meta-analysis of 19,731 patients. *Cancers*, 14(1):180.
Sampson JR et al. (2003), Autosomal recessive colorectal adenomatous polyposis due to inherited mutations of MYH. *Lancet*, 362:39–41.
Samuels, Y., et al. (2005). Mutant PIK3CA promotes cell growth and invasion of human cancer cells. *Cancer Cell*, 7(6):561–573.
Siebenhofer, A., et al. (2015). Positive family history of colorectal cancer in a general practice setting [FRIDA.Frankfurt]: study protocol of a cross-sectional study. *BMC Cancer*, 15: 605. https://doi.org/10.1186/s12885-015-1600-7
Sierra, J., et al. (2006). The APC tumor suppressor counteracts beta-catenin activation and H3K4 methylation at Wnt target genes. *Genes Develop*, 20(5):586–600.
Slattery, M. L. and Kerber, R. A. (1994). Family history of cancer and colon cancer risk: the Utah Population Database. *J Natl Cancer Inst*, 86(21):1618–1626.
Søndergaard, J. O., et al. (1991). Cancer incidence among parents of patients with colorectal cancer. *Int J Cancer*, 47(2):202–206.

St John, D. J., et al. (1993). Cancer risk in relatives of patients with common colorectal cancer. *Ann Intern Med*, 118(10):785–790.

Stintzing S. (2014). Management of colorectal cancer. *F1000prime Reports*, 6:108.

Stoffel, E. M., et al. (2015). Hereditary CRC syndromes: American Society of Clinical Oncology Clinical Practice Guideline endorsement of the familial risk-CRC: European Society for Medical Oncology Clinical Practice Guidelines. *J Clin Oncol*, 33:209–17.

Tomlinson, I. P., et al. (2011). Multiple common susceptibility variants near BMP pathway loci GREM1, BMP4, and BMP2 explain part of the missing heritability of colorectal cancer. *PLoS Genet*, 7(6):e1002105.

Tomlinson, I., et al. (2007). A genome-wide association scan of tag SNPs identifies a susceptibility variant for colorectal cancer at 8q24.21. *Nature Genet*, 39(8):984–988.

Tomlinson, I. P., et al. (2008). A genome-wide association study identifies colorectal cancer susceptibility loci on chromosomes 10p14 and 8q23.3. *Nat Genet.*, 40(5):623–630.

Utsunomiya J et al. (1975). Peutz-Jeghers syndrome: its natural course and management. *Johns Hopkins Med J*, 136(2):71–82.

Vieira, A. R., et al. (2017). Foods and beverages and colorectal cancer risk: a systematic review and meta-analysis of cohort studies, an update of the evidence of the WCRF-AICR Continuous Update Project. *Ann Oncol*, 28(8):1788–1802.

Wang, H., et al. (2021). Variation rs9929218 and risk of the colorectal cancer and adenomas: a meta-analysis. *BMC Cancer*, 21:190.

Wang, J. Y., et al. (2004). Molecular detection of APC, K- ras, and p53 mutations in the serum of colorectal cancer patients as circulating biomarkers. *World J Surg*, 28(7):721–726.

Weisenberger, D. J., et al. (2006). CpG island methylator phenotype underlies sporadic microsatellite instability and is tightly associated with BRAF mutation in colorectal cancer. *Nature Genet*, 38(7):787–793.

Weiss, J. M., et al. (2021). NCCN Guidelines® insights: genetic/familial high-risk assessment: colorectal, Version 1.2021. *J Nat Comp Cancer Netw*, 19(10), 1122–1132.

Weissman, S. M., et al. (2011). Genetic counseling considerations in the evaluation of families for Lynch syndrome: a review. *J Genet Counsel*, 20(1):5–19.

Will, O. C., et al. (2009). Adrenal incidentaloma in familial adenomatous polyposis: a long-term follow-up study and schema for management. *Dis Colon Rectum*, 52(9):1637–1644.

Xie, YH., et al. (2020). Comprehensive review of targeted therapy for colorectal cancer. *Sig Transduct Target Ther*, 5:22.

Yang, H., et al. (2014). Meta-analysis of the rs4779584 polymorphism and colorectal cancer risk. *PloS One*, 9(2):e89736.

5 Screening and Early Detection Strategies

Mohammad Taghi Ahady, Marjan Assefi, Ananya Chakraborty, Manar G. Shalabi and Wael Hafez

5.1 INTRODUCTION

In the United States, colorectal cancer (CRC) is the second most common cause of cancer-related death, according to Lieberman (2009). It ranks third among men and second among women as the fourth leading cause of disease-related death overall. The expression "screening" alludes to the most common way of utilizing a demonstrative test to distinguish irregularities or precancerous or beginning phase malignant growth in a populace that gives off an impression of being solid yet is in danger of fostering a specific sickness. Interestingly, observation includes rehashing the previously mentioned assessment on people who have been determined to have disease or precancerous sores. The adenoma-carcinoma sequence is the process by which a significant number of colorectal cancers are formed from adenomas (Burnett-Hartman et al., 2021). Most adenomas with increasing dysplasia grow from small to large and eventually develop into invasive cancer. The development of the adenoma-carcinoma group in sporadic cases takes more than ten years, so screening measures are taken to stop this progression (Maida et al., 2017). Screening has demonstrated its viability by lowering death rates. About 60%–70% of cases of CRC are discovered at a later stage in unexamined cohorts. However, CRC screening procedures like colonoscopy and the fecal immunochemical test (FIT) increase the likelihood of five-year survival, speed up detection, and reduce mortality rates. The precise identification of colorectal lesions is essential for maximizing outcomes due to the central role that colonoscopy plays in CRC screening and monitoring (Maida et al., 2017). Proposals from the United States for colorectal malignant growth evaluation in the United States The Habitats for Infectious Prevention and Countermeasures (CDC) and the Preventive Administrations Team recommend screening with sigmoidoscopy, colonoscopy, or high-awareness waste mysterious blood tests (FOBT) for individuals between the ages of 50 and 75. This is essential for lead assessing for precancerous colorectal polyps or starting stage sickness before the start of secondary effects. Numerous challenges persist despite the significant reduction in CRC cases achieved through screening. Due to the complexity of primary prevention efforts, the focus remains on screening even though screening has not eliminated CRC (Ladabaum et al., 2020). A great number of qualified Americans have not gone through screening; it is guessed that somewhere in the range of 46% and 63% of CRC-related passings in the US are brought about by flawed screening entryways. Colorectal cancer is typically diagnosed using one of two methods: invasive or non-invasive. Intrusive systems include colonoscopy, adaptable sigmoidoscopy, virtual colonoscopy (CT colonography), twofold differentiation barium douche, and biopsies, anomalies, and direct perception of the colorectal location. In contrast, non-invasive tests like the stool DNA test (MT-sDNA), the fecal immunochemical test (FIT), and the fecal occult blood test (FOBT) look for colorectal cancer-associated blood, genetic markers, or DNA changes in the stool. The choice between intrusive and nonintrusive methods is influenced by individual gambling variables, side effects, and recommendations from medical service providers (Shaukat & Levin, 2022). In the diagnosis and treatment of colorectal sickness, both prominent and effortless methods expect a critical part, in achieving predominant outcomes and changing therapy decisions (Beniwal et al., 2023) (Figure 5.1).

5.2 ENDOSCOPY, COLONOSCOPY, AND SIGMOIDOSCOPY

Endoscopy is a common screening procedure for CRC and the subject of extensive research. It permits tissue biopsies and polyp removal simultaneously, enabling representation of the distal colon. There are several advantages to endoscopy over colonoscopy, including less preparation of the bowel, faster examination, fewer complications, and lower costs (Jain et al., 2022). However, its effectiveness in preventing proximal CRC is limited by its inability to investigate the proximal colon (Maida et al., 2017). The standard is still colonoscopy because of its direct visualization. Colonoscopy is the most common endoscopic procedure in the United States, and it is generally regarded as the standard for CRC observation and screening (Leung et al., 2023). The American Cancer Society recommended beginning CRC screening at age 45 in 2018 to improve prognosis and lessen the disease's impact on society, as stated by Thomas et al. (2023). It has been demonstrated that polypectomy and colonoscopy reduce the mortality and incidence of CRC, making

DOI: 10.1201/9781032704517-5

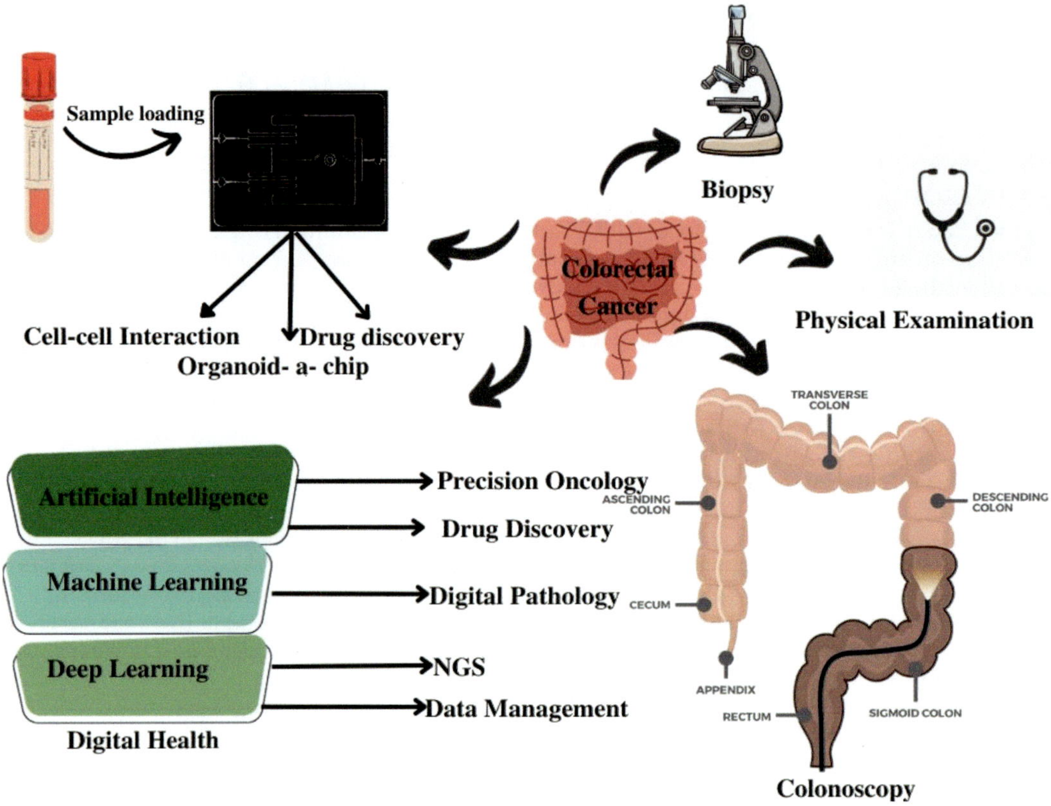

Figure 5.1 Diagnostic strategies in colorectal cancer.

them an essential part of effective preventative measures. According to the National Polyp Study, CRC mortality can be reduced by half compared to the general population's control groups when adenomas are removed during a colonoscopy (Kaltenbach et al., 2020). The planning of a colonoscopy after a positive stool-based screening has been entirely investigated due to the significance of this strategy following positive outcomes from stool tests (Forbes et al., 2021). The risk of malignant growth-related mortality is reduced by delaying a colonoscopy by more than a year after a positive stool test, according to American concentrations (Zhang et al., 2023a). Along these lines, Taiwanese exploration uncovers an expanded gamble of colorectal disease and high-level colon malignant growth stages III-IV on the off chance that a positive FIT is not circled back to a colonoscopy in six months or less. A microsimulation model study indicates that the overall benefits of screening will decrease by 10% if follow-up colonoscopy is delayed by a year (Rutter et al., 2022). Patients who do not show up for their appointments and issues with the provider or system, like not ordering a colonoscopy after a positive FIT result or not having a comprehensive tracking system for positive FIT results, are two reasons people do not get a colonoscopy after a positive stool screening. The USPSTF and the USMSTF have accepted circuitous visual screening tests for a five-year evaluation period as a semi-intrusive method for assessing CRC. These tests are like CT colonography (CTC) or virtual colonoscopy. Injuries bigger than 10 mm have awareness levels going from 67% to 94%. Despite this, there are clear limitations to this procedure, such as increased costs, coincidental sores requiring additional resources for evaluation with negligible benefit, and a lack of framework for performing a colonoscopy at the same time as CTC in most medical offices. Assuming polyps are found, extra entrail planning is expected for the resulting colonoscopy. The painless procedure known as colon container endoscopy (CCE) has been irrefutable since its new turn of events. In the wake of setting up the entrail, this original strategy includes taking a little case, which makes it conceivable to take various pictures of the colon (Iordache et al., 2023; Perlo, 2021). 2009 saw the presentation of Pillcam-2, a second-generation CCE. Its responsiveness and specificity of 88% and 82%, respectively, for injuries larger than 6 mm define its indicative exactness. The Food and Drug Administration (FDA) has approved the use of CCE in patients who have had a limited colonoscopy for specialized or other reasons (Jalayeri Nia et al., 2023; Lee & Schnoll-Sussman, 2022). However it is important to remember that none of the national societies have supported using CCE as a screening method. It is fundamental to recognize explicit constraints of CCE, like the potential for significant expenses, absence of repayment, and diminished awareness in distinguishing sessile serrated sores. The analysis and checking of colorectal malignant growth depend vigorously on colonoscopy. Direct visualization and a thorough examination of the entire colon and rectum are made possible by this procedure, which is the gold standard for screening and diagnosing CRC lesions in their early stages (Thattakath, 2023). Using a flexible colonoscope and an advanced imaging system, healthcare professionals can visually examine the mucosal lining of the rectum and colon (Patel et al., 2024). They can identify abnormal tissue, polyps, or tumours that are a sign of CRC more easily because of this. Furthermore, colonoscopy makes it conceivable to gather tissue tests (biopsies) for histopathological assessment, which helps with exact organizing and deciding the infection's forcefulness (Sanduleanu et al., 2010). Notwithstanding its demonstrative worth, colonoscopy is fundamental for post-treatment reconnaissance since it empowers the ID of new polyps or possible repeats in people with a CRC history. The overall efficacy of colonoscopy, which ranges from 93% to 94% and can be as high as 99.8%, emphasizes the significance of the procedure as a fundamental pillar in colorectal cancer prevention, early detection, and continuous monitoring.

However, a colonoscopy comes with many drawbacks and limitations, such as the possibility of perforation and adverse reactions to contrast agents (Bhagatwala et al., 2015). Additionally, the preparation of the bowel is required for the procedure, which can be costly, drain resources from the medical system, and cause additional stress to patients, particularly the elderly and those who are more vulnerable. Many precancerous polyps can be discovered during a colonoscopy, however only a small percentage of these lesions have the potential to develop into cancer (Winawer et al., 1990). Patients with polyps must have shorter surveillance intervals since definitive evidence about which individual polyps are at higher risk is lacking (Gupta & East, 2020). Without clearly defined benefits, this results in an increased demand for healthcare services and additional strain on the healthcare system (Beniwal et al., 2023). A colonoscopy is the final phase of evaluation in each program to screen for colorectal cancer. Numerous large-scale cohort studies have demonstrated that a colonoscopy is a primary screening test that is both safe and effective (Shaukat & Levin, 2022). These studies indicate that 0.5 to 1.0% of people with an average risk profile are diagnosed with colon cancer and 5 to 10% have advanced neoplasia that can be removed. Colonoscopy has been connected to a diminishing in colorectal disease frequency and death rates on the off chance of control studies. In any case, no randomized controlled preliminary studies have directly compared the outcomes of a colonoscopy to those of other types of screening. As per various examinations, colorectal malignant growth might be created in 0.3 to 0.9 percent of patients who had an adenoma found and eliminated during a screening colonoscopy. Some of the cancers that were discovered during the subsequent colonoscopy may have been caused by missed lesions. Lesions with a diameter of 10 mm or greater may be overlooked in 2 to 12% of patients. Identifying flat adenomas can be particularly difficult and may necessitate specialized techniques. The painstakingness of the assessment and the total expulsion of all polyps are essential for colonoscopy to lessen the gamble of proximal colon malignant growth. Measurement tools for evaluating quality are now readily available, and these concerns highlight the significance of monitoring and improving the quality of colonoscopy (Lieberman, 2009). In case-control studies, the use of sigmoidoscopy has been linked to a reduction in colorectal cancer mortality, particularly in the section of the colon being examined. There was no evidence that participants in screening sigmoidoscopy had a lower risk of colorectal cancer in a large, randomized, and controlled trial. I addition, an assessment observing the assumption to-treat guideline showed an immaterial verifiable drop in mortality keeping six years in these individuals diverged from the benchmark bunch. However, screening colonoscopy has shown that more than 30% of patients with innovative neoplasia have proximal injuries that sigmoidoscopy would miss. This happens more frequently to women than men, and it also happens more frequently to people over 60 who look different when compared to younger partners. A clinical visit and planning of the entrail are expected for the method, which regularly causes some distress. Because of the chance of sedation, which empowers an exhaustive colon assessment and polypectomy, colonoscopy arises as a greater choice than sigmoidoscopy for most medical services suppliers and patients (Thorne, 2008). From a financial standpoint, sigmoidoscopy reimbursement is relatively modest given the resources involved. These various restrictions have led to underutilization in the United States (Lieberman, 2009).

5.3 BLOOD TESTING

Patients who are more likely to develop CRC can be distinguished by using specific demographic and hematologic data. One method for recognizing individuals in danger is to utilize the total blood count (CBC), particularly if microcytic iron lacks pallor is available. In a retrospective case-control study, anemia and elevated red cell distribution width (RDW) were found to be common in colon cancer patients. People with right-sided colon cancer had more RDW. Moreover, an elevated RDW in colorectal illness patients is related to a raised setback risk (Ye et al., 2016). The speculation sets that steady depleting associated with dangerous development in CRC patients prompts the absence of iron slightness and positive baffling blood test achieves stools, a quirk more prevalent in additional carefully prepared patients. CRC risk factors in the general population can be identified by utilizing age, CBC, and demographic data from artificial intelligence (AI). Goyal et al. (2020) state that research on this novel approach to CRC risk assessment has yielded promising results (Goyal et al., 2020). Precancerous polyps and early-stage colorectal cancer can only be detected through blood tests, which are necessary to increase screening acceptance, reduce risks, and reduce costs. An option for individuals who would rather not utilize other screening techniques is a blood test that has been endorsed by the FDA and focuses on the methylation of septin 9 DNA. A recent meta-analysis of 19 trials in the screening population found a combined sensitivity and specificity of 69% to 92% for the diagnosis of colorectal cancer, as stated by Kanth and Inadomi (2021). In 2008, Lofton-Day developed the concept of a plasma Septin9 methylation assay. Epigenomics promoted it as the Epi proColon 1.0 following its underlying turn of events, encouraging comparative tests on various substances. The Epi 455 proColon 2.0 has since been introduced, showing additionally created responsiveness and distinction in arranged case-control and sidekick studies, including the PRESEPT study. Despite this, several studies have shown that the Septin9 assay's low sensitivity casts doubt on its current efficacy in detecting adenomas. The Multi-Society Task Force on Colorectal Cancer says that US regulations specifically say that it should not be used for colorectal cancer screening (Jones et al., 2010). Despite not being suitable as a primary screening tool, this assay may still have a limited role, albeit cautiously, in patients who reject all other screening methods (Maida et al., 2017).

5.4 STOOL TESTING

Numerous investigations have demonstrated a connection between the development of gastrointestinal adenoma, a precursor to colorectal malignant growth, and the microbiota in the stomach. A healthy gut microbiota, on the other hand, is closely linked to a lower risk of advanced adenoma. There is substantial evidence to suggest that the design and activity of the stomach microbiota are fundamentally influenced by numerous perceived risk factors for CRC. These modifications have an impact on the onset of CRC, immune responses, host metabolism, and cancer-related genomics and epigenomics. A diet high in red and processed meat is known to be bad for health because it causes the production of short-chain fatty acids (SCFAs) through bacterial fermentation. This is like how more fiber in the diet makes beneficial bacteria like Bifidobacterium and Lactobacillus spp. grow. This is because these meats contain both organic and sulfur-containing amino acids. As a result, there are a lot of bacteria that can digest these protein-rich foods. Polyamines, hydrogen sulfide, and ammonia, among other metabolites, have been shown to have genotoxic effects on the gut. Thus, the movement of colonic carcinogenesis is helped by this DNA harm in digestive epithelial cells. The gut microbiota of people with CRC consists of several distinct bacterial species, including fusobacterium, solo bacterium, bacteroides, pepto streptococcus, and parvimonas. Strikingly, their stomach microbiota varies fundamentally from healthy people concerning assortment and construction. These microbial compositions can be found in people of all races and ethnicities, according to some studies. Additionally, it may be possible to use certain microbial genes as early-stage CRC diagnostic indicators due to their prevalence. The immune system is unavoidably triggered by the gut microbiota and its byproducts due to their proximity to the colonic mucosa, possibly resulting in persistent low-grade inflammation. Persistent inflammation is a known risk factor for numerous cancers, including CRC. Several additional studies, on the other hand, have demonstrated a link between various kinds of cancer and persistent inflammation. Under normal physiological conditions, the host and the bacteria in the gut coexist in equilibrium, which must be emphasized. Regardless, when the equilibrium of the stomach microbiome is upset by components like alcohol use, unequivocal dietary penchants, or even the use of counter agents' poisons, outrageous intricacies could result. According to Olovo, Huang, Zheng, and Xu (2021), dysbiosis, a disturbance in the gut, is a significant risk factor for several diseases, including CRC, type 2 diabetes, cardiovascular disease, and inflammatory bowel disease (Olovo et al., 2021). Various specialists have connected dysbiosis, a condition brought about by unevenness in the gastrointestinal microbiota, to the beginning of adenomas and CRC. Unrelenting irritability is speculated to be a significant force to be reckoned with, even though the specific mechanism by which dysbiosis may influence CRC pathogenesis remains largely undefined. Zeller and colleagues discovered a large cluster of provocative and pathogenic cycles originating from a wide variety of gram-negative microorganisms simultaneously with the observed metabolic changes in the microbiome of CRC patients (Olovo et al., 2021). The evaluation of waste samples has revealed flaws in the composition and abundance of the stomach microbiota in healthy individuals as well as those with adenoma or CRC (Nistal et al., 2015). This demonstrates how the waste microbiome might be useful for quickly and safely determining CRC. Different research methods have been utilized to examine the profiles of the gut microbiome in CRC. The shotgun metagenome sequencing of excrement and the enhancement and sequencing of different variable districts (V1, V2, V4) of the 16S rRNA acquired from waste DNA are remembered for these techniques. qPCR has also been used in some studies to compare the levels of microbial gene markers in the specimens of interest to certain controls (Olovo et al., 2021).

5.4.1 Molecular Biomarkers in Stool

The most crucial rule of waste testing is the possibility that both early and high-level precancerous irregularities can cause death. Using existing technology, this can be determined by discharging platelets into the digestive lumen. Due to the gastrointestinal parcel's soluble environment, using excellent separated DNA has advantages. The methylation detection method of the gene in feces is becoming increasingly recognized as a noninvasive method for diagnosing colorectal cancer because it can identify abnormal methylation in the early stages of the disease. In cases of cancer, feces may also contain a significant amount of cancer cells and tumor DNA. A less invasive and highly effective colorectal cancer screening method was developed as a result of the discovery of abnormal gene methylation in feces. Single genes such as BMP3, family member 4, and secreted frizzled-related protein 2 (SFRP2) can be utilized in fecal analysis. There are also tests based on WNT inhibitory factor 1 (WIF-1) and SFRP2 (Wu et al., 2020). RNA particles, particularly microRNAs, are RNA particles with solitary strands that are under 22 nucleotides in length and can impede courier RNA interpretation and work with mRNA corruption. Quality articulation of miRNAs is a crucial transcriptional administrative component and a crucial modulator in tumorigenesis, according to ongoing research. Due to their tissue-explicit articulation, microRNAs, which can be found in various organic examples like blood, pee, and tissue liquid, have a great deal of potential as biomarkers. They intricately regulate the signaling pathways of cancer cells as well as various physiological and pathological processes. Research has demonstrated the way that miRNAs can be dependably recognized in squander tests. Patients with colorectal cancer and healthy controls have distinct differences in the articulation of their miRNAs that can be reliably separated. To decide if colorectal cancer is present, RNA from shed development cells on the stomach-related lumen walls is assessed as a feature of the evaluation. Colorectal malignant growth patients express miR-92a, miR-20a, miR-21, miR-135, and miR-144* in various ways, according to important examinations. However, sample sizes are small and the methods used in the current studies on waste DNA are still in their infancy. Regardless, waste miRNA is a promising marker for the conclusion of colorectal malignant growth because miRNA can definitively demonstrate the movement and improvement of cancers (Wu et al., 2020). According to a protein point of view, the guaiac-based squander puzzling blood test (gFOBT) and squander immunochemical test (FIT) can recognize hemoglobin in excrement. However, some districts have chosen not to use gFOBT for colorectal malignant growth detection or have restricted its use to a useful symptomatic tool due to its limited explicitness. In contrast to popular belief, FIT is thought to be a novel guaiac-based FOBT cycle that avoids cross-reactivity with beef hemoglobin and uses a globulin immune response unaffected by food sources with increased peroxidase activity.

The fecal hemoglobin (f-Hb) concentration can be estimated numerically using quantitative fitting. Notwithstanding having moderately low mathematical qualities for f-Hb, FIT recognizes colorectal disease with a serious level of responsiveness. Wu et al. state that when symptoms are particularly severe, rapid colorectal cancer screening with FIT is a viable option (Wu et al., 2020). When it comes to identifying advanced-stage colorectal cancers, both the fecal immunochemical tests (FITs) and the fecal immunochemical test-DNA (Cologuard) assay, which are based on stools, have a relatively higher level of sensitivity; however, their capacity to identify cancers that are still in the early stages of the disease is relatively limited. Waste immunochemical tests (FITs) and the

FIT-DNA (Cologuard) test, both based on stools, frequently distinguish between advanced-stage colorectal tumors. However, they are only able to identify 28% (FIT) to 40% (FIT-DNA) of advanced adenomas, which limits their ability to identify cancers in their early stages. Patient adherence is essential to the successful implementation of these assessments. The noninvasive stool examination known as FIT is currently the method that is utilized the most frequently in two-phase population-based screening initiatives. The tolerably low cost per evaluation accounts for this tendency. The US Multi-society Team (USMSTF) on Colorectal Malignancy currently supports FIT, which is the standard for most tests designed to approve alternative painless markers. A single FIT assessment for colorectal cancer may have different responsiveness and specificity depending on the point of inspiration. However, regardless of the threshold chosen, the sensitivity to identify advanced adenomas remains low. Because it focuses on the biological changes that underlie the neoplastic process, a molecular strategy for colorectal cancer screening is appealing. A panel of molecular markers in a multitarget assay demonstrated that altered DNA can be found in stool. The DeeP-C study, which included close to 10,000 people who were at average risk, found that a panel that included a variety of DNA markers in addition to FIT had a sensitivity of 92.3%, in comparison to 73.8% for FIT alone (with colonoscopy serving as the standard). The panel had a sensitivity of 42.4 percent when it came to identifying advanced precancerous lesions with a minimum size of 1 cm, such as advanced adenomas or sessile serrated adenomas (SSA), whereas the FIT alone had a sensitivity of 23.8%. Individually, members with nonadvanced or negative colonoscopy findings had specificities of 86.6% and 94.9% for the DNA and FIT tests, respectively. In accordance with the USMSTF guidelines (recommended predictably; level 2 proposition), the multi-assigned stool FIT-DNA test (Cologuard, Unequivocal Sciences) has received support from the FDA, the Center for Governmental Clinical Consideration, and Medicaid Organizations (CMS). Due to the high number of colonoscopies required for every life-year gained, the modeling analysis performed for the American Cancer Society (ACS) did not recommend it as a test. (Bresalier et al., 2020) assert that the earlier GLNE 10 trial sought to contrast FIT-DNA-discovered biomarkers with those discovered by EDRN.

5.5 NANOPARTICLES

The role of nanotechnology in advancing colorectal cancer diagnosis through various innovative approaches. It showcases different types of nanoparticles, such as carbon-based, lipid-based, and quantum dots, along with gold nanoparticles, nanoprobes, and nano-bioconjugates (Figure 5.2). These nanoparticles enable improved techniques like nanoprobe-based colonoscopy, advanced nano-imaging, nano-sensors for early detection, biomarker detection, and liquid biopsies. Collectively, these nano-innovations enhance the precision, sensitivity, and efficiency of colorectal cancer detection, leading to earlier diagnosis and better patient outcomes.

Recently, the topic of nanotechnology has emerged as one that is developing at a rapid pace in terms of the development of new diagnostics for CRC. Because of the special qualities and abilities of nanomaterials, there are new avenues for the precise and targeted identification of CRC biomarkers, which can lead to early detection and customized treatment plans. When nanotechnology is included in CRC diagnostics, it offers unmatched accuracy and effectiveness that surpasses traditional techniques. This is explained by the unique physicochemical properties of nanomaterials, which facilitate the rapid entrapment and improvement of biomarkers due to their large surface area-to-volume ratio. Moreover, nanotechnology-based approaches have the potential to allow for the simultaneous evaluation of multiple targets in a single experiment.

Furthermore, these methods can be tailored to accommodate a wide range of specimen types, including blood, tissue, and feces, thus increasing their applicability in CRC diagnosis. In this regard, we primarily focus on nanoparticles, including gold, magnetic, and quantum dots; dendrimers have also been thoroughly investigated for their potential use in CRC diagnosis (Beniwal et al., 2023).

5.5.1 Quantum Dots (QDs)

Quantum dots (QDs) have gained recognition as fluorescent semiconductor crystals at the nano-scale level, primarily due to their unique optical characteristics, including their impressive brightness, stability against light-induced degradation, and ability to adjust emission wavelengths. Through the attachment of ligands, QDs can be bound to specific target molecules such as cancer cells or tumor vasculature, allowing for precise and targeted visualization of molecular markers. Consequently, the utilization of QDs presents a promising opportunity for the advancement of sophisticated imaging methodologies in the realm of CRC research and diagnostics (Beniwal et al., 2023).

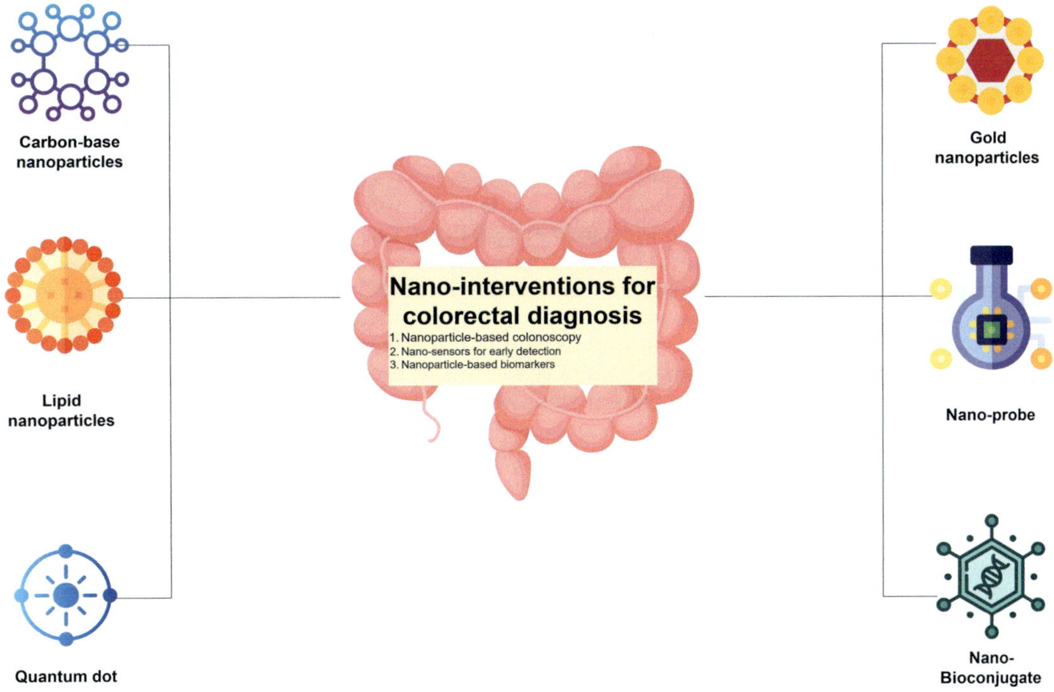

Figure 5.2 Nanoparticles for diagnosis of colorectal cancer.

5.6 CARBON-BASED NANOPARTICLES

There are several advantages of using carbonaceous nanoparticles, such as carbon nanotubes, graphene, and carbon dots, in the diagnosis of CRC. When compared to gold and quantum dots (QDs), carbon nanoparticles exhibit exceptional biocompatibility, which guarantees their safe application in vivo with little negative health effects. These nanoparticles also have a large surface area and good electrical conductivity, which makes them excellent candidates for functionalization with particular targeted ligands or biomolecules. Furthermore, they work with electrically based sensors, which are frequently used in point-of-care systems. Examples of these sensors are impedance and electrochemical sensors. Yan et al., for instance, have published a study that evaluates the effectiveness of carbon nanoparticles in identifying lymph node metastases in early-stage (T1-2) colorectal cancer.

The trial cohort consisted of people who had undergone submucosal injections of carbon nanoparticles around the tumor site following surgical resection for T1-2 CRC. By using near-infrared fluorescence imaging, the results demonstrated how well the nanoparticles performed in tracking lymphatic drainage and accurately locating sentinel lymph nodes in T1-2 CRC. Significantly, compared to conventional histological analysis, the carbon nanoparticle-guided sentinel lymph node biopsy demonstrated a higher detection rate for lymph node metastasis, with a heightened sensitivity of 91.67% and a 100% specificity. In a different study, Cai et al. investigated the efficacy of methylene blue and activated carbon nanoparticle dispersion for lymph node staining in colorectal cancer.

In an in vivo study, mice with colorectal cancer were given a suspension of activated carbon nanoparticles via injection into a lymphatic conduit. This effectively stained the lymph nodes in the area and made them easier to see. Moreover, an in vitro study that stained lymph nodes taken from CRC patients showed that the staining efficacy of activated carbon nanoparticle suspension was similar to that of methylene blue. This suggests that activated carbon nanoparticles may be used as a substitute for methylene blue when it comes to staining lymph nodes in CRC patients. In a different trial, Wang et al. used endoscopic assessment prior to surgery to inject carbon nanoparticles into the submucosal layer surrounding the tumor site. The carbon nanoparticles served as a navigational aid for locating and following lymph nodes during laparoscopic surgery.

The outcomes showcased the successful preoperative endoscopic localization of CRC utilizing carbon nanoparticles in all enrolled patients during surgery. Collectively, the integration of carbon nanoparticles into laparoscopic procedures enabled real-time visualization and enhanced precision in lymph node dissection, offering substantial promise for optimizing surgical outcomes, decreasing complications, and improving overall patient care in CRC (Beniwal et al., 2023).

5.7 LIPID-BASED NANOPARTICLES

The use of lipid-based nanoparticles, which are made of phospholipids or cholesterol, has several benefits for the detection of CRC. Lipid-based nanoparticles, like other nanoparticles, can be modified with certain ligands or antibodies to identify CRC biomarkers or antigens associated with tumors in a tumor-selective manner. Moreover, lipid-based nanoparticles can encapsulate contrast agents or fluorescent dyes, which helps improve the visibility of colorectal cancer lesions using a variety of imaging techniques, such as MRI or fluorescence imaging. Lipid-based nanoparticles are special because of their stability, ability to transfer imaging payloads, and compatibility with biological systems. These characteristics make them ideal for accurate and non-invasive colorectal cancer diagnostics.

Nanotechnology-driven advancements in CRC detection bring about significant limitations that necessitate careful scientific consideration. Of these limitations, the possibility of problems with biocompatibility and toxicity stands out the most. When used in diagnostic applications, nanoparticles may cause unanticipated interactions in biological systems, which could lead to unfavorable consequences. Therefore, it is crucial to guarantee these nanoparticles' safety for in vivo application by conducting in-depth toxicity assessments and meticulous empirical inspection. Furthermore, there may be substantial financial and technological barriers that prevent the large-scale development and manufacture of nanoscale diagnostic techniques, thereby impeding their widespread use in clinical settings. Moreover, the long-term stability and robustness of nanoscale diagnostic systems require scrutiny to confirm their reliable performance over time.

The complex regulatory processes that regulate the application of nanotechnology-based diagnostic tools provide significant temporal barriers, potentially delaying their instant integration into standard clinical practice. While there is hope that nanotechnology will improve CRC detection, these issues must be carefully addressed to fully realize the potential of nanotechnology to improve clinical management paradigms and early cancer identification (Beniwal et al., 2023).

5.8 BIOMARKERS

Biomarkers, according to the definition provided by the National Cancer Institute, are biological entities identifiable in blood, bodily fluids, or tissues, serving as indicators of normal or abnormal processes, conditions, or diseases, notably cancer. They play a pivotal role in distinguishing between patients with a particular condition and those who are healthy, with variations potentially arising from diverse factors such as post-translational modifications, genetic mutations, and transcriptional changes, among others. This broad category of biomarkers encompasses diverse molecules including proteins (e.g., enzymes, receptors), nucleic acids (e.g., microRNAs, non-coding RNAs), antibodies, and peptides, with identifiable changes including alterations in gene expression, proteomic profiles, and metabolomic patterns. For non-invasive and continuous monitoring, biomarkers can be detected in the circulation (e.g., whole blood, serum, plasma), excretions or secretions (e.g., stool, urine, sputum, nipple discharge), or obtained from tissues, sometimes necessitating a biopsy or specialized imaging techniques. Genetic biomarkers, both inherited and somatic, can be identified through sequence variations in DNA from different sources, such as whole blood, sputum, or buccal cells for inherited biomarkers, and tumor tissue for somatic biomarkers. These biomarkers possess the potential to provide insights into a patient's prognosis, representing the likelihood of disease recurrence post-cancer diagnosis, with modern methodologies contributing to more accurate prognostic predictions for specific malignancies. Moreover, biomarkers have the capacity to influence treatment response customization, acting as "predictive factors" that guide the selection of optimal therapeutic strategies, as illustrated by the predictive role of KRAS mutations in colorectal cancer treatment response to EGFR-targeted therapy (Rashid et al., 2023).

Biomarkers, characterized by molecular signatures, play a crucial role in cancer detection and the tailoring of treatment strategies, falling into diagnostic, prognostic, and predictive categories. These biomarkers offer a range of advantages across different disease stages, facilitating the assessment of disease progression, recurrence risk, and personalized treatment effectiveness evaluation. The primary goal is early colorectal cancer detection in asymptomatic individuals and the identification of high-risk lesions like pre-malignant polyps to enhance patient survival outcomes. Presently, the sensitivity of colorectal cancer detection using the fecal immunochemical test (FIT) at a threshold of 100ng/mL stands at 73.8%, contrasting with the 92.3% sensitivity of stool-based DNA assays targeting KRAS, NDRG4 aberrations, and BMP3 methylation.

Moreover, the sensitivity of FIT testing for advanced precancerous lesions is 23.8%, whereas stool DNA testing demonstrates a sensitivity of 42.4%. Furthermore, the detection rate of polyps with high-grade dysplasia is 46.2% using FIT testing, in contrast to 69.2% with stool DNA testing. Additionally, in the case of serrated sessile polyps larger than 1 cm, FIT testing achieves a detection rate of only 5.1%, while stool DNA sampling shows a significantly higher rate of 42.4%. These results highlight the constraints of current diagnostic screening approaches and the difficulties in identifying suitable surrogate markers for early disease detection. Present non-invasive screening methods for colorectal cancer exhibit inadequate sensitivity in detecting pre-cancerous lesions and may miss important cases of early-stage colorectal cancer. As a result, it is crucial to maintain a low threshold for more invasive colonoscopy in these individuals, and the creation of additional tools is vital to improving the recognition of early-stage colorectal cancer (Ogunwobi et al., 2020).

Second, the application of prognostic biomarkers is promising for predicting the course of the disease, including early recurrence and death. KRAS promotes cell proliferation and is a member of the TPases RAS proto-oncogene family. KRAS mutations have been linked to a lower overall survival rate following hepatic mastectomy in metastatic CRC and a higher likelihood of recurring metastatic CRC after curative resection. Furthermore, the RAS-RAF-MEK-ERK pathway, which controls cell transcription, is how the BRAF proto-oncogene functions. When compared to individuals with the BRAF wild type, the presence of the BRAF V600E mutation is associated with a worse survival rate, including progression-free survival and overall survival rates that can be as much as 50% lower.

When compared to each modality alone, a combination of radiological and genetic features may improve sensitivity for prognostic reasons in the developing field of radio genomics. Last but not least, although having low sensitivity and specificity, the high molecular weight glycoprotein carcinoembryonic antigen (CEA) is used as a biomarker to predict early recurrence in post-operative patients. Therefore, the use of prognostic markers may be able to modify the cutoff points for additional research on recurrent illness and offer chances for early intervention. They may also have an effect on the cutoff points for patients receiving more intensive forms of care. Predictive biomarkers are also used to customize therapy according to molecular subtypes. A restricted response to anti-EGFR receptor therapy, including cetuximab and panitumumab, has been associated with KRAS mutations.

When treated with FOLFIRI plus cetuximab, the overall response rate increased by 16% in KRAS wildtype patients and decreased by 4% in KRAS mutant patients. Considering that up to 40% of patients have KRAS mutations, a sizable percentage of patients may be able to avoid expensive anti-EGFR treatment. Additionally, diphosphate-glucuronosyltransferase 1A (UGT1A) metabolizes irinotecan, a topoisomerase inhibitor in the FOLFIRI regimen. When compared to the UGT1A1*1 genotype, homozygosity for the UGT1A1*28 allele has been associated with a dose-dependent increase in toxicity. Furthermore, more than 80% of 5-FU is metabolized via dihydropyridine dehydrogenase (DPD). Because DYPD*2A and DPYD*13 mutations are linked to higher toxicity, it is possible to reduce toxicity by 25–50% by lowering the 5-FU dose.

These therapies have the potential to improve therapeutic responsiveness and decrease treatment-related toxicity. They can also help with dosage adjustments so that a chosen regimen yields the greatest possible advantages. National guidelines that advocate using just KRAS, NRAS, BRAF, and MSI status for predicting therapy response and outcome in colorectal cancer highlight the need for discovering additional biomarkers. However, research is still ongoing in many possible biomarker areas (Ogunwobi et al., 2020).

5.9 BIOSENSORS

According to the definition provided by the International Union of Pure and Applied Chemistry (IUPAC), a biosensor is described as an autonomous integrated device capable of providing specific quantitative or semi-quantitative analytical information. This is achieved through the utilization of a biological recognition element, also referred to as a biochemical receptor, which is in direct spatial proximity to a transducer component. Various types of biochemical receptors, such as enzymes, proteins, antibodies, nucleic acids, cells, tissues, or receptor molecules, have been utilized in the development of biosensors. The biological receptor selectively interacts with a particular analyte, leading to a biochemical reaction that is then transformed into a detectable signal by the transducer. The intensity of this signal is dependent on the concentration of the analyte. Subsequently, the signal is processed by a signal processor, amplified, and finally exhibited on an electronic display system. Biosensors can be categorized based on the physicochemical mechanism employed by the transducer to convert the information obtained from the interaction between the biological receptor and the analyte. The transducer can take various forms, including electrochemical, optical, piezoelectric, and thermoelectric, among others. Electrochemical biosensors offer distinct advantages due to their durability and specificity, resulting from the fusion of the transducer platform and the biological element. This integration can yield highly sensitive devices with a wide dynamic linear range and a low limit of detection. Furthermore, electrochemical biosensors are cost-efficient to produce, user-friendly, and exhibit rapid response times. Additionally, they can be miniaturized, enabling the development of portable devices that require minimal sample volumes for operation. Electrochemical biosensors can also be classified based on the electrochemical method employed to record the signal. The field of biosensing commonly employs various electroanalytical methods such as cyclic voltammetry, differential pulse voltammetry, square wave voltammetry, amperometry, potentiometry, and electrochemical impedance spectroscopy. In voltammetric approaches, a time-varying potential is applied in an electrochemical cell, and the resulting current is measured relative to the applied potential. Amperometric and voltammetric biosensors entail the application of a potential between a working electrode and a reference electrode, generating a current due to the oxidation/reduction processes of electroactive species at the electrode surface. A counter electrode is utilized to complete the circuit and facilitate electron flow. Voltammetric biosensors monitor the current while sweeping the potential at a constant rate within a specific range, with the resulting peak often being proportional to the analyte concentration. Conversely, amperometric biosensors detect variations in current at a fixed potential, providing selectivity as the oxidation/reduction potential is unique to the species under analysis. Differential pulse voltammetry and square wave voltammetry are applied in biosensing to reduce capacitive current and enhance sensitivity. A potentiometric biosensor integrates a biological sensing element with an electrochemical transducer, where the analytical signal is an electrical potential. The potential of the electrochemical cell is measured under static conditions with minimal current flow (Quinchia Cardona et al., 2020).

With the progress made in the fields of nanoscience and nanotechnology, a diverse range of nanomaterials with exceptional chemical, physical, and mechanical properties have been developed and applied in biosensing platforms. These encompass various types of nanomaterials, such as noble metals, metal oxides, metal chalcogenides, magnetic nanoparticles (NPs), carbon-based nanomaterials, and conductive polymers, among others. Particularly noteworthy are the metal nanostructures, which have demonstrated significant appeal in the creation of highly sensitive electrochemical biosensors. Noble metal nanostructures, like Au, Ag, Pt, Pd, Ni, and others, are crucial in advancing biosensors to meet the increasing need for diagnostic devices with both high specificity and sensitivity. The outstanding attributes of nanostructured noble metals greatly enhance electrochemical biosensors, primarily due to their increased surface energy, reduced dimensions, and rapid electron transfer capabilities, all of which facilitate electrochemical reactions at the electrode-solution interface. Additionally, nanostructured metal oxides have gained attention as labeling agents and surface modifiers, effectively boosting the electroactive surface area and facilitating electron transfer at the transducer surface (Quinchia Cardona et al., 2020).

Target molecule analytical signals can be transformed into readable electrical signals by electrochemical detection techniques like electrochemistry, electrochemiluminescence (ECL), and photoelectrochemistry (PEC). High sensitivity and selectivity, ease of use, quick detection, affordability, the capacity to identify several biomarkers at once, and the possibility of miniaturization are among the intrinsic advantages of these techniques. Based on the signals that they produce, electrochemical techniques can be divided into five groups: conductometry, voltammetry, resistance methods, amperometry, and potentiometry. Among them, the development of electrochemical biosensors, sensor characterization, and quantitative analysis frequently make use of voltammetry and electrochemical impedance spectroscopy (EIS). Voltammetry has many benefits because of its broad use, easy data interpretation, high sensitivity, and simple operation.

However, in contrast to other electrochemical techniques, EIS leaves the system undisturbed during measurements and provides greater information about electrode dynamics and interface structure. Analyte detection with electrochemical technology is accomplished by highly sensitive devices known as electrochemical sensors. These sensors usually consist of conversion circuits, transducers, and sensitive elements. Target molecules can be used to categorize electrochemical sensors, which include ion, gas, and biosensors. The biometric element is an essential component in the design of biosensors because it functions as a sensitive element with molecular recognition capabilities. Biosensors can be further classified as enzyme sensors, immunological sensors, aptamer sensors, and so on, based on the biometric elements that are used. sensibility. Moreover, the electrode reaction in the sensor takes place at the interface where the electrode and solution meet, and the characteristics of this interface have a big impact on the reaction. Consequently, strategically changing the interface is essential to improving sensing effectiveness (Zhang et al., 2023b).

To improve sensor detection capabilities, nanomaterials and nanocomposites such as carbon-based materials, conductive polymers (CPs), metal oxide nanomaterials, metal nanomaterials, and silicon nanomaterials are frequently used in biosensors. These materials increase electrode surface electrical conductivity, stabilize biometric components, improve surface area, and speed up electrochemical reactions. Among the various nanomaterial options, CPs have attracted significant attention and are often known as "synthetic metals" due to their remarkable electrical, optical, and magnetic properties. Notably, CPs like PPy, polyaniline (PANI), polythiophene (PTh), and poly (3,4-ethylene dioxythiophene) (PEDOT) have gained considerable interest for their practical use in biomedicine, electronics, energy equipment, and other fields, owing to their biocompatibility, large surface area, environmental stability, electrical conductivity, and other physical characteristics. These CPs demonstrate effectiveness in a wide range of electrochemical biosensor applications (Zhang et al., 2023b).

5.10 ARTIFICIAL INTELLIGENCE

Since the origination of the term artificial intelligence (AI) in 1956, it has developed into a thriving field with significant implications across various sectors, particularly in the medical domain (Figure 5.3). The concept of AI revolves around the application of advanced technologies that empower computational systems to perform tasks that conventionally require human expertise. The AI domain encompasses a broad array of subdisciplines, including robotics, computer vision, natural language processing, and machine learning. Often, these subfields converge to provide more intricate functionalities and capabilities. In the medical field, robotic instruments are increasingly utilized in minimally invasive surgical interventions, such as robotic-assisted procedures for patients suffering from CRC. Moreover, natural language processing plays a crucial role in AI, enabling machines to comprehend, analyze, and interpret human language. In the context of CRC management, natural language processing has demonstrated its worth by extracting relevant clinical data from scanned colonoscopy and pathology reports, a task that would otherwise require manual intervention. Furthermore, computer vision and image processing have shown their utility in exploring colonoscopy and enhancing the visualization of abnormal tissues. Nevertheless, among the various AI subfields, machine learning stands out as a predominant force, extensively applied in three key areas of medicine: early detection and diagnosis, treatment strategies, and forecasting outcomes and prognosis (Viscaino et al., 2021).

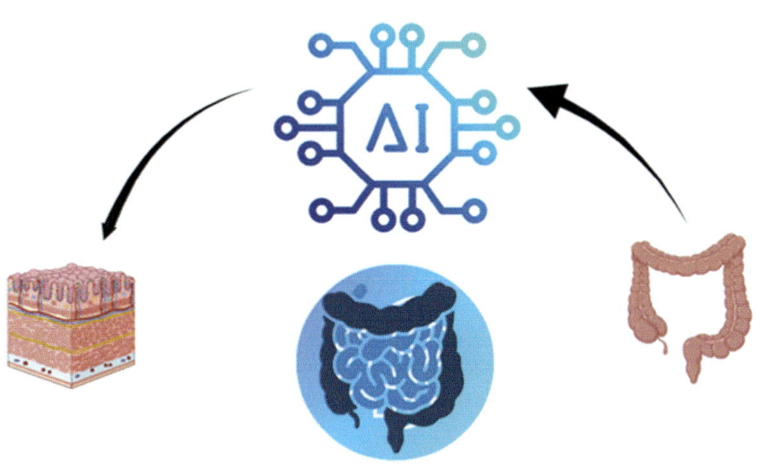

Figure 5.3 Artificial intelligence (AI) can help colorectal cancer screening.

An assortment of programming structures that simplify it to gather and examine different and multifaceted datasets is one method for characterizing computerized reasoning (man-made intelligence). To replicate or simulate the fundamental cognitive processes that make up human intelligence, AI is a disciplinary field that combines formal logic, cognition, and computer science, despite differences in methodology. PC Upheld Area (PC supported plan) structures emerge as an imaginative development that might potentially begin a have a significant impact on context in the field of suggestive colonoscopy by utilizing cutting-edge PC-based knowledge techniques (Maida et al., 2017). Regarding CRC screening, PC-based knowledge can be used to update the repeat of early screening, resulting in a significant reduction in both the event rate and the passing rate among CRC patients. The unmistakable proof and screening of a greater extent of CRC biomarkers can be streamlined by integrating bioinformatics instruments into recreated insight. Pathologists may be able to reduce the number of cases that are either misdiagnosed or not diagnosed at all by making use of AI-supported pathology recognition technology. Image recognition using AI relies heavily on machine learning (ML). It has the potential to help doctors make more accurate clinical judgments, reduce errors, provide objective, dependable reference opinions, and significantly improve the interpretability of medical images. The blend of man-made comprehension (mechanized dissuading) colonoscopy has shown expansive obligation in raising the adenoma affirmation rate (ADR). In addition to lowering the risk of CRC, this strategy ensures precise resection, thereby reducing the clinical burden of removing many benign polyps. Real-time polyp detection has been demonstrated to be extremely effective when a YOLOV3 AI algorithm is applied via media. Moreover, its moderateness and transparency make it ideal for all-over use in less advanced areas. Notwithstanding, the consistent PC upheld area (CADe) structure infrequently can recognize harmless and little adenomas and polyps that ordinary colonoscopies could miss. Thus, the ascent in CRC cases might be ascribed to some extent to the CADe framework. The diagnosis of CRC with AI-assisted colonoscopy has been supported by numerous studies. Additionally, clinical applications of express optical biopsy techniques that rely on manual capture of consistent images have demonstrated fundamental responsibility. The adenoma misses rate (AMR), which highlights the variation in lesions detected across consecutive endoscopies, is another widely accepted metric for evaluating the efficacy of colonoscopy (Yin et al., 2023).

REFERENCES

Beniwal, S. S., et al. (2023). Current status and emerging trends in colorectal cancer screening and diagnostics. *Biosensors, 13*(10):926.

Bhagatwala, J., et al., (2015). Colonoscopy—indications and contraindications. In: Ettahr R., ed., *Screening for colorectal cancer with colonoscopy*, InTech Open, 35–47.

Bresalier, R. S., et al. (2020). Biomarkers for early detection of colorectal cancer: the early detection research network, a framework for clinical translation. *Cancer Epidemiol Biomark Prevent 29*(12):2431–2440.

Burnett-Hartman, AN et al. (2021). An update on the epidemiology, molecular characterization, diagnosis, and screening strategies for early-onset colorectal cancer. *Gastroenterol 160*(4):1041–1049.

Forbes, N., et al. (2021). Association between time to colonoscopy after positive fecal testing and colorectal cancer outcomes: a systematic review. *Clini Gastroenterol Hepatol, 19*(7):1344–1354. e1348.

Goyal, H., et al. (2020). Scope of artificial intelligence in screening and diagnosis of colorectal cancer. *J Clin Med, 9*(10):3313.

Gupta, V., East, J. E. (2020). Optimal endoscopic treatment and surveillance of serrated polyps. *Gut Liver, 14*(4):423.

Iordache, M. M., et al. (2023). Calprotectin, biomarker of depression in patients with inflammatory bowel disease? *Medicina, 59*(7):1240.

Jain, S., Maque, J., et al. (2022). Optimal strategies for colorectal cancer screening. *Current Treatm Options Oncol 23*(4):474–493.

Jalayeri Nia, G., et al. (2023). Clinical utility of colon capsule endoscopy: a moving target? *Therap Advances Gastroenterol, 16*:17562848231195680.

Jones, R. M., et al. (2010). Patient-reported barriers to colorectal cancer screening: a mixed-methods analysis. *Am J Prevent Med, 38*(5):508–516.

Kaltenbach, T., et al. (2020). Endoscopic removal of colorectal lesions: recommendations by the US Multi-Society Task Force on Colorectal Cancer. *Am J Gastroenterol, 115*(3):435–464.

Kanth, P., Inadomi, J. M. (2021). Screening and prevention of colorectal cancer. *BMJ, 374*:n1855.

Ladabaum, U., et al. (2020). Strategies for colorectal cancer screening. *Gastroenterol, 158*(2):418–432.

Lee, M. W., Schnoll-Sussman, F. (2022). Capsule endoscopy. In: Cohen J, ed., *Successful training in gastrointestinal endoscopy*, Wiley, 113–124.

Leung, L. J., et al. (2023). Post-colonoscopy colorectal cancer etiologies in a large integrated United States healthcare setting. *Gastroenterol, 164*(3):470.

Lieberman, D. A. (2009). Screening for colorectal cancer. *New Engl J Med 361*(12):1179–1187.

Maida, M., et al. (2017). Screening of colorectal cancer: present and future. *Expert Rev Anticancer Ther, 17*(12):1131–1146.

Nistal, E., et al. (2015). Factors determining colorectal cancer: the role of the intestinal microbiota. *Frontiers Oncol 5*:220.

Ogunwobi, O. O., et al. (2020). Biomarkers in colorectal cancer: current research and future prospects. *Int J Molec Sci, 21*(15):5311.

Olovo, C. V., et al. (2021). Fecal microbial biomarkers in early diagnosis of colorectal cancer. *J Cellul Molec Med, 25*(23):10783–10797.

Patel, R., et al. (2024). The evolving landscape of colonoscopy: recent developments and complication management. In: Rodrigo L, ed., *Colonoscopy-diagnostic and therapeutic advances*. Intech Open.

Perlo, D. (2021). Deep Learning for medical imaging: perspectives from real use cases. Thesis, University of Turin.

Quinchia Cardona, J. P., et al (2020). Electrochemical biosensors for determination of colorectal tumor biomarkers. *Micromachines* 11(4):411.

Rashid, G., et al. (2023). Non-steroidal anti-inflammatory drugs and biomarkers: a new paradigm in colorectal cancer. *Frontiers Med 10*:130710.

Rutter, C. M., et al. (2022). The impact of cumulative colorectal cancer screening delays: A simulation study. *J Med Screen, 29*(2):92–98.

Sanduleanu, S., et al. (2010). In vivo diagnosis and classification of colorectal neoplasia by chromoendoscopy-guided confocal laser endomicroscopy. *Clin Gastroenterol Hepatol, 8*(4):371–378.

Shaukat, A., Levin, T. R. (2022). Current and future colorectal cancer screening strategies. *Nature Rev Gastroenterol Hepatol 19*(8):521–531.

Thattakath, C. W. (2023). *Impact of colorectal cancer screening program on mortality in lithuania. Thesis,* Vilniaus University.

Thomas, J., et al. (2023). Advancing colorectal cancer screening: a comprehensive systematic review of artificial intelligence (AI)-assisted versus routine colonoscopy. *Cureus, 15*(9).

Thorne, K. (2008). *An independent evaluation of the NHS Modernisation Agency's' Modernising Endoscopy Services' project using routinely collected, service-related endoscopy data*. Thesis, Swansea University.

Viscaino, M., et al. (2021). Artificial intelligence for the early detection of colorectal cancer: A comprehensive review of its advantages and misconceptions. *World J Gastroenterol, 27*(38):6399.

Winawer, S., et al. (1990). Risk and surveillance of individuals with colorectal polyps. Who Collaborating Centre for the Prevention of Colorectal Cancer. *Bulletin WHO, 68*(6):89.

Wu, Z., et al (2020). Colorectal cancer screening methods and molecular markers for early detection. *Technol Cancer Res Treatm, 19*:1533033820980426.

Ye, X., et al. (2016). Association of neutrophil-to-lymphocyte ratio, neutrophil count, and VEGF on metastasis in patients with nasopharyngeal carcinoma. *Res Square*, 1–16, https://doi.org/10.21203/rs.3.rs-26726/v1

Yin, Z., et al. (2023). Application of artificial intelligence in diagnosis and treatment of colorectal cancer: A novel Prospect. *Frontiers Med 10*:1128084.

Zhang, X., et al. (2023a). Application of polypyrrole-based electrochemical biosensor for the early diagnosis of colorectal cancer. *Nanomat, 13*(4), 674.

Zhang, Z., et al. (2023b). Research Progress of Intestinal Microecology in the Pathogenesis of Colorectal Adenoma and Carcinogenesis. *Technol Cancer Res Treatm, 22*:15330338221135938.

6 Diagnostic Techniques and Imaging Modalities

Mohd Amir Shafeeque, Dilshad Ahmad, Satya N. Sankwar, Noorin Zaidi and Zainab Siddiqui

6.1 INTRODUCTION

Colorectal cancer (CRC) is a significant health concern, and the use of diagnostic techniques and imaging modalities have a crucial role in its management. For decades, various imaging modalities have been developed to aid in the detection, surveillance, and treatment of CRC. These modalities include computed-tomography (CT), magnetic resonance imaging (MRI), and positron emission tomography (PET), narrow band imaging, and computer-aided detection devices. Additionally, the use of hybrid imaging techniques, such as PET-CT or PET-MRI, has enhanced the detection of tumor recurrence and distant metastases in treated CRC cases (Revheim et al., 2015). Narrow band imaging and other equipment-based image-enhanced endoscopy methods have been used to differentiate colorectal polyps, aiding in the observation of mucosal epithelium and capillaries to detect and differentially diagnose colorectal tumors (Tischendorf et al., 2007; Zeng et al., 2020; González-Suárez et al., 2020; Plekhanov et al., 2023).

In this chapter, we discuss the integration of advanced imaging modalities, such as the use of deep learning methods in precision diagnosis, and different immunohistochemical biomarkers that have significantly improved the diagnostic capabilities for CRC.

6.2 HISTOPATHOLOGICAL ANALYSIS OF COLORECTAL CANCER

CRC histological variants include mucinous, adenosquamous, spindle cell, medullary, signet ring, serrated, cribriform, micropapillary, and undifferentiated types (Fleming et al., 2012). Adenocarcinomas comprise 90% of CRC cases (Hamilton 2000; Fleming et al., 2012), classified based on glandular formation: well-differentiated (>95% glandular structures), moderately differentiated (50-95%), and poorly differentiated (<50%) (Fleming et al., 2012). Tumors are categorized as low-grade (well and moderately differentiated) and high-grade (poorly differentiated) (Figure 6.1) (Hamilton 2000; Fleming et al., 2012). Invasive CRC may invade the mucosal layer, termed carcinoma in situ or intramucosal carcinoma (Fleming et al., 2012). Mucinous adenocarcinomas (10–20% of cases) show upregulation of mucin 2 and 5AC, contributing to a protective mucosal barrier, with downregulation linked to poor prognosis (Luo et al., 2019). Microsatellite instability is common in mucinous adenocarcinomas (Fleming et al., 2012; Luo et al., 2019), while signet ring carcinoma is rare (<1%) and associated with specific gene mutations and downregulation of proteins (Farraj et al., 2019).

Tumor-budding, a feature of the tumor stroma, arises from cell response to the extracellular matrix (Lugli et al., 2017). Tumor buds, derived from epithelial to mesenchymal cells, can be classified by their origin, including peri-tumoral (PTB) at the tumor front and intra-tumor (ITB) at the core, often showing a strong correlation (Lugli et al., 2017; Fujiyoshi et al., 2020). A study noted that tumor buds at the invasive front were primarily found in the ascending and transverse colon, comprising 33% of cases (Fujiyoshi et al., 2020).

The concept of tumor budding was introduced at the ITBCC in April 2016 (Lugli et al., 2017), defined as single cells or clusters of up to four at the invasive margin. Tumor budding correlates with higher TNM stage (tumour, node, and metastasis), grade, and lymph node metastasis and can influence treatment plans. High tumor budding is found to be associated with reduced overall, disease-free, and progression-free survival rates (Grigore et al., 2016). Recently, a Delphi consensus proposed categorizing pseudobuds for small clusters of tumor cells, i.e. up to 4 (Haddad et al., 2021).

Tumor buds are often confused with poorly differentiated clusters (PDCs), which consist of ≥5 cells lacking glandular differentiation and can be identified in stained slides without special markers (Reggiani Bonetti et al., 2016). PDC counts inform colorectal tumor grading: <5 as grade 1, 5–9 as grade 2, and ≥10 as grade 3 (Ueno et al., 2012; Reggiani Bonetti et al., 2016). Tumor-infiltrating lymphocytes do not correlate with PDC density but are inversely related to tumor budding (Fujiyoshi et al., 2020). Patients with PDCs show significantly lower disease-free survival; grade 1 tumors have a better five-year survival rate (95.9%) compared to grade 2 (84.6%) and grade 3 (59.3%) tumors (Ueno et al., 2012). Thus, the presence of PDCs has an impact on the prognosis of CRC.

 DOI: 10.1201/9781032704517-6

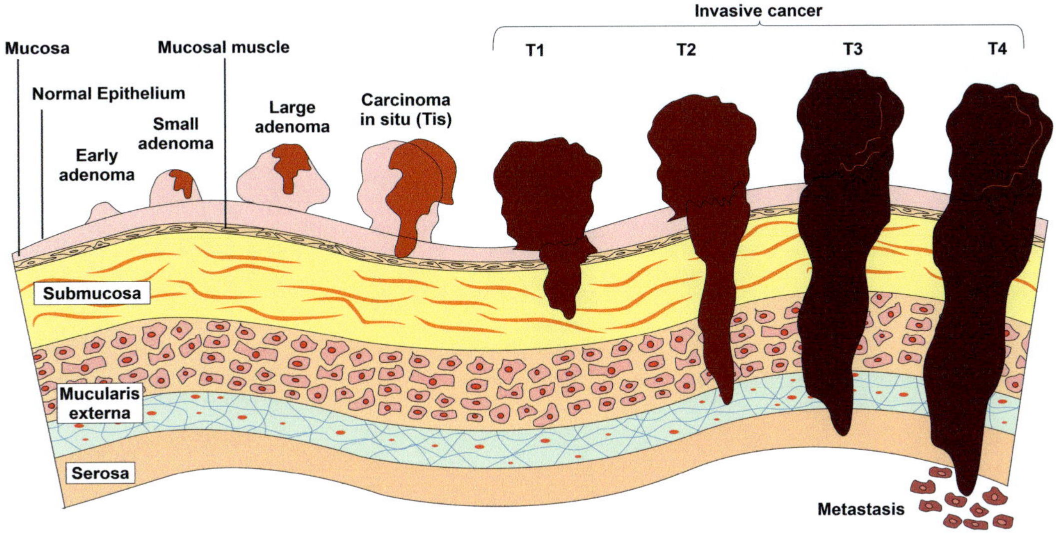

Figure 6.1 The progression of adenomas to carcinomas and the corresponding primary tumor T stages. The process starts with the pro-carcinogenic events that lead to the formation of a benign tumor, which later on progresses to higher stages if not diagnosed earlier or left untreated. Cancer is at its earliest stage and occurs only in the mucosa while the T1 staged tumor is present only in the inner layer of the submucosa of the bowel. The T2 staged tumors grow into the muscle layer of the bowel wall and the T3 staged tumors grow into the outer lining of the bowel wall but don't spread beyond it. In contrast, the T4 staged tumors spread into the tissue layer covering the organs in the abdomen. The Tis, carcinoma in situ refers to the invasive carcinomas.

6.2.1 Immunohistochemical Analysis

The immunohistochemistry technique in the field of oncology is employed for tissue-based diagnosis and is accepted worldwide. Table 6.1 presents an overview of immunohistochemical markers utilized in the context of CRC, including their functions, sensitivity, specificity, and accuracy percentages or area under the curve (AUC) values.

CEAg is a cell surface glycoprotein marker associated with metastasis, chemotherapy resistance, and anoikis initiation in CRC, as depicted in Figure 6.2 (Samara et al., 2007; Jansen et al., 2023). The CEAg family is found to be upregulated in approximately 70% of all cancers (Li et al., 2010). Anoikis is mediated by CEAg through the activation of the TRAIL-R2- DR5-caspase 8 signaling pathway (Samara et al., 2007). A study found that PELPK amino acids trigger the binding of CEAg to the hnRNP M4 receptor, activating Kupffer cells and promoting liver metastasis (Bajenova et al., 2001; Palermo et al., 2012). This activation leads to the release of cytokines, increased levels of matrix metalloproteinases, reactive oxygen and nitrogen species as shown in Figure 6.3 (Lee and Lee, 2017). CEAg also binds to the TGF-beta-RI receptor, and activates c-Myc, a downstream target of SMAD3, suggesting its role in promoting cancer cell proliferation (Figure 6.3) (Li et al., 2010).

Calretinin, also known as calbindin 2 (CALB2), plays a significant role in colon cancer. Stevenson et al. examined the effects of CALB2 gene silencing in colon cancer cells treated with 5-fluorouracil, demonstrating that CALB2 is crucial for inducing pro-apoptotic factors such as caspase 3/7, poly (ADP-ribose) polymerase cleavage, and loss of mitochondrial membrane potential (Figure 6.2). Additionally, low CALB2 levels correlate with poor overall survival in CRC patients (Stevenson et al., 2011).

The epidermal growth factor receptor (EGFR), part of the receptor tyrosine kinases family, is upregulated in colon cancer, promoting cell survival and proliferation (Wieduwilt and Moasser, 2008). Growth factors binding to EGFR activate the PI3K-AKT and KRAS-BRAF signaling pathways (Figure 6.3) (Hagland et al., 2013). Anti-EGFR therapies, including monoclonal antibodies (cetuximab and C225), gefitinib, CI1033, PKI166, GW572016, and erlotinib, have shown positive effects in responders (Wieduwilt and Moasser, 2008; Bardelli and Siena, 2010; Spano et al., 2005).

Folate receptor-α expression is associated with metastatic tumor behavior and poor response to chemotherapy (Chen et al., 2022). It is overexpressed in many epithelial cancers but in CRC it is

Table 6.1: List of immunohistochemical markers in colorectal cancer

List of immunohistochemical markers in colorectal cancer

Markers	Function in CRC	Study type	Sample type	Sample size	Sensitivity	Specificity	References
Carcinoembryonic antigen (CEA)	Carcinoembryonic antigen comprises a diverse cohort of proteins, like CEACAM5, which function as biomarkers in the context of colorectal cancer.	Retrospective study	Tissue samples	Cases=280; control=280	93.70%	96.10%	Tiernan et al., 2013
Tumour-associated glycoprotein-72 (TAG-72)	This glycoprotein complex is found in numerous adenocarcinomas. It is a potential therapeutic target in colorectal cancer as well as breast and lung cancers.	Retrospective study	Tissue samples	Cases=280; control=280	70.30%	78.50%	Tiernan et al., 2013
Folate receptor- α	The folate receptor-alpha is a membrane-bound protein that binds and transports folic acid and has emerged as a potential tumor biomarker useful for targeted cancer therapies and cancer detection.	Retrospective study	Tissue samples	Cases=280; control=280	39.30%	99.20%	Tiernan et al., 2013
Endothelial growth factor receptor (EGFR)	Overexpression of EGFR can lead to the development of colorectal cancer.	Retrospective study	Tissue samples	Cases=280; control=280	21.70%	90.50%	Tiernan et al., 2013
Caudal type homeobox-2	Caudal type homeobox-2 is a critical nuclear transcription factor for intestinal development and is expressed in intestinal epithelium and adenocarcinomas. It is a highly sensitive marker of colorectal origin and can be used as an adjunct for the diagnosis of intestinal adenocarcinomas.	Retrospective study	Tissue samples	Colorectal adenocarcinomas=118; gastric adenocarcinomas=59; pancreatic adenocarcinomas=32	96.6% (for cut-off values > 5%) and 78% (for cut-off values > 50%)	54.9%(for cut-off values > 5%) and 84.6% (for cut-off values > 50%)	Bayrak et al., 2012

Cytokeratin 20	Cytokeratin 20 expression is more common in rectal carcinomas and low-grade colorectal carcinomas.	Retrospective study	Tissue samples	Colorectal adenocarcinomas=118; gastric adenocarcinomas=59; pancreatic adenocarcinomas=32	63.60%	96.70%	Bayrak et al., 2012
Keratin 7	CK7 expression is observed in a subset of colon, gastric, and pancreatic adenocarcinomas.	Retrospective study	Tissue samples	Colorectal adenocarcinomas=118; gastric adenocarcinomas=59; pancreatic adenocarcinomas=32	63.60%	96.70%	Bayrak et al., 2012
BRAF (V600E)	The BRAF V600E mutation assumes a crucial role in the initiation and advancement of colorectal cancer. This mutation, situated within the BRAF gene, entails the substitution of valine (V) by glutamic acid (E) at position 600 in the protein sequence. Accurate identification of the BRAF V600E mutation in CRC holds significance for diagnostic and therapeutic considerations.	Retrospective study	Formalin-fixed paraffin-embedded tissue (FFPE)	Positive Genotype =48; Negative Genotype=69	93.70%	95.60%	Vakiani et al., 2015
Cadherin-17	In colorectal cancer, CDH17 is exclusively expressed in the membrane and cytoplasm of epithelial cells in the small intestine and colon, both during embryogenesis and in adulthood.	Retrospective study	Tissue samples	Colorectal adenocarcinoma=35; Non-colorectal adenocarcinoma=30	100.00%	50%	Abouelkhair et al., 2021
Special AT-rich sequence-binding protein 2 (SATB2)	SATB2 has been evaluated as a diagnostic marker for identifying the colonic origin of tumors.	Retrospective study	Tissue samples	Primary colorectal cancer=70; Non-colorectal cancer=80	94.30%	97.5%	Elnady et al., 2022

(Continued)

Table 6.1: (Continued) List of immunohistochemical markers in colorectal cancer

List of immunohistochemical markers in colorectal cancer							
Markers	**Function in CRC**	**Study type**	**Sample type**	**Sample size**	**Sensitivity**	**Specificity**	**References**
Cyclooxygenase- 2	Cyclooxygenase- 2 activity leads to the upregulation of certain genes, such as mPGES-1, DUSP4, Programmed cell death 4, Trop2, and genes from the TGFb and p53 pathways, which play a role in colorectal cancer.	Meta-analysis	Tissue samples	Cases= 30; control=30	90.10%	89.90%	Hidalgo-Estévez et al., 2020, Nanda and Dhawan 2020
Glycoprotein A33	It is a tumor associated antigen glycoproteinwhich is found to be significantly expressed in colorectal cancer.	Pilot study	Tissue samples	51 cases of liver metastatic carcinoma of different origins have been studied- colorectum (n=32), breast (n=3), oesophagogastric tract (n=4), lung (n=3), pancreas (n=8), and prostate (n=1)	95.90%	85.40%	Wong et al., 2017a
Calretinin	Calretinin is a widely used immunohistochemical marker for mesothelial cells and malignant mesothelioma. It is expressed in a small percentage of colorectal tumor tissues.	Retrospective study	Tissue samples	Colorectal mucinous adenocarcinoma (MA)=75; Non-mucinous adenocarcinoma (NMA)=75	MA= 98.7 % NMA= 90.7 %	MA= 1.3% NMA= 9.3%	Foda et al., 2016, Yahya et al., 2013

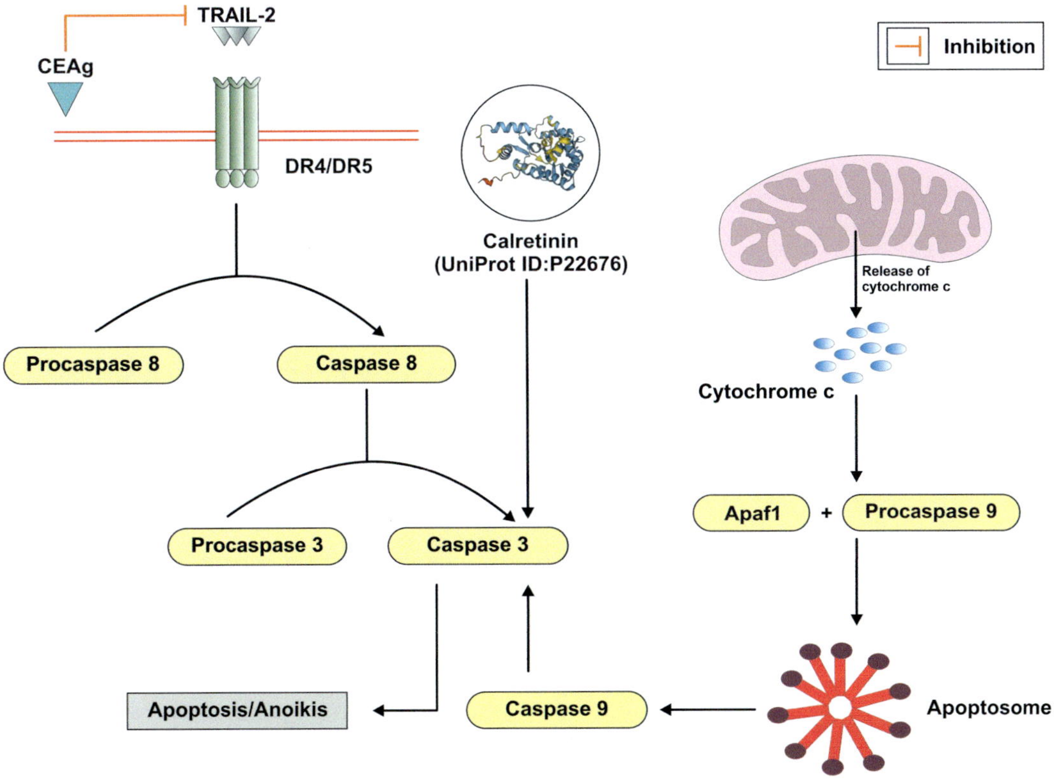

Figure 6.2 The carcinoembryonic antigen regulates apoptosis/anokis by inhibiting TRAIL2. TRAIL 2 binds to the DR4/DR5 receptor, which cleaves procaspase-8 into caspase-8. Cleaved caspase-8 further cleaves procaspase-3 into caspase-3, which ultimately leads to Anokis. Calretinin is also involved in the activation of caspase-3. The release of cytochrome C from the mitochondria is the alarming sign that the cell must undergo anokis through the formation apoptosome.

considered a promising target for therapeutic and imaging agents (Chen et al., 2022). Folic acid is the known activator of the folate receptor-α, which in turn activates the downstream signaling cascade involving cSrc-ERK1&2- nuclear factor kappa beta-protein 53 intermediates as shown in Figure 6.3 (Kuo and Lee, 2016).

BRAF (V600E) mutation is a constitutive kinase that activates the RAS-RAF-MEK-ERK pathway (Figure 6.3), causing transcription of genes involved in colon cancer progression (Guerrero et al., 2022). This mutation is found to be associated with 8% to 10% of metastatic CRC cases (Hosseinalizadeh et al., 2024). CRC can also develop resistance to BRAF inhibitors by upregulating the WNT-beta catenin-mediated pathway (Chen et al., 2018). It has also been reported that only a very few percentage of CRC patients respond to BRAF inhibitor therapy (Hosseinalizadeh et al., 2024). Therefore, targeting both the RAS/RAF/MEK/ ERK and the WNT-beta catenin pathway might prove to be beneficial for treating BRAF (V600E) mutated CRC.

Cadherin-17 also known as liver intestine cadherin plays an important role in mediating adhesion between the cells in tumor tissues (Ng et al., 2023; Yui et al., 2023). Colorectal tumors with increased cadherin-17 expression are found to be significantly associated with distant metastasis and tumor aggressiveness (Ng et al., 2023). The Glu739 mutation of the cadherin-17 gene promotes more weakened cell aggregation than the Lys115 mutation. Therefore, weakened cell-cell adhesion contributes to an increased risk of lymph node metastases in CRC (Yui et al., 2023).

Glycoprotein A33 is a protein found exclusively in the intestinal epithelium and human colon cancers (Mao et al., 2003). The activation of peroxisome proliferator-activated receptor-gamma has been observed to induce the expression of glycoprotein A33 in colon cancer cells (Rageul et al., 2009). Its expression is regulated by two transcription factors i.e. CDX1 and Kruppel-like factor 4 as

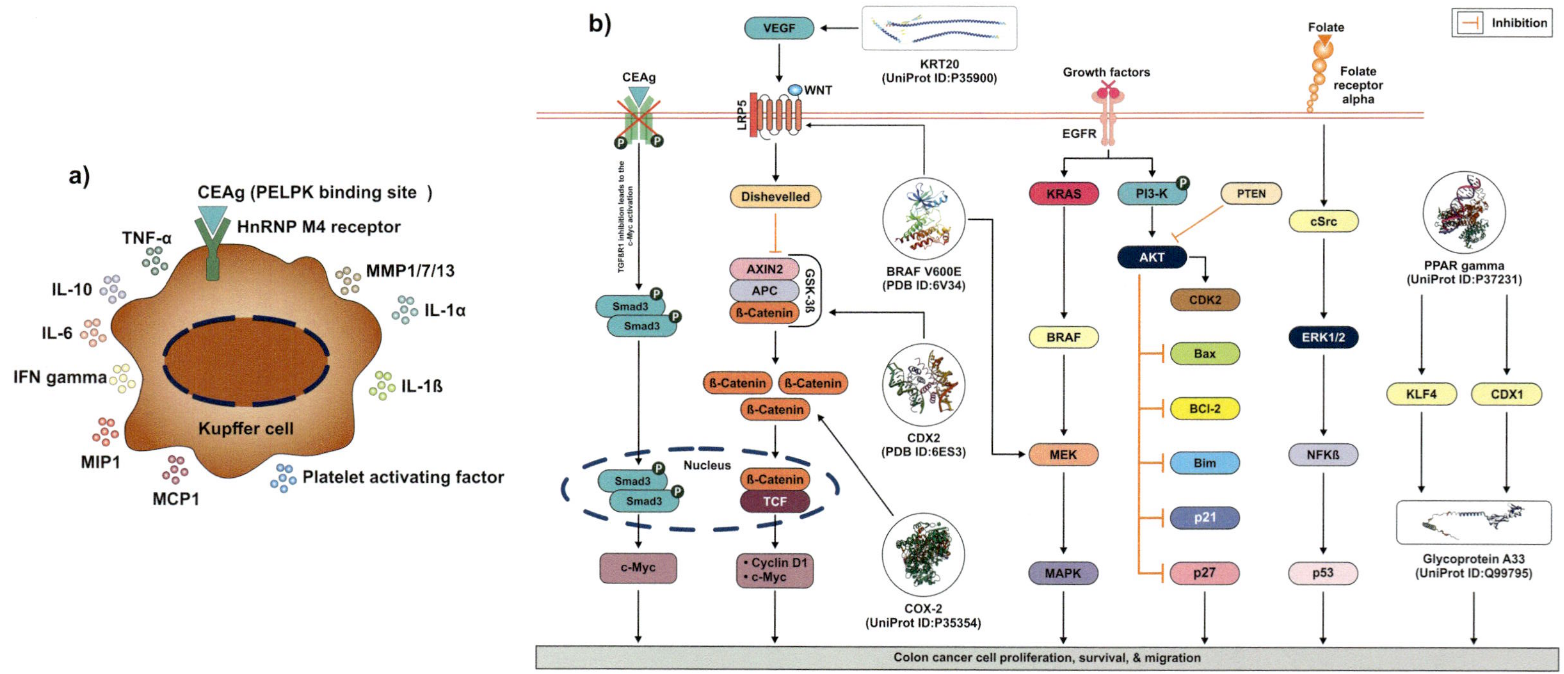

Figure 6.3 The role of molecular markers in the pathogenesis of colorectal cancer (a) shows that the binding of the carcinoembryonic antigen to its receptor hnRNP M4, leads to the activation of Kupffer cells. Many cytokines, including interleukin (IL)-1alpha, IL-1beta, IL-6, IL-10, interferon-gamma, transforming growth factor-beta, tumor necrosis factor-alpha, platelet-activating factor, monocyte chemotactic protein 1, and macrophage inflammatory protein 1, are secreted in response to Kupffer cell activation. Moreover, certain matrix metalloproteinases (MMPs), namely MMP-1, MMP-7, and MMP-13, are also released and are involved in the extracellular re-modelling, hence leading to liver metastasis. (b) The above-mentioned signaling pathways summarizes the colorectal cancer cells survival, proliferative, and migration. Proteins, including peroxisome proliferator-activated receptor gamma and cytokeratin-20, stimulate the production of glycoproteins A33 and cyclooxygenase-2, respectively. Cyclooxygenase-2, in turn, amplifies beta catenin through Wnt signaling, and the folate receptor alpha triggers nuclear factor kappa beta, which promotes colorectal cancer cell proliferation. Furthermore, carcinoembryonic antigen disrupts the transforming growth factor-beta RI signaling cascade, increases c-Myc activation, and accelerates cancer cell progression.

shown in Figure 6.3 (Rageul et al., 2009; Mao et al., 2003). Additionally, peroxisome proliferator-activated receptor gamma activation increases Kruppel-like factor 4 expression (Rageul et al., 2009).

Cyclooxygenase-2 plays a key role in producing various prostaglandins, such as E2, D2, F2alpha, and I2, which serve as substrates for specific synthases (Dixon et al., 2013). In colon cancer cells, F2alpha activation by cyclooxygenase-2 enhances early growth response 1 transcription, increasing prostaglandin E2 synthase 1 mRNA (Stamatakis et al., 2015). The PTGS-2 gene transcription is influenced by signaling mediators like NF-kappaB, beta-catenin, and CCAAT enhancer-binding proteins (Figure 6.3). Cyclooxygenase-2 overexpression upregulates 16 genes and downregulates 4, predominantly via TGF-beta and p53 pathways (Hidalgo-Estévez et al., 2020). Thus, cyclooxygenase-2 is a crucial marker for metastatic identification.

Cytokeratin-20 is a valuable diagnostic marker for CRC (Figure 6.3). Its potential role in promoting angiogenesis is suggested by its positive correlation with vascular endothelial growth factor (VEGF) levels (Auf et al., 2011). The interplay between VEGF and WNT components facilitates angiogenesis (Kasprzak, 2020). Other research indicates that mRNA expression of VEGFA and chromodomain helicase DNA binding protein 4 is elevated in cytokeratin-20 positive tumor-free lymph nodes. Thus, cytokeratin-20 expression can help identify non-metastatic CRC patients at higher risk for tumor recurrence (Wong et al., 2017b). The expression of cytokeratin-20 is regulated by CDX1, which binds to elements 246 bp upstream of the transcription start site (Chan et al., 2009). Additional factors significantly linked to high cytokeratin-20 levels in CRC patients include advanced tumor stages, liver metastasis, lung metastasis, and elevated CEAg levels (Auf et al., 2011).

Most metastatic CRC patients (94%) are keratin 7 negative and cytokeratin-20 positive. Those with the BRAF (V600E) mutation show increased keratin-7 expression, especially in microsatellite stable cases (39%) compared to MSI cases (6%), correlating with poor survival (Figure 6.4) (Hosseinalizadeh et al., 2024). Gurzu et al. examined the relationship between keratin-7, cytokeratin-20, protein 53, MLH-1, and KRAS mutations, finding mucinous adenocarcinomas with diffuse keratin-7 expression, 50% of which were cytokeratin-20 negative. Keratin-7 positive KRAS mutated samples in the distal colon and rectum had low protein 53 and MLH-1 positivity (Gurzu et al., 2013). Micropapillary adenocarcinomas showed higher keratin-7 expression, but this was not seen in colorectal adenocarcinomas (Hosseinalizadeh et al., 2024).

TAG-72, a glycoprotein complex linked to membranes is also analyzed for the detection of CRC immunohistochemically (Cho et al., 2019). Its upregulation is found to be significantly associated with lymphatic invasion, and venous invasion in CRC patients. However, TAG-72 downregulation was found to be associated with high MSI status (Figure 6.4), perineural invasion, and mucinous histological type (Cho et al., 2019). This marker was recently utilized in CRC surgeries by using it as a target. During surgery TAG-72 targeting helps in the precise identification and removal of the malignant tissue (Hitchcock et al., 2021). The TAG-72-directed approaches involve the utilization of specific antibodies or ligands like ^{125}I-labelled murine monoclonal antibody that recognizes and binds to the TAG-72 on the cancer cells (Hitchcock et al., 2021).

The SATB2 low expression is found to be associated with tumor progression and metastasis (Wang et al., 2019). Furthermore, the authors reported that SATB2 regulates the expression of

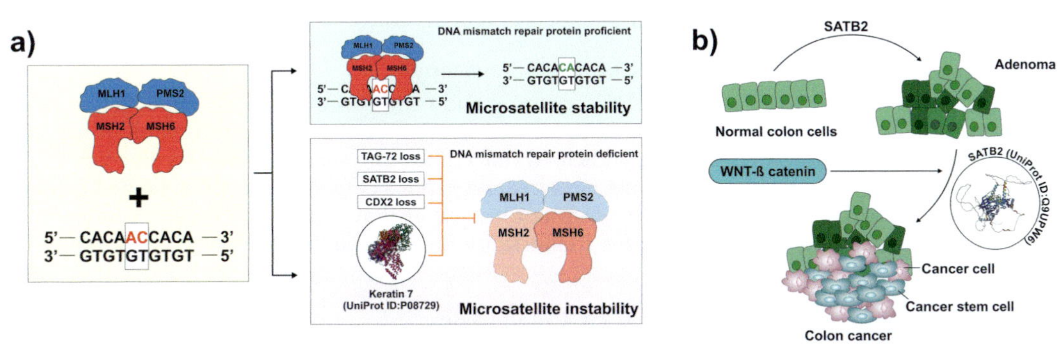

Figure 6.4 (a) MLH1, MSH2, MSH6, and PMS2, the components of mismatch repair system. The mismatch repair system deficiencies lead to microsatellite instability (MSI), which is distinguished by unstable microsatellites, a specific type of simple DNA sequence repeat. (b) An overview of the normal colon cells transformation to cancer stem cells due to the overexpression of SATB2.

beta-catenin which further induces T cell factor/transducin-like enhancer family transcription factors. This causes the switch-on of various downstream genes such as Bcl-2, CD44, CD133, Survivin, c-Myc, Nanog, Cyclin D1, Zeb1, and XIAP (Yu et al., 2017). Hence, it can be concluded that SATB2 overexpression may lead to cancer stemness in CRC (Figure 6.4). SATB2 also recruits HDAC1 to the Snail promoter and represses Snail transcription and inhibits epithelial-to-mesenchymal transition (Wang et al., 2019).

Loss of caudal-type homeobox-2 expression in CRC cases is often associated with deficiencies in mismatch repair genes (Figure 6.4) (Lugli et al., 2008; Vlahović et al. 2023). This loss correlates with reduced overall and progression-free survival (Singh et al., 2022). Additionally, caudal-type homeobox-2 expression loss is more prevalent in advanced CRC and high-grade tumors (Singh et al., 2022). The tumorigenic effects of caudal-type homeobox-2 loss involve activation of WNT-beta-catenin signaling, which is mediated by the inhibition of axin 2 and glycogen synthase kinase 3 beta. Key downstream effectors of WNT signaling include c-Myc, beta-catenin, and cyclin D1 (Yu et al., 2019).

6.3 IMAGING OF COLORECTAL CANCER

Over the years, various criteria for diagnosing CRC by different imaging techniques have been proposed and routinely used. Majorly, the imaging techniques involved in its detection are endoscopic ultrasound, elastography, virtual colonoscopy, CT, MRI, and PET scan. Several onco-radiologists support the combinations of these imaging techniques to formulate a conclusive disease interpretation. Radiological imaging modalities also play a crucial role in the management of CRC (Table 6.2). These modalities are used to detect primary tumors, assess disease extent, evaluate local invasion, and detect distant metastases. Tracers also play an important role in imaging which has been outlined in Table 6.3.

6.3.1 Ultrasonography

Ultrasound is often used in the evaluation of CRC, particularly for assessing liver metastases. It is non-invasive, cost-effective, and readily available. While it may not be as sensitive as other imaging modalities for detecting small liver lesions, it can be useful for guiding interventions such as biopsy or drainage procedures. But overall it has high sensitivity (79.6%) and specificity (92.1%) for the detection of CRC (Martínez-Ares et al., 2005).

6.3.1.1 Ultrasound Elastography

CRC is characterized by tumor stiffness, which drives epithelial to mesenchymal transition and contributes to tumor progression (Liu, 2020; Qiu et al., 2023). This stiffness results from extracellular matrix remodeling, primarily through collagen overexpression and cross-linking, mediated by the enzyme lysyl oxidase (Liu, 2020). Therefore, measuring tissue stiffness is crucial and can be achieved with ultrasound elastography, which determines tissue elasticity by applying force and measuring the resulting shear deformation (tumor tissue displacement). Ultrasound elastography is categorized into quasi-static, vibroacoustic, and shear-wave types, based on different stressors (Qiu et al., 2023; Loft et al., 2022). Ultrasonic elastography can differentiate adenomas from early-stage rectal cancer with 79% sensitivity and 95% specificity and is also used to evaluate treatment responses in colorectal cancer patients (Loft et al., 2022).

6.3.2 Sigmoidoscopy

Sigmoidoscopy is commonly used for CRC screening and tumor recurrence monitoring. It is an effective tool for polyp detection compared to other screening methods (Issa and Noureddine 2017; Wang et al., 2024). A meta-analysis indicated that combining sigmoidoscopy with colonoscopy reduced mortality and incidence rates in screened groups compared to non-screened groups (Wang et al., 2024). However, a limitation of sigmoidoscopy is that it primarily detects tumors in the distal colon, leading to a 46% reduction in mortality for those tumors (Issa and Noureddine 2017).

6.3.3 Colonoscopy

Colonoscopy is the gold standard for CRC screening and diagnosis, as it allows direct visualization of the colorectal mucosa. It enables polyp removal and tissue sample collection for histopathological analysis (Issa and Noureddine, 2017). Research has shown that colonoscopic polyp detection significantly reduces mortality rates in screened populations, by 68% to 88% (Issa and Noureddine, 2017).

Table 6.2: List of imaging technologies used for the detection of colorectal cancer

S.No.	Technique	Study type	Sample size	Discrimination between	Sensitivity	Specificity	Accuracy%/ AUC	Reference
1	**Positron-emission tomography (PET)/Computed tomography (CT)**	Retrospective study	Adenocarcinomas=38; Adenomas=67	The colonoscopy findings were categorized into two groups: with and without cancer possibility.	65%	75.40%	72.70%	Nowicki, 2019
2	**Colonoscopy**	Retrospective study	Adenocarcinomas=38; Adenomas=67	The colonoscopy findings were categorized into two groups: with and without cancer possibility.	80%	68.40%	71.4%.	Nowicki 2019
3	**Endorectal ultrasound**	Meta-analysis	Adenoma cases=62; Adenocarcinoma cases=24	Adenoma and adenocarcinoma (T1 & T2)	79% (95% CI=0.58–0.93)	95% (95% CI=0.87–0.99)	86.00%	Loft et al., 2022
4	**Magnetic resonance imaging (MRI)**	Meta-analysis	Adenoma cases=62; Adenocarcinoma cases=24	Adenoma and adenocarcinoma (T1 & T2)	73% (95% CI=0.50–0.89)	90% (95% CI=0.80–0.96)	78.00%	Loft et al., 2022
5	**Shear-wave elastography**	Meta-analysis	Adenoma cases=62; Adenocarcinoma cases=24	Adenoma and adenocarcinoma (T1 & T2)	75% (for cut off value ≥ 41.3)	90% (for cut off value ≥ 41.3)	88.31%	Loft et al., 2022
6	**Magnifying chromoendoscopy (MC)**	Prospective study	A group(MC→EUS)=35; B group (EUS→MC)=31	Patients were randomly assigned in to 2 groups: A group= primary MC followed by secondary EUS; B group=primary EUS followed by secondary MC	74.20%	68.60%	71.20%	Shimura et al., 2014
7	**Endoscopic ultrasonography (EUS)**	Prospective study	A group(MC→EUS)=35; B group (EUS→MC)=32	Patients were randomly assigned in to 2 groups: A group= primary MC followed by secondary EUS; B group=primary EUS followed by secondary MC	67.70%	74.30%	71.20%	Shimura et al., 2014

(Continued)

Table 6.2: (Continued) List of imaging technologies used for the detection of colorectal cancer

S.No.	Technique	Study type	Sample size	Discrimination between	Sensitivity	Specificity	Accuracy%/ AUC	Reference
8	**Colon capsule endoscopy (CCE)**	Prospective study	CCE group=173; CTC group=176	Comparing the efficacy of CCE and CTC for the detection of individuals with colorectal neoplasia among FIT-positive participants in a population-based, organized CRC screening program, which represent an enhanced population for colorectal neoplasms.	98.1% (95% CI=94.1; 99.4)	76.6% (95% CI=68.8; 82.9)	93.30%	González-Suárez et al., 2020
9	**Endoscopic ultrasound elastography**	Meta-analysis	Derivation study=33; validation study=55	Differentiating intramucosal and shallow submucosal neoplasms from deep submucosal and advanced colorectal cancers. Assessment of the depth of submucosal or deeper invasion in the derivation and validation groups.	Derivative study=100 (95% CI=84.1–100); Validation study=77.8 (95% CI=60.4–89.0)	Derivative study=88.2 (95% CI=76.1–88.2); Validation study=86.1 (95% CI=77.4–91.7)	Derivative study=93.3 (95% CI=79.5–93.3); Validation study=83.3 (95% CI=71.7–90.8)	Esaki et al., 2021
10	**Magnifying chromoendoscopy**	Meta-analysis	Derivation study=33; validation study=55	Differentiating intramucosal and shallow submucosal neoplasms from deep submucosal (dSM) and advanced colorectal cancers. Assessment of the depth of submucosal or deeper invasion in the derivation and validation groups.	Derivative study=66.7 (95% CI= 50.1–72.1); Validation study=84.2 (95% CI=68.8–92.6)	Derivative study=94.4 (95% CI= 80.6–99.0); Validation study=91.4 (95% CI=83.1–96.0)	Derivative study=81.8 (95% CI=66.8–86.8); Validation study=88.9 (95% CI=78.0–94.8)	Esaki et al., 2021

11	**Endoscopic ultrasonography**	Meta-analysis	Derivation study=33; Validation study=55	Differentiating intramucosal and shallow submucosal neoplasms from deep submucosal (dSM) and advanced colorectal cancers. Assessment of the depth of submucosal or deeper invasion in the derivation and validation groups.	Derivative study=93.3 (95% CI= 76.6–98.8); Validation study=89.5 (95% CI=72.8–96.9)	Derivative study=77.8 (95% CI=63.9–82.3); Validation study=65.7 (95% CI=56.7–69.8)	Derivative study=84.8 (95% CI=69.7–89.8); Validation study=74.1 (95% CI=62.3–79.3)	Esaki et al., 2021
12	**Computed tomography colonography (CTC)**	Retrospective observational study	Chronic diverticular disease=53; CRC=42	CTC was used to differentiate between CRC and chronic diverticular disease based on morphological and textural parameters.	47%	98%		Valletta et al., 2022
13	**Computer aided detection device (CAD)**	Prospective study	No Polyps(6 mm or larger)=464; Polyps(6-9 mm)=52	The CAD-assisted findings were categorized into no polyps or polyps	90.9% (95% CI=88.0%, 93.4%)	76.9% (95% CI=63.2%, 87.5%)		Regge et al., 2013
14	**Dual-energy spectral computed tomography imaging**	Retrospective study	Low-grade=108; High-grade=65		86.20%	70.40%	88.60%	Cao et al., 2021
15	**Diffusion-weighted imaging**	Prospective study	Cases=33; control=15		90%	100%		Leufkens et al., 2013
16	**Chromoendoscopy**	Prospective study	Group 1 (aberrant crypt foci)=310;Group 2 (control group)=357		96%	99%		Kowalczyk et al., 2016
17	**Dynamic contrast-enhanced magnetic resonance imaging**	Retrospective study	Hepatocellular carcinoma =41; Hepatic metastasis of rectal cancer=34		90.24%	73.53%	80.00%	Li et al., 2020

Table 6.3 Tracer/contrast agents used in imaging techniques for cancer diagnosis

S.No.	Tracer name/ contrast used	Name of imaging technique	Approved by or not approved	Advantages	Disadvantages	Side effects	References
1.	Fluorodeoxyglucose (FDG)	Positron emission tomography (PET)	FDA approved	- Detects metabolic activity in tumors - Provides whole-body imaging	- Limited spatial resolution - False positives in inflammatory lesions	Minimal, such as allergic reactions	de Geus-Oei et al., 2006, Németh et al., 2019, Fletcher et al., 2008, Votrubova et al., 2006
2.	Gadolinium-based contrast agents	Magnetic resonance imaging (MRI)	FDA approved	- Excellent soft tissue contrast - Multi-parametric imaging capabilities	- Risk of nephrogenic systemic fibrosis in patients with kidney issues - Gadolinium deposition in the brain	Generally well-tolerated, but allergic reactions are possible	Zhou and Lu 2013, Asadollahzade et al., 2022
3.	Superparamagnetic Iron Oxide (SPIO)	Magnetic resonance imaging (MRI)	Investigational	- Enhanced contrast for colorectal liver metastases imaging - High sensitivity for detecting liver lesions	- Limited availability - Long imaging time	Generally well-tolerated, rare adverse events	Hekimoglu et al., 2011, Kumano et al., 2003
4.	18F-fluorothymidine (FLT)	Positron emission tomography (PET)	Investigational	- Measures cellular proliferation - Potential for early cancer detection	- Limited clinical validation - Higher cost compared to FDG-PET	Limited, mild reactions at injection site	Kim et al., 2019
5.	Manganese-based contrast agents	Magnetic resonance imaging (MRI)	Investigational	- Potential for improved brain imaging Manganese's essential role in cellular processes	- Limited research and validation - Manganese toxicity concerns	Limited data, potential for neurotoxicity	Henoumont et al., 2023, Chabanova et al., 2011
6.	CUB Domain-Containing Protein 1 (CDCP1)	Positron emission tomography (PET)	Investigational	- Helps in early detection-can be used to select cancer patients for personalised treatment with CDCP1-targeting agents, including antibody-drug conjugates	- Limited to cancers with sufficient CDCP1 expression - Not all cancer types show high CDCP1 expression.		Khan et al., 2022
7.	Anti-CCSP-2 scFv-FITC	Real-time fluorescence colonoscopy	Investigational	- Enables real-time detection of CRCa lesions - Provides high sensitivity in detecting CRCa lesions	- More extensive assessment is needed for the clinical use of anti-CCSP-2 scFv-FITC		Kim et al., 2021

Virtual colonoscopy, or CT colonography, is a minimally invasive method that utilizes CT imaging for a 3D view of the colon. Though it is less sensitive than traditional colonoscopy in detecting small polyps, it serves as an alternative for patients who cannot or refuse the procedure (Pickhardt, 2017). It is also valuable for preoperative T staging (Porté et al., 2017). In CRC patients, anastomotic recurrences occur when tumors grow post-treatment, related to inadequate margins or missed lesions. CT colonography exhibits 95% sensitivity and 100% specificity for diagnosing these recurrences (Porté et al., 2017). It accurately identifies extra-colonic findings such as liver metastases, with a 93% accuracy rate (Elnaggar et al., 2023). Recently, Wesp et al. introduced two deep learning models utilizing 3D CT colonography; the segmentation model achieved 69% specificity and 80% sensitivity for distinguishing premalignant from benign polyps, while the non-segmentation model had 44% specificity but maintained 80% sensitivity (Wesp et al., 2022). Overall, it is a cost-effective alternative to conventional colonoscopy and CT scans (Porté et al., 2017).

6.3.3.1 Recent Developments in Colonoscopy

6.3.3.1.1 Colon Capsule Endoscopy

An alternative to the conventional colonoscopy is the colon capsule endoscopy, which is usually recommended for the screening of CRC patients. It is a minimally invasive technique that provides a customizable frame rate and a panoramic image of the colon. Hausmann et al. included 153 patients to study the efficacy of the second-generation colon capsule endoscopy and found it to be effective in 111 patients (Hausmann et al., 2021). The colon capsule endoscopy method has higher sensitivity (2.0; 95% CI = 1.34-2.98) to detect polyps of size ≥6mm than CT colonography in patients with incomplete colonoscopy (Spada et al., 2015). A single-center, randomized trial study conducted by González-Suárez and colleagues to evaluate the efficacy of colon capsule endoscopy over CT colonography showed higher sensitivity for colorectal neoplasia (98.1% versus 64.9%), neoplastic lesions (96.1% versus 79.3%) and advanced colorectal neoplasm (100% versus 93.1%) in patients with fecal immunochemical positive test (González-Suárez et al., 2020). This procedure is a choice of many CRC patients in comparison to colonoscopy.

6.3.3.1.2 Chromoendoscopy

Incorporating chromoendoscopy enhances conventional endoscopy's sensitivity by aiding lesion detection and providing insights into colonic inflammation (Shah et al., 2014). This method uses dyes, like crystal violet and indigo carmine, to improve lesion visualization (Trivedi and Braden 2013). Crystal violet effectively differentiates colonic mucosal subtypes (Singh et al., 2020), while indigo carmine significantly boosts adenoma detection rates (Tischendorf et al., 2007; Singh et al., 2020).

Kudo and colleagues introduced the "pit-pattern" concept for classifying colorectal tumor lesions into types I to V, where types I and II indicate non-neoplastic lesions, while types III to V indicate neoplastic ones (Kudo et al., 1996) (Figure 6.5). Narrow-band imaging employs a specific light spectrum for in vivo visualization of colonic mucosal patterns (Tischendorf et al., 2007). Studies show chromoendoscopy has a notable diagnostic advantage (sensitivity=91.7%, specificity=90%) over magnifying endoscopy, and it shares comparable accuracy with narrow band imaging for differentiating polyp types. Vascular pattern analysis identifies abnormal and normal vascularity, with chromoendoscopy showing 66.7% sensitivity and 95% specificity (Tischendorf et al., 2007). Recently, Venkatayogi et al. developed the HySenSe system combined with machine learning, achieving 100% sensitivity and 99.3% specificity for non-neoplastic lesion detection (Venkatayogi et al., 2022), facilitating early diagnosis of non-neoplastic polyps.

6.3.3.1.3 Compression Optical Coherence Elastography

The era of precision diagnosis emphasizes the diagnosis of CRC beyond tumor grading and staging parameters. Morphological categorization of colorectal tumors is an effective strategy to know more about tumor characteristics. It makes the treatment planning more accurate. Hence, compression optical coherence elastography can be used for the colorectal tumor localization, differentiation, and characterization of its morphological subtypes (Plekhanov et al., 2023; Zeng et al., 2020). Plekhanov et al. employed optical coherence elastography to measure the tissue stiffness of the normal colon, adenoma, and CRC. The authors reported the highest stiffness in CRC, i.e. >500 kPa in comparison to the adenoma and the normal colon mucosa. Moreover, the high-grade colorectal tumors showed higher tumor stiffness than the low-grade tumors (Plekhanov et al., 2023). Zeng and co-workers designed a deep learning-based pattern recognition

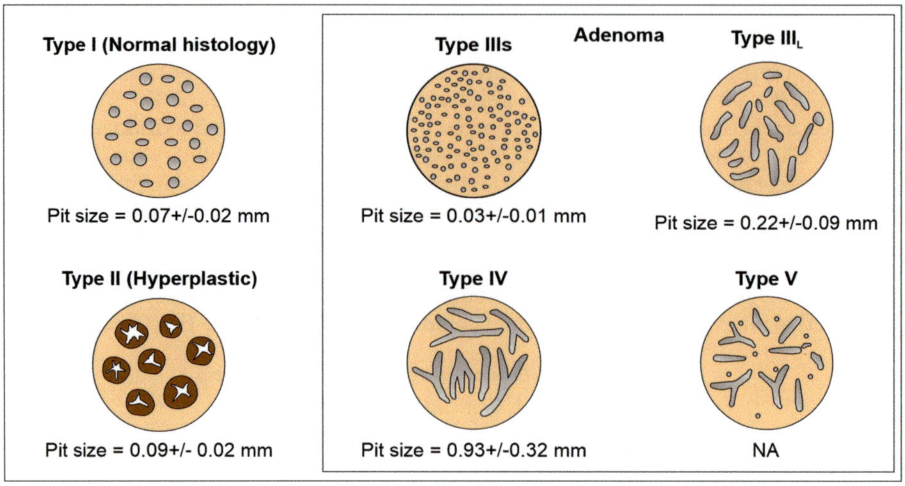

Figure 6.5 The Kudo's pit pattern colorectal tumor classification system, which uses the pattern and size of the pits as seen during a colonoscopy to classify colorectal tumors into type I, type II, and type IIIs, IIIL, type IV, and type V.

optical coherence elastography system that was able to identify normal and neoplastic colorectal tissue with a sensitivity of 100% and a specificity of 99.7% (Zeng et al., 2020).

6.3.4 Computed Tomography (CT) Scan

CRC is typically assessed using CT scans, which yield detailed images of the abdomen and pelvis, aiding in the detection and characterization of colorectal tumors. These scans are crucial for evaluating disease extent, lymph node involvement, and distant metastases, especially in the liver and lungs (Spirov et al., 2021; Yang et al., 2022). 2D and 3D colon imaging from volumetric CT data enhances visualization of the proximal colon and assessment of surrounding organs (Plumb and Halligan, 2015). However, challenges like patient preparation and standardization of imaging parameters exist. 3D scans offer precise lesion measurements and assist in identifying metastases (Plumb and Halligan, 2015). In colon and rectal cancer, PET/CT is valuable for detecting lung metastasis and characterizing nodules larger than 1 cm, although its effectiveness diminishes for smaller nodules (Lee et al., 2023). Perfusion CT imaging aids in assessing CRC angiogenesis and provides insights into tumor perfusion and microvascular function, essential for tumor characterization and treatment decisions (Petralia et al., 2010). This technique can be combined with PET for comprehensive evaluations (Kinahan and Fletcher, 2010). Tracer kinetics in perfusion CT can be influenced by blood flow, vessel permeability, and technical factors like image noise (Miles et al., 1991). Contrast agent administration poses risks for some patients, and interpretation variability may arise (Hayano et al., 2014). Although perfusion CT is an affordable screening alternative, further research is needed for its optimization in CRC screening. CT may not be the best choice for CRC diagnosis, as PET/CT shows advantages in certain circumstances, particularly in restaging and detecting extra-hepatic metastases (Çınar et al., 2013). While essential for CRC management, CT's limitations highlight the need for a multimodal imaging approach, with PET/CT serving as a valuable support tool.

6.3.5 Magnetic Resonance Imaging (MRI)

MRI is vital for assessing colorectal tumor invasion, particularly in the rectum, providing excellent soft tissue contrast for evaluating tumor size, invasion depth, and nearby organ involvement (Horvat et al., 2019b). This often makes MRI preferable to CT for staging rectal cancer and preoperative planning. Its capability to outline tumor margins and lymph node involvement facilitates accurate staging and treatment planning for CRC patients (Horvat et al., 2019b; Torkzad et al., 2010).

The use of MRI hepatobiliary contrast agents has improved the detection of colorectal liver metastases, reducing the need for further assessments (Lee et al., 2023). In a randomized trial with 360 cases, CT with intravenous contrast required further evaluation in 39.3% of studies, while MRI with extracellular contrast did so in only 17% (Lee et al., 2023). A meta-analysis of 17 studies found MRI's sensitivity and specificity for detecting liver metastases at 89% and 87%, respectively (Tsili et al. 2021; Mao et al., 2020). Diffusion-weighted imaging MRI aids in evaluating neoadjuvant treatment response and restaging (Beets-Tan et al., 2018). However, MRI has limitations, particularly in identifying metastases in lymph nodes and distant organs due to its anatomical imaging focus, which restricts spatial resolution (Matthews and Choi, 2016). Despite advancements in MRI's role in CRC imaging, challenges persist in surveillance, diagnosis, and staging. MRI cannot fully assess tumor biology and treatment response, whereas functional imaging techniques like PET/CT and PET/MRI provide a more comprehensive view of tumor behavior and therapy response (van Cutsem et al., 2016; Yeoh and Miles 2019). Additionally, contrast-enhanced MRI is invasive and time-intensive, limiting its use (Sabry et al., 2021). Routine postoperative MRI after curative resection has not significantly improved local recurrence detection. While whole-body MRI shows potential for efficient staging, its cost-effectiveness and utility for all CRC patients remain concerns (Miles et al., 2019). In summary, MRI is a valuable CRC imaging tool but has limitations in comprehensively assessing tumor biology and spatial resolution for metastases. Integrating functional imaging modalities is essential for enhancing CRC treatment.

6.3.6 Positron Emission Tomography (PET) Scan

PET scans are effective in identifying distant metastases, particularly in the liver, lungs, and lymph nodes. They utilize radiotracers like fluorodeoxyglucose (FDG) to highlight areas with elevated metabolic activity indicative of cancer. The 18F-FDG PET approach is pivotal for evaluating cancers such as CRC, which is characterized by increased glucose metabolism (Vikram and Iyer, 2008). This modality is essential for detecting CRC recurrence, especially when clinical doubts remain despite negative CT results (Revheim et al., 2015). The PET/CT combination provides both functional and anatomical data, aiding in CRC diagnosis, staging, and management (Kinahan and Fletcher 2010). Research indicates that PET/CT with 18F-FDG is beneficial for initial CRC staging, evaluating operability, ruling out distant diseases, and guiding treatment decisions (Rodríguez-Fraile et al., 2020; Spirov et al., 2021). One study found that PET-CT could detect lymph node metastases with 78.4% sensitivity and 100.0% specificity. It also plays a significant role in therapy response assessment in oligometastatic CRC patients (Nasr et al., 2023). A meta-analysis reported pooled sensitivity and specificity of 96% and 97% respectively for CRC liver metastases (Tsili et al., 2021), and another study highlighted its effectiveness in detecting loco-regional recurrence in postoperative CRC with raised CEAg levels (Ragheb and Sharara, 2020). PET/CT and PET/MRI are advised for various CRC stages, including initial staging for operability assessment and identifying high-risk metastases (Revheim et al., 2015). Despite challenges in detecting small lesions, PET remains valuable for CRC diagnosis and staging.

6.3.7 Radiomics and Radiogenomics

Radiomics is an emerging area in the field of biomedical imaging. Imaging data obtained by CT, MRI, or PET scans are now becoming the source for the formulation of imaging biomarkers with the help of artificial intelligence techniques such as deep learning and machine learning algorithms (Horvat et al., 2019a; Inchingolo et al., 2023). The identification and characterization of the degree of tumor heterogeneity through radiomics will unravel many unanswered questions such as how the treatment response will be, what the clinical outcome and survival of the patients (Horvat et al., 2019a). The radiomics technology comprises several steps to finally select imaging biomarkers (Figure 6.6).

Radiogenomics, integrating radiomics with genomics, provides a non-invasive technique to understand tumor biology (Figure 6.6). The abdominal contrast enhancement CT-based images were utilized by Porto-Álvarez and co-workers to predict the presence of KRAS mutation in patients with CRC. Among all the classifiers such as neural network, decision tree, support vector machine, and random forest used for KRAS mutation presence and absence the proposed AdaBoost exhibited the highest accuracy of 76.8%. Furthermore, it also showed the highest kappa value (53.7%) upon Cohen's kappa analysis, which was done to analyze the performance of the classifiers and AdaBoost (Porto-Álvarez et al., 2023). Zhong et al. conducted a multi-institutional study whereby they used radiogenomic analysis on 1601 CRC samples from five datasets to identify radiogenomic markers for prognosis prediction and to uncover intra-tumoral heterogeneity. The latter is recognized as a contributor to aggressive disease development and therapy resistance in various solid malignancies. The authors formulated a risk prediction model based on cell cycle and extracellular matrix genes and further divided the patients into two risk groups,

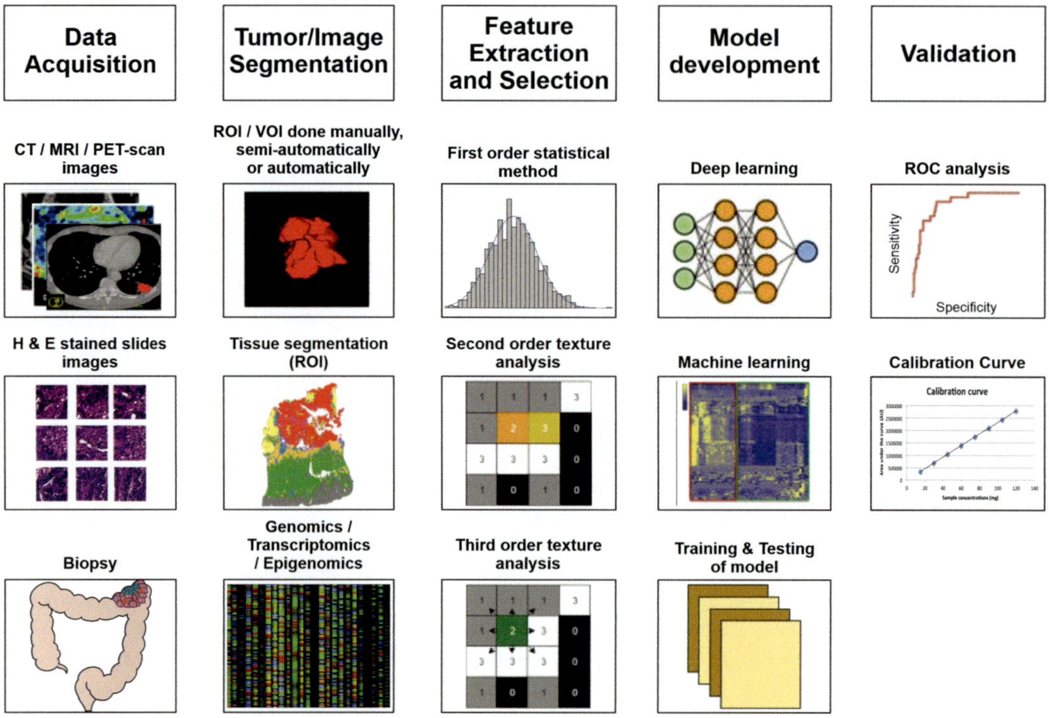

Figure 6.6 The several steps of the radiomics, radiopathomics and radiogenomics workflow. The first step involves the collection of raw data, i.e. the acquisition of either genomics/transcriptomics/proteomics data, medical images or H&E stained slide images. Subsequently, a dataset is compiled by incorporating the relevant clinical and radiological information. Furthermore, the images are segmented to identify the regions of interest or volume of interest. Image segmentation can be performed manually, semi-automatically, or automatically. Features are then extracted from the segmented images. Feature extraction and selection, is a crucial step, which involves the selection of the most relevant features to construct a model that can predict the desired endpoint. Finally, the developed model is validated on the internal and/or external datasets using several methods such as receiver operating curve analysis and a calibration curve.

i.e. high and low. The high-risk group had a 2.2-fold reduced survival rate than low-risk group individuals (Zhong et al., 2022).

Radiomics extracts numerous quantitative features from digital images and stores them in databases for hypothesis generation and testing. While it extends computer-aided diagnosis, it faces challenges limiting clinical application, such as lack of standardization, validation, and multicenter studies (Inchingolo et al., 2023). These issues affect the generalization and reproducibility of models, alongside concerns about data quality and model overfitting.

6.3.8 Radiopathomics

The integration of radiological and pathological data through advanced methods, termed radiopathomics, shows promise in predicting tumor characteristics, treatment responses, and patient outcomes in CRC (Figure 6.6). Tsai et al. developed the Multi-omics and Multi-cohort Assessment (MOMA) software, which analyzes whole slide histopathology images from four databases to predict genetic alterations, gene expression, and MSI status. MOMA achieved an AUC of 88% for TCGA and 76% for HPFS datasets in predicting MSI. The system also classified BECN1 gene expression into high and low categories and linked histopathological image patterns with these expression levels, along with predicting BRAF mutation status with an AUC of 71% (Tsai et al., 2023). The TCGA-Colon Adenocarcinoma and TCGA-Rectum Adenocarcinoma datasets were used to retrieve Whole Slide Histopathological Images (WSI) for the semi-quantitative calculation of the gland formation (GF) ratio. ROI was identified by tissue segmentation, and it was then further classified into four groups: GF1 (tubular morphology), GF2 (irregular tubular structures), GF3 (glands-forming cells), and X (normal colon and rectal epithelial cells). The GF classifiers showed a high discriminating percentage, as expected given the AUC values for GF1=98.3%, GF2 & GF3=96.3%, and X=98.1% (Chen, 2022). The WSI-Survival Analysis framework is also based on a cluster-level Deep Convolutional Survival algorithm and is used to predict a patient's survival status (Zhu et al., 2017). The authors evaluated the significance of WSI-Survival Analysis in three datasets (TCGA- Lung Squamous Cell Carcinoma, NLST, and TCGA- Glioblastoma Multiforme) and compared it with the ROI-based experiments. Several regression methods were used to determine the C-index values, and it was found that WSI-Survival Analysis was a better predictor of survival than ROI-based (Zhu et al., 2017). Using WSI-Survival Analysis, [the] pathomics signature was developed in another study conducted by Wang and colleagues. The researchers also developed a CT image-based radiomics model by using a support vector machine classifier and synthetic minority oversampling techniques. This further led to the categorization of CRC patients with lung metastasis into high-radiomic scores and low-radionics scores. Similarly, the immunoscores were calculated based on CD3+ and CD8+ T-lymphocyte count and high immunoscores were found to be associated with improved survival rates. Next, a combined nomogram of radiomics, pathomics, and immunoscore was developed and evaluated to predict the overall and disease-free survival in such patients. In comparison to the approaches being utilized individually, the combined nomogram revealed a greater AUC for both overall survival (86%) and disease-free survival (87%) (Wang et al., 2022). In this line of medical technology advancement, an observational study was done by Feng et al. in 2022 to develop and validate the radiopathomics model RAdioPathomics Integrated Prediction System (RAPIDS). This model uses advanced machine learning and artificial intelligence and provides an interpretation of tumor regression or residual after treatment. Analysis of this study demonstrated that RAPIDS exhibited superior performance in predicting pathological complete response status. The results of the AUC curve (81.2%) and statistical data showed significant improvements over the MRI, pathomics nucleus, and pathomics microenvironment models. Moreover, Kaplan-Meier survival analysis demonstrated that, when outcomes such as overall survival, disease-free survival, local recurrence-free survival, and distant metastasis-free survival were examined, the patients predicted by the RAPIDS model exhibited a better prognosis. In conclusion, multivariate analysis revealed RAPIDS to be an independent predictor of both overall and disease-free survival in patients with locally advanced rectal cancer (Feng et al., 2022).

Despite its potential, radiopathomics faces challenges, including lack of standardization and validation, which complicate clinical application. Automated systems are necessary to improve tumor delineation efficiency and accuracy. Tian et al. (2020) and Huang et al. (2023) discussed enhancing tumor representation and addressing data imbalances in deep learning models. The study by Brancato et al. faced limitations related to cohort size, missing localization information, and lack of shared reference standards in radiomics and pathomics research (Brancato et al., 2022).

6.4 CONCLUSION

The combination of various diagnostic techniques and imaging tools plays a crucial role in the effective management of CRC. It ranges from the traditional approaches like per-rectal examination to imaging modalities such as CT, MRI, and PET. Histopathological examination, especially in deciphering tumor variants like mucinous adenocarcinomas and tumor budding, remains critical for predicting outcomes and planning treatments. Immunohistochemical analyses add further accuracy by revealing molecular changes and tumor microenvironment dynamics. Future advancements aim to enhance the current diagnostic methods with the incorporation of advanced computational analyses such as radiomics, radiogenomics, and radiopathomics which may aid in precision diagnosis. As research progresses and validation studies continue, the primary goal is to move towards precision diagnosis and treatment efficacy in individuals with CRC.

REFERENCES

Abouelkhair, M. B., et al. (2021). The diagnostic value of cadherin 17 and CDX2 expression as immunohistochemical markers in colorectal adenocarcinoma. *J Gastrointest Cancer*, *52*:960–969.

Asadollahzade, E., et al. (2022). The benefits and side effects of gadolinium-based contrast agents in multiple sclerosis patients. *Revista da Associação Médica Brasileira*, *68*(8):979–981.

Auf, F. A., et al. (2011). Cytokeratin 20 and vascular endothelial growth factor as molecular markers in Egyptian patients with colorectal cancer. *J Oncol Pharma Pract*, *17*(3):160–167.

Bajenova, O. V., et al. (2001). Heterogeneous RNA-binding protein M4 is a receptor for carcinoembryonic antigen in Kupffer cells. *J Biolog Chem*, *276*(33):31067–31073.
Bardelli, A., Siena, S. (2010). Molecular mechanisms of resistance to cetuximab and panitumumab in colorectal cancer. *J Clin Oncol*, *28*(7):1254–1261.
Bayrak, R., et al. (2012). The value of CDX2 and cytokeratins 7 and 20 expression in differentiating colorectal adenocarcinomas from extraintestinal gastrointestinal adenocarcinomas: cytokeratin 7-/20+ phenotype is more specific than CDX2 antibody. *Diagnost Pathol*, *7*:1–11.
Beets-Tan, R. G., et al. (2018). Magnetic resonance imaging for clinical management of rectal cancer: updated recommendations from the 2016 European Society of Gastrointestinal and Abdominal Radiology (ESGAR) consensus meeting. *Eur Radiol*, *28*:1465–1475.
Brancato, V., et al. (2022). The relationship between radiomics and pathomics in glioblastoma patients: preliminary results from a cross-scale association study. *Frontiers Oncol*, *12*:1005805.
Cao, Y., et al. (2021). Development of a dual-energy spectral computed tomography-based nomogram for the preoperative discrimination of histological grade in colorectal adenocarcinoma patients. *J Gastrointest Oncol*, *12*(2):544.
Chabanova, E, et al. (2011). Manganese based MR contrast agents: formulation and clinical applications. *Open Drug Safety J*, *2*(1):29–38.
Chan, C. W. et al. (2009). Gastrointestinal differentiation marker Cytokeratin 20 is regulated by homeobox gene CDX1. *Proc Nat Acad Sci*, *106*(6):1936–1941.
Chen, C. I., et al. (2022). High expression of folate receptor alpha (Folr1) is associated with aggressive tumor behavior, poor response to chemoradiotherapy, and worse survival in rectal cancer. *Technol Cancer Res Treatm*, *21*:15330338221141795.
Chen, G., et al. (2018). Wnt/β-catenin pathway activation mediates adaptive resistance to BRAF inhibition in colorectal cancer. *Molec cancer Therapeut*, *17*(4):806–813.
Chen, S., (2022). Automatic tumor grading on colorectal cancer whole-slide images: Semi-quantitative gland formation percentage and new indicator exploration. *Frontiers Oncol*, *12*:833978.
Cho, J., et al. (2019). The prognostic role of tumor associated glycoprotein 72 (TAG-72) in stage II and III colorectal adenocarcinoma. *Pathol Res Pract*, *215*(1):171–176.
Çınar, A. et al. (2013). Restaging of colorectal cancer and PET/CT. *Turkish J Surg*, *29*(2):76.
de Geus-Oei, L. F., et al. (2006). FDG-PET in colorectal cancer. *Cancer Imag 6* (Spec No A), S71.
Dixon, D. A., et al. (2013). Mechanistic aspects of COX-2 expression in colorectal neoplasia. *Recent Results Cancer Res*, *191*:7–37.
Elnady, M. S., et al. (2022). Diagnostic accuracy of SATB2 in identifying primary and metastatic colorectal carcinoma: a comparative immunohistochemical study. *ecancermedicalscience*, *16*:1491.
Elnaggar, M., et al. (2023). The accuracy of different modalities used for preoperative primary tumour localisation in operated colorectal cancer patients. *Cureus*, *15*(3):.
Esaki, M., et al. (2021). Endoscopic ultrasound elastography as a novel diagnostic method for the assessment of hardness and depth of invasion in colorectal neoplasms. *Digestion*, *102*(5):701–713.
Farraj, F. A., et al. (2019). Signet ring cell carcinoma of the colon in young adults: a case report and literature review. *Case Reports in Oncol Med*, 3092674.
Feng, L., et al. (2022). Development and validation of a radiopathomics model to predict pathological complete response to neoadjuvant chemoradiotherapy in locally advanced rectal cancer: a multicentre observational study. *Lancet Digital Health*, *4*(1):e8–e17.
Fleming, M., et al. (2012). Colorectal carcinoma: Pathologic aspects. *J Gastrointest Oncol*, *3*(3):153.
Fletcher, J. W., et al. (2008). Recommendations on the use of 18F-FDG PET in oncology. *J Nucl Med*, *49*(3):480–508.
Foda, A. A., et al. (2016). Aberrant Expression of Calretinin, D2-40 and Mesothelin in Mucinous and Non-Mucinous Colorectal Carcinomas and Relation to Clinicopathological Features and Prognosis. *Pathol Oncol Res*, *22*(4):725–732.
Fujiyoshi, K., et al. (2020). Tumour budding, poorly differentiated clusters, and T-cell response in colorectal cancer. *EBioMedicine*, *57*: 2352–3954.
González-Suárez, B., et al. (2020). Colon capsule endoscopy versus CT colonography in FIT-positive colorectal cancer screening subjects: a prospective randomised trial—the VICOCA study. *BMC Med*, *18*:1–11.
Grigore, A. D., et al. (2016). Tumor budding: the name is EMT. Partial EMT. *J Clin Med*, *5*(5), 51.
Guerrero, R. M., et al. (2022). Targeting BRAF V600E in metastatic colorectal cancer: where are we today?. *ecancermedicalscience*, *16*:1489.

Gurzu, S., et al. (2013). Serrated pathway adenocarcinomas: molecular and immunohistochemical insights into their recognition. *PLoS One*, *8*(3):e57699.

Haddad, T. S., et al. (2021). Improving tumor budding reporting in colorectal cancer: a Delphi consensus study. *Virchows Archiv*, *479*(3): 1–11.

Hagland, H. R., et al. (2013). Molecular pathways and cellular metabolism in colorectal cancer. *Digestive Surg*, *30*(1):12–25.

Hamilton, S. R. (2000). Carcinoma of the colon and rectum. In: Hamilton SR, Aaltonen LA, eds, *Pathology and genetics of tumors of digestive system (WHO Classification of Tumours Vol 2)*, third edn, IARC.

Hausmann, J., et al. (2021). Colon capsule endoscopy: Indications, findings, and complications—data from a prospective german colon capsule registry trial (DEKOR). *Clin. Endosc*, *54*:92–99.

Hayano, K., et al. (2014). Computed tomography perfusion imaging as a potential imaging biomarker of colorectal cancer. *World J Gastroenterol*, *20*(46):17345.

Hekimoglu, K., et al. (2011). Small colorectal liver metastases: detection with SPIO-enhanced MRI in comparison with gadobenate dimeglumine-enhanced MRI and CT imaging. *Eur J Radiol* *77*(3):468–472.

Henoumont, C., et al. (2023). Mn-based MRI contrast agents: an overview. *Molecules*, *28*(21):7275.

Hidalgo-Estévez, A. M., et al. (2020). Cyclooxygenase 2-regulated genes an alternative avenue to the development of new therapeutic drugs for colorectal cancer. *Frontiers Pharmacol*, *11*:529720.

Hitchcock, C. L., et al. (2021). Survival advantage following TAG-72 antigen-directed cancer surgery in patients with colorectal carcinoma: Proposed mechanisms of action. *Frontiers Oncol*, *11*:731350.

Horvat, N., et al. (2019a). Novel imaging techniques of rectal cancer: what do radiomics and radiogenomics have to offer? A literature review. *Abdom Radiol*, *44*:3764–3774.

Horvat, N., et al. (2019b). MRI of rectal cancer: tumor staging, imaging techniques, and management. *Radiograph*, *39*(2):367–387.

Hosseinalizadeh, H., et al. (2024). Emerging insights into keratin 7 roles in tumor progression and metastasis of cancers. *Frontiers Oncol*, *13*:1243871.

Huang, Y., et al. (2023). Deep learning radiopathomics based on preoperative US images and biopsy whole slide images can distinguish between luminal and non-luminal tumors in early-stage breast cancers. *EBioMedicine*, *94*:104706.

Inchingolo, R., et al. (2023). Radiomics in colorectal cancer patients. *World J Gastroenterol*, *29*(19):2888.

Issa, I. A., Noureddine, M. (2017). Colorectal cancer screening: An updated review of the available options. *World J Gastroenterol*, *23*(28):5086.

Jansen, K., et al. (2023). 2355P Carcinoembryonic antigen (CEA) expression in human tumors: A tissue microarray study on 15,413 tumors. *Ann Oncol* *34*:S1197.

Kasprzak, A. (2020). Angiogenesis-related functions of Wnt signaling in colorectal carcinogenesis. *Cancers*, *12*(12):3601.

Khan, T., et al. (2022). CUB Domain-Containing Protein 1 (CDCP1) is a rational target for the development of imaging tracers and antibody-drug conjugates for cancer detection and therapy. *Theranost*, *12*(16):6915.

Kim, H. I., et al. (2021). Biomolecular imaging of colorectal tumor lesions using a FITC-labeled scFv-Cκ fragment antibody. *Scientific Rep*, *11*(1):17155.

Kim, J. E., et al. (2019). 3′-Deoxy-3′-18 F-Fluorothymidine and 18 F-Fluorodeoxyglucose positron emission tomography for the early prediction of response to Regorafenib in patients with metastatic colorectal cancer refractory to all standard therapies. *Eur J Nucl Med Molec Imag*, *46*:1713–1722.

Kinahan, P. E., Fletcher, J. W. (2010). Positron emission tomography-computed tomography standardized uptake values in clinical practice and assessing response to therapy. *Seminars in Ultrasound, CT and MRI*, 31/6: 496–505.

Kowalczyk, M., et al. (2016). Chromoendoscopy with a standard-resolution Colonoscope for evaluation of rectal aberrant crypt foci. *PLoS One*, *11*(2):e0148286.

Kudo, S. E., et al. (1996). Diagnosis of colorectal tumorous lesions by magnifying endoscopy. *Gastrointest Endosc*, *44*(1):8–14.

Kumano, S., et al. (2003). Using superparamagnetic iron oxide–enhanced MRI to differentiate metastatic hepatic tumors and nonsolid benign lesions. *Am J Roentgen*, *181*(5):1335–1339.

Kuo, C. T., Lee, W. S. (2016). Progesterone receptor activation is required for folic acid-induced antiproliferation in colorectal cancer cell lines. *Cancer Letters*, *378*(2):104–110.

Lee, J. H., Lee, S. W. (2017). The roles of carcinoembryonic antigen in liver metastasis and therapeutic approaches. *Gastroenterol Res Pract*, 7521987.
Lee, S., et al. (2023). Imaging of colon and rectal cancer. *Curr Problems Cancer*, *47*(2):100970.
Leufkens, A. M., et al. (2013). Diffusion-weighted MRI for the detection of colorectal polyps: feasibility study. *Magnetic Resonance Imag*, *31*(1):28–35.
Li, J., et al. (2020). Dynamic contrast-enhanced MRI differentiates hepatocellular carcinoma from hepatic metastasis of rectal cancer by extracting pharmacokinetic parameters and radiomic features. *Experiment Therap Med*, *20*(4):3643–3652.
Li, Y., et al. (2010). Carcinoembryonic antigen interacts with TGF-β receptor and inhibits TGF-β signaling in colorectal cancers. *Cancer Res*, *70*(20):8159–8168.
Liu, C., (2020). Matrix stiffness and colorectal cancer. *OncoTargets Ther*, 2747–2755.
Loft, M. K., et al. (2022). Endorectal ultrasound shear-wave elastography of complex rectal adenoma and early rectal cancer. *Diagnost*, *12*(9):2166.
Lugli, A., et al. (2017). Recommendations for reporting tumor budding in colorectal cancer based on the International Tumor Budding Consensus Conference (ITBCC) 2016. *Modern Pathol*, *30*(9):1299–1311.
Lugli, A., et al. (2008). Differential diagnostic and functional role of the multi-marker phenotype CDX2/CK20/CK7 in colorectal cancer stratified by mismatch repair status. *Modern Pathol*, *21*(11):1403–1412.
Luo, C., et al. (2019). Mucinous colorectal adenocarcinoma: clinical pathology and treatment options. *Cancer Communicat*, *39*:1–13.
Mao, Y., et al. (2020). Diagnostic performance of magnetic resonance imaging for colorectal liver metastasis: A systematic review and meta-analysis. *Scientific Rep*, *10*(1):1969.
Mao, Z., et al. (2003). Transcriptional regulation of A33 antigen expression by gut-enriched Krüppel-like factor. *Oncogene*, *22*(28):4434–4443.
Martínez-Ares, D., et al. (2005). The value of abdominal ultrasound in the diagnosis of colon cancer. *Revista Espanola de Enfermedades Digestivas*, *97*(12):877–886.
Matthews, R., Choi, M. (2016). Clinical utility of positron emission tomography magnetic resonance imaging (PET-MRI) in gastrointestinal cancers. *Diagnost*, *6*(3):35.
Miles, A., et al. (2019). Patient preferences for whole-body MRI or conventional staging pathways in lung and colorectal cancer: a discrete choice experiment. *Eur Radiol*, *29*:3889–3900.
Miles, K. A., et al. (1991). Colour perfusion imaging: a new application of computed tomography. *Lancet*, *337*(8742):643–645.
Nanda, N., Dhawan, D. K. (2020). Role of cyclooxygenase-2 in colorectal cancer. *Frontiers Bioscience-Landmark*, *26*(4):706–716.
Nasr, I. M. et al. (2023). 18F-FDG PET/CT in therapy response assessment: oligometastatic colorectal cancer. *Egypt J Radiol Nucl Med*, *54*(1):33.
Németh, Z., et al. (2019). Advantages of 18 F FDG-PET/CT over Conventional Staging for Sarcoma Patients. *Pathol Oncol Res*, *25*:131–136.
Ng, L., et al. (2023). Tissue cadherin 17 (CDH17): A favorable prognostic determinant of colorectal cancer using digital image analysis. *J Clin Oncol*, *41*(16_suppl):e14651–e14651.
Nowicki, A., (2019). Wartość kliniczna badań kolonoskopii i pozytonowej tomografii emisyjnej z tomografią komputerową w rozpoznawaniu raka jelita grubego. *Polish J Surg*, *91*:6–9.
Palermo, N. Y., et al. (2012). Hexapeptide fragment of carcinoembryonic antigen which acts as an agonist of heterogeneous ribonucleoprotein M. *J Peptide Sci*, *18*(4):252–260.
Petralia, G., et al. (2010). CT perfusion in oncology: how to do it. *Cancer Imag*, *10*(1):8.
Pickhardt, P. J. (2017). Imaging and screening for colorectal cancer with CT colonography. *Radiol Clin*, *55*(6):1183–1196.
Plekhanov, A. A., et al. (2023). Towards targeted colorectal cancer biopsy based on tissue morphology assessment by compression optical coherence elastography. *Frontiers Oncol*, *13*:1121838.
Plumb, A. A., Halligan, S. (2015). Colorectal cancer screening. *Seminars Roentgenol*, 50/2:101–110.
Porté, F., et al. (2017). CT colonography for surveillance of patients with colorectal cancer: systematic review and meta-analysis of diagnostic efficacy. *Eur Radiol*, *27*:51–60.
Porto-Álvarez, J., et al. (2023). CT-Based Radiomics to Predict KRAS Mutation in CRC Patients Using a Machine Learning Algorithm: A Retrospective Study. *Biomed*, *11*(8):2144.
Qiu, Z., et al. (2023). Clinical Usefulness of Ultrasound Elastography in Colonic Diseases: A Narrative Review. *Curr Med Imag*, *20*:e300323215222.
Rageul, J., et al. (2009). KLF4-dependent, PPARγ-induced expression of GPA33 in colon cancer cell lines. *Int J Cancer*, *125*(12):2802–2809.

Ragheb, S. R., Sharara, S. M. (2020). Can PET/CT detect recurrence in post-operative colorectal carcinoma patients with elevated CEA level?. *Egypt J Radiol Nucl Med*, *51*:1–9.
Regge, D., et al. (2013). Efficacy of computer-aided detection as a second reader for 6–9-mm lesions at CT colonography: multicenter prospective trial. *Radiol*, *266*(1):168–176.
Reggiani Bonetti, L., et al. (2016). Poorly differentiated clusters (PDC) in colorectal cancer: what is and ought to be known. *Diagnost Pathol*, *11*:1–4.
Revheim, M. E., et al. (2015). Respiratory gated and prolonged acquisition 18F-FDG PET improve preoperative assessment of colorectal liver metastases. *Acta Radiolog*, *56*(4):397–403.
Rodríguez-Fraile, M., et al. (2020). FDG PET/CT in colorectal cancer. *Revista Española de Medicina Nuclear e Imagen Molecular*, *39*(1):57–66.
Sabry, M. S., et al. (2021). Role of diffusion-weighted MRI in diagnosis and post therapeutic follow-up of colorectal cancer. *Egypt J Radiol Nucl Med*, *52*:1–11.
Samara, R. N., et al. (2007). Carcinoembryonic antigen inhibits anoikis in colorectal carcinoma cells by interfering with TRAIL-R2 (DR5) signaling. *Cancer Res*, *67*(10):4774–4782.
Shah, S. A., et al. (2014). Chromoendoscopy for colorectal cancer surveillance in patients with inflammatory bowel disease. *Curr Gastroenterol Rep*, *16*:1–10.
Shimura, T., et al. (2014). Magnifying chromoendoscopy and endoscopic ultrasonography measure invasion depth of early stage colorectal cancer with equal accuracy on the basis of a prospective trial. *Clin Gastroenterol Hepatol*, *12*(4):662–668.
Singh, J., et al. (2022). Pattern of expression of CDX2 in colorectal cancer and its role in prognosis. *J Cancer Res Therapeut*, *18*(Suppl 2):S420–S427.
Singh, R., et al. (2020). Chromoendoscopy: role in modern endoscopic imaging. *Transl Gastroenterol Hepatol*, *5*:39.
Spada, C., et al. (2015). Colon capsule versus CT colonography in patients with incomplete colonoscopy: a prospective, comparative trial. *Gut*, *64*(2):272–281.
Spano, J. P., et al. (2005). Epidermal growth factor receptor signaling in colorectal cancer: preclinical data and therapeutic perspectives. *Ann Oncol*, *16*(2):189–194.
Spirov, G., et al. (2021). The value of PET/CT in detecting colorectal cancer recurrence in patients with negative CT findings. *J Morphol Sci*, *4*(2):57–65.
Stamatakis, K., et al. (2015). Prostaglandins induce early growth response 1 transcription factor mediated microsomal prostaglandin E2 synthase up-regulation for colorectal cancer progression. *Oncotarget*, *6*(37):39941.
Stevenson, L., et al. (2011). Calbindin 2 (CALB2) regulates 5-fluorouracil sensitivity in colorectal cancer by modulating the intrinsic apoptotic pathway. *PLoS One*, *6*(5):e20276.
Tian, J., et al. (2020). ASO author reflections: radiopathomics strategy of combing multi-scale tumor information on pretreatment to predict the pathologic response to neoadjuvant therapy. *Ann Surg Oncol*, *27*:4307–4308.
Tiernan, J. P., et al. (2013). Carcinoembryonic antigen is the preferred biomarker for in vivo colorectal cancer targeting. *Br J Cancer*, *108*(3):662–667.
Tischendorf, J. J. W., et al. (2007). Value of magnifying chromoendoscopy and narrow band imaging (NBI) in classifying colorectal polyps: a prospective controlled study. *Endosc*, *39*(12):1092–1096.
Torkzad, M. R., et al. (2010). Magnetic resonance imaging (MRI) in rectal cancer: a comprehensive review. *Insights Imag*, *1*:245–267.
Trivedi, P. J., Braden, B. (2013). Indications, stains and techniques in chromoendoscopy. *QJM*, *106*(2):117–131.
Tsai, P. C., et al. (2023). Histopathology images predict multi-omics aberrations and prognoses in colorectal cancer patients. *Nature Communicat*, *14*(1):2102.
Tsili, A. C., et al. (2021). Imaging of colorectal cancer liver metastases using contrast-enhanced US, multidetector CT, MRI, and FDG PET/CT: a meta-analysis. *Acta Radiolog*, *62*(3):302–312.
Ueno, H., et al. (2012). New criteria for histologic grading of colorectal cancer. *Am J Surg Pathol*, *36*(2):193–201.
Vakiani, E., et al. (2015). Immunohistochemical detection of the BRAF V600E mutant protein in colorectal neoplasms. *Applied Immunohistochem Molec Morphol*, *23*(6):438–443.
Valletta, R., et al. (2022). Role of CT colonography in differentiating sigmoid cancer from chronic diverticular disease. *Japan J Radiol*, *40*:48–55.
Van Cutsem, E., et al. (2016). Imaging in colorectal cancer: progress and challenges for the clinicians. *Cancers*, *8*(9):81.
Venkatayogi, N., et al. (2022). Pit-pattern classification of colorectal cancer polyps using a hyper sensitive vision-based tactile sensor and dilated residual networks. *arXiv:2211.06814*.

Vikram, R., Iyer, R. B. (2008). PET/CT imaging in the diagnosis, staging, and follow-up of colorectal cancer. *Cancer Imag, 8*(Spec Iss A):S46.

Vlahović, I., et al. (2023). Potential predictors for CDX2 expression loss and mismatch repair deficiency in colorectal cancer. *Pathol Oncol Res, 29*:1610908.

Votrubova, J., et al. (2006). The role of FDG-PET/CT in the detection of recurrent colorectal cancer. *Eur J Nucl Med Molec Imag, 33*:779–784.

Wang, D., et al. (2024). Effects of sigmoidoscopy screening (including colonoscopy) on colorectal cancer: A meta-analysis based on randomized controlled trials. *Preventive Med Rep, 39*:102636.

Wang, R., et al. (2022). Development of a novel combined nomogram model integrating deep learning-pathomics, radiomics and immunoscore to predict postoperative outcome of colorectal cancer lung metastasis patients. *J Hematol Oncol, 15*(1):11.

Wang, Y. Q., et al. (2019). SATB2-AS1 suppresses colorectal carcinoma aggressiveness by inhibiting SATB2-dependent Snail transcription and epithelial–mesenchymal transition. *Cancer Res, 79*(14):3542–3556.

Wesp, P., et al. (2022). Deep learning in CT colonography: differentiating premalignant from benign colorectal polyps. *Eur Radiol, 32*(7):4749–4759.

Wieduwilt, M. J., Moasser, M. (2008). The epidermal growth factor receptor family: biology driving targeted therapeutics. *Cellular and molecular life sciences, 65*, 1566–1584.

Wong, N. A., et al. (2017a). A33 shows similar sensitivity to but is more specific than CDX2 as an immunomarker of colorectal carcinoma. *Histopathol, 71*(1):34–41.

Wong, S. C. C., et al. (2017b). Prognostic significance of Cytokeratin 20-positive lymph node vascular endothelial growth factor A mRNA and chromodomain helicase DNA binding protein 4 in pN0 colorectal cancer patients. *Oncotarget, 9*(6):6737.

Yang, T., et al. (2022). FAPI PET/CT in the diagnosis of abdominal and pelvic tumors. *Frontiers Oncol, 11*:797960.

Yeoh, E., Miles, K. (2019). Simultaneous positron emission tomography and magnetic resonance imaging for the detection and characterisation of liver lesions in patients with colorectal cancer: A pictorial review. *J Med Imag Radiat Oncol, 63*(5):624–629.

Yu, J., et al. (2019). CDX2 inhibits the proliferation and tumor formation of colon cancer cells by suppressing Wnt/β-catenin signaling via transactivation of GSK-3β and Axin2 expression. *Cell Death Dis, 10*(1):26.

Yu, W., et al. (2017). SATB2/β-catenin/TCF-LEF pathway induces cellular transformation by generating cancer stem cells in colorectal cancer. *Scientific Rep, 7*(1):10939.

Yui, A., et al. (2023). Molecular mechanism underlying the increased risk of colorectal cancer metastasis caused by single nucleotide polymorphisms in LI-cadherin gene. *Scientific Rep, 13*(1):6493.

Zeng, Y., et al. (2020). Real-time colorectal cancer diagnosis using PR-OCT with deep learning. *Theranostics*, 10(6):2587–2596.

Zhong, M. E., et al. (2022). CT-based radiogenomic analysis dissects intratumor heterogeneity and predicts prognosis of colorectal cancer: a multi-institutional retrospective study. *J Translat Med, 20*(1):574.

Zhou, Z., Lu, Z. R. (2013). Gadolinium-based contrast agents for magnetic resonance cancer imaging. *Wiley Interdisciplinary Rev: Nanomed Nanobiotechnol, 5*(1):1–18.

Zhu, X., et al. (2017). Wsisa: Making survival prediction from whole slide histopathological images. *IEEE Conference on Computer Vision and Pattern Recognition (CVPR)*, Honolulu: 6855–6863.

Yahya, Z. M., et al. (2013). Evaluation of the sensitivity and specificity of immunohistochemical markers in the differential diagnosis of effusion cytology. *Oman Med J, 28*(6):410–416.

7 Adjuvants and Neoadjuvant Therapies in the Treatment of Colorectal Cancer

Subhajit Karmakar, Atharva Anand Mahajan and Subhrojyoti Ghosh

7.1 INTRODUCTION

One of the most prevalent diseases in the world, colon cancer is a kind of cancer that begins in the colon, which is part of the large intestine. Depending on the tumor's features and stage, treatment usually consists of a mix of radiation therapy, chemotherapy, and surgery. Although colon cancer is a serious health risk, results can be substantially enhanced with early identification and treatment. The foundation of early detection and prevention is a healthy lifestyle, symptom awareness, and routine screening. Treatment advances, such as tailored medicines and novel surgical methods, are improving the prognosis for colon cancer patients. The majority of patients with colon cancer are diagnosed at an early stage (stages I to III), which provides an opportunity for cure (Chakrabarti et al., 2020). To achieve a cure for early-stage colon cancer, the current treatment paradigm involves surgery and, in a small number of patients, adjuvant chemotherapy. The goal of this combination of treatments is to eradicate minimal residual disease.

For the majority of colon cancer patients, surgery alone can be curative. However, adjuvant and neoadjuvant therapies are essential in the treatment of colorectal cancer (CRC) as they aim to enhance outcomes by targeting microscopic disease that might not be visible or treatable through surgery alone (PDQ Adult Treatment Editorial Board, 2002; Smith et al., 2024).

An estimated 1,926,425 new cases in 2022 and 10 million deaths each year are attributed to CRC. It is among the world's leading causes of mortality from cancer (Colorectal Cancer Statistics | WCRF International, n.d.). The incidence of colorectal cancer (CRC) is expected to rise by 60% by 2030 (Vekic et al., 2020). Numerous variables, including aging populations, lifestyle changes, and improved detection techniques in some areas, are responsible for this considerable increase. There is a worrying trend toward a higher incidence of colorectal cancer (CRC) in those under 50, according to recent statistics from Western nations. Because CRC has always been thought to be a disease that mostly affects elderly persons, although the rise in younger patients is especially concerning (Sifaki-Pistolla et al., 2022). Although the exact causes of this rise are unknown, potential causes include dietary modifications, obesity, sedentary lifestyles, and genetic predispositions. Of those with newly diagnosed colorectal cancer, around 75% had early-stage, non-metastatic illness upon diagnosis. However, flat or depressed precancerous lesions account for around 25% of false-negative outcomes from traditional colonoscopies. A thorough review of studies, including colonoscopy data, found a combined miss rate of 22% for all polyps and 26% for polyps smaller than 5 mm (Hong et al., 2012; Ikematsu et al., 2022). This is an important topic since there is a large rise in the likelihood of effective treatment and even a cure with early identification.

Adjuvant therapy, including chemotherapy, in conjunction with surgery can be curative for patients whose illness is still in its early stages. However, avoiding illness recurrence is still a major issue, even with therapy. Individuals (5–30%) with CRC have recurrence after the first treatment, despite advancements in surgery and adjuvant therapy (Osterman et al., 2020; Xie et al., 2020). This broad variation is due to several factors, including the initial treatment's thoroughness, tumor biology, and the stage upon diagnosis. Improving preventative efforts is important, especially for those at higher risk, including younger people, and should involve lifestyle changes, more knowledge, and routine screening. Colonoscopy and other screening techniques are essential for early and more manageable detection of colorectal cancer. Sustained investigation is necessary to enhance comprehension of recurrence reasons and to create better adjuvant medicines (Triantafillidis et al., 2015). Younger populations are seeing increased CRC, necessitating focused study into the underlying causes and effective treatment approaches. Since the effects of CRC vary depending on where one lives, addressing the global burden of CRC also necessitates focusing on inequities in healthcare access. Reducing the worldwide burden of this illness requires ensuring access to screening, early diagnosis, and treatment in low-resource settings.

However, this chapter will discuss the novel adjuvant and neoadjuvant-based therapies that can increase treatment regimes and open some windows toward a cure. This chapter also addresses the current standard of care, changing paradigms for treatment, and newly discovered biomarkers that may help with patient selection and therapy personalization for better results.

DOI: 10.1201/9781032704517-7

7.2 ADJUVANT THERAPY

After the initial treatment, typically surgery, adjuvant therapy (AT) is further treatment administered to eradicate any remaining cancer cells and lower the chance of recurrence (Figure 7.1). These medications are continually evolving because of ongoing research aimed at improving the timing, combination, and sequencing of therapies to boost survival rates and minimize side effects in patients with colorectal cancer. Adjuvant treatment is crucial for managing colon cancer, especially when the disease has spread to nearby lymph nodes (stage III) or when there is a high risk of recurrence following surgery (Taieb & Gallois, 2020; Yeom et al., 2019). Patients with resected stage III colon cancer should consistently receive adjuvant therapy after surgery, as numerous clinical studies have demonstrated a clear survival benefit with both 5-fluorouracil (5-FU) monotherapy and oxaliplatin-based regimens. (Tomita et al., 2019). It has also been shown that different groups of stage I patients have different results. Others have demonstrated that in patients in stage I, specific immunologic or genetic markers may correlate with worse outcomes. Despite these findings, risk stratification for directing therapy in stage I patients is not routinely performed, unlike in stage II patients. This may be partly due to concerns that the potential negative outcomes and side effects of chemotherapy could outweigh the possible survival benefits.

7.2.1 Common Chemotherapeutic Drugs in Adjuvant-Based Therapy

The advantages of adjuvant chemotherapy were first demonstrated in a landmark 1990 study by Moertel et al., which revealed that administering adjuvant chemotherapy with 5-FU and levamisole for 12 months after surgery significantly improved survival rates and reduced recurrence compared to surgery alone. When evaluating chemotherapy options, it is important to consider side effects such as oxaliplatin-induced neuropathy against the potential benefits, especially for stage II cancer patients, where the therapeutic gains might be less evident. Disease-free survival (DFS) at two and three years are established benchmarks used in clinical trials to assess the efficacy of adjuvant treatments for stage III colon cancer. (Sargent et al., 2011). This is supported by the Adjuvant Colon Cancer End Points (ACCENT) joint group study, which found that in colon cancer patients receiving adjuvant chemotherapy with 5-FU, DFS at two and three years correlated with overall survival (OS) at five and six years, respectively (Chan & Chee, 2019; Yeom et al., 2019).

After five years, the recurrence rate was less than 1.5% years, according to an update of this investigation, which also revealed that most relapses happen early, usually within two years following surgery. When it came to adjuvant chemotherapy scheduling, patients with resected colorectal cancer who received adjuvant chemotherapy later in life fared poorly. Meta-analyses indicate a reduction in OS when adjuvant chemotherapy is initiated more than eight weeks after surgery, and

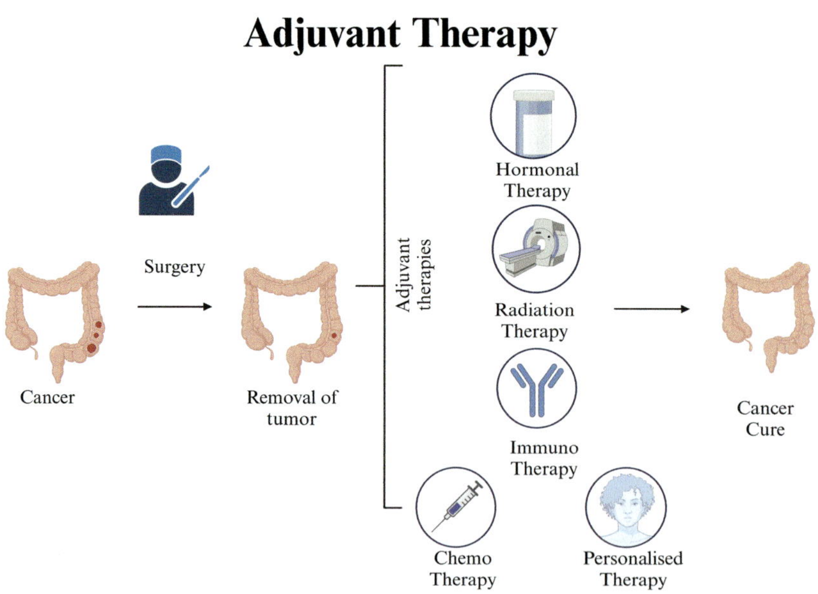

Figure 7.1 Adjuvant therapy is given after tumor surgery to reduce the risk of cancer recurrence.

particularly if it is delayed beyond four weeks. Therefore, adjuvant chemotherapy should be started within 12 weeks of receiving medical clearance, whenever possible (Zhan et al., 2018).

FOLFOX (5-FU, leucovorin, and oxaliplatin) and CAPOX (capecitabine and oxaliplatin) are typical regimens (Jones et al., 2023). Furthermore, chemotherapy may be an option for medically fit, high-risk stage II patients due to the potential little absolute advantage of the single drug 5-fluorouracil (5-FU). Consideration may also be given to a combination of chemotherapy and oxaliplatin in some high-risk situations. It is important to note that the current definition of high-risk stage II colon cancer may be inadequate, as many patients deemed at medium risk could still experience recurrence, while those with high-risk characteristics might not. For stage III colon cancer, combination chemotherapy based on oxaliplatin and fluoropyrimidine (e.g., FLOX, FOLFOX, XELOX) is the standard treatment. This approach is more effective than single-agent 5-FU in terms of OS and DFS (Iveson et al., 2018; Tsai et al., 2016). It is generally known that adjuvant therapy in CC lowers the probability of mortality in stage III by 10–15% when fluoropyrimidines alone are used and by an additional 4–5% when combinations incorporating oxaliplatin are used. The benefits of oxaliplatin as adjuvant therapy for stage III cancer are sustained for both DFS and OS with a 10-year follow-up, as demonstrated in the Multicentre International Study of oxaliplatin/5-FU/leucovorin (LV) in the Adjuvant Treatment of Colon Cancer (MOSAIC). Researchers look forward to similar OS and DFS data for patients with stage II colon cancer (André et al., 2015).

7.2.1.1 5-fluorouracil (5-FU)-Based Chemotherapy

Chemotherapy using 5-FU is a standard treatment for colorectal cancer, and it is used both for advanced or metastatic illness and as an adjuvant (post-surgery) measure (Varghese, 2015). For many years, 5-FU has been one of the most often used medications to treat colorectal carcinoma, which includes tumors of the colon and the rectum. The body breaks down 5-FU into 5-fluoro-2'-deoxyuridine-5'-monophosphate (FdUMP), which combines with folic acid and thymidylate synthase to generate a stable complex that stops deoxyuridine monophosphate (dUMP) from becoming deoxythymidine monophosphate (dTMP) (Kurasaka et al., 2022; Zhang et al., 2008). This inhibition of DNA synthesis brings on cell death. Moreover, 5-FU can be broken down into 5-fluorouridine triphosphate (FUTP), which is incorporated into RNA and interferes with processing (Noordhuis et al., 2004). 5-FU-based chemotherapy is frequently used to eradicate a microscopic illness that remains after surgically resecting colon tumors, therefore lowering the likelihood of recurrence of cancer. Patients with stage III (lymph node-positive) colon cancer and certain high-risk stage II cases benefit most from it. An oral prodrug of 5-FU, especially in the CAPOX (Capecitabine and Oxaliplatin) regimen, is occasionally used in place of intravenous 5-FU (Allegra et al., 2015; Pectasides et al., 2015). Chemotherapy based on 5-FU is still a mainstay of the treatment of colorectal cancer; it is important in managing metastatic disease as well as in adjuvant situations. Its usage is still being optimized by ongoing studies, especially when combined with other chemotherapeutic drugs and targeted therapies. The combination of 5-FU and LV was shown to be better in subsequent studies, and six months of treatment was sufficient to produce OS advantages comparable to those of 12 months (Chau et al., 2005). Survival benefits have been observed with both the Roswell Park regimen, which involves a weekly bolus of 5-FU for six weeks within an eight-week cycle over six months, and the Mayo Clinic regimen, which consists of a daily bolus of 5-FU for five days in a 28-day cycle for six months (Morse, 2005; Patel et al., 2004). Infusional 5-FU regimens have proven to be as effective as bolus regimens but with fewer side effects. Moreover, the X-ACT study demonstrated that oral capecitabine is a practical, safe, and equally effective alternative to the Mayo Clinic regimen for treating stage III colon cancer.

7.2.1.2 Oxaliplatin-Based Therapies

A common platinum-based chemotherapy drug for CRC, especially in advanced and metastatic stages, is oxaliplatin. It is frequently used in regimens like FOLFOX (folinic acid, 5-FU, and oxaliplatin) together with other chemotherapy medications, most frequently fluoropyrimidines like 5-fluorouracil (5-FU) (Jonker et al., 2006). Oxaliplatin creates platinum-DNA adducts, which cause DNA crosslinking, which is how it works against cancer. This stops transcription and DNA replication, which eventually causes cancer cells to undergo apoptosis (Virag et al., 2013). For certain people, oxaliplatin is a better alternative than older platinum medications like cisplatin since it has a distinct adverse effect profile and is less nephrotoxic. In 2004, oxaliplatin was approved for use in adjuvant therapy for stage III colon cancer, based on its impact on achieving a three-year DFS. Combining oxaliplatin with 5-FU or capecitabine led to a 20% relative risk reduction for DFS, which also corresponds to similar improvements in overall survival. (Chan &

Chee, 2019; Marshall et al., 2007). The goals of ongoing research are reducing side effects, enhancing patient selection using biomarkers, and conquering oxaliplatin resistance. Combination treatments involving immunotherapies and targeted medicines are also being researched to improve oxaliplatin-based regimens' efficacy.

7.2.2 Radiation Therapy

In the multimodal treatment of colorectal cancer, radiation therapy plays a critical role, especially for rectal cancer. It is strategically applied to enhance surgical results, lower recurrence, and control symptoms in more advanced instances. While attempting to lower the hazards involved, technological and scientific advancements in radiation treatment continue to improve its accuracy and efficacy. External beam radiation therapy (EBRT), the most used kind, uses radiation to target the tumor from outside the body. It is given across many sessions, frequently every day for a few weeks (Calvo et al., 2013; Koka et al., 2022). It is less typically used to prevent local recurrence in rectal cancer but is often utilized in colon cancer. However, patients with a pathologic stage of pT3-4 or N+ who initially undergo surgery are often also treated with postoperative radiation (Peng et al., 2019; Skóra et al., 2015). The current standard of care for resectable rectal cancer involves aggressive surgery based on the principle of complete mesorectal excision, which has significantly reduced the rate of local recurrence. Research findings indicate that among pT3N0 rectal patients undergoing mesorectal excision surgery alone, local recurrence rates varied between 4.1 and 6.5% (Peng et al., 2019). The effectiveness of further radiation therapy in this category of rectal cancer is debatable because of the low probability of local failure. They proposed that individuals with pT3N0 stages might not require post-operative adjuvant radiotherapy and that it might constitute overtreatment.

7.2.3 Targeted Therapy

This comprises immune checkpoint inhibitors, BRAF inhibitors, anti-VEGF treatment, and anti-EGFR therapy (Kasi et al., 2023; Li et al., 2021). Drugs such as cetuximab (anti-EGFR) or bevacizumab (anti-VEGF) may be added in some circumstances, particularly in high-risk situations. These monoclonal antibodies target the epidermal growth factor receptor (EGFR) on the cancer cell surface. EGFR is involved in signaling pathways that promote cell survival and proliferation. Additionally, drugs like bevacizumab, aflibercept, and ramucirumab inhibit the production of vascular endothelial growth factor (VEGF), a protein that facilitates angiogenesis, or the formation of new blood vessels within tumors (Lu et al., 2023). Through VEGF inhibition, these treatments deprive the tumor of the oxygen and nutrition it needs to thrive. Patients with metastatic colorectal cancer (CRC) carrying a BRAF V600E mutation often receive additional treatments, such as MEK inhibitors and anti-EGFR antibodies, in conjunction with other therapies (Kopetz et al., 2019). On the other hand, immune checkpoint inhibitors have demonstrated robust responses and increased longevity in this patient population, frequently resulting in long-term disease management. These treatments can potentially lessen the negative effects of conventional chemotherapy while also improving results by concentrating on certain molecular abnormalities in cancer cells. To provide CRC patients with more individualized and efficient treatment choices, ongoing research endeavors to enhance and broaden the application of targeted treatments.

7.2.3.1 Immunotherapy

Immunotherapy is increasingly being explored as an adjuvant treatment for colorectal cancer, particularly for patients with specific molecular features (Sahin et al., 2019). Checkpoint inhibitors, such as pembrolizumab and nivolumab, are a major focus, especially for tumors with mismatch repair deficiency (dMMR) or microsatellite instability-high (MSI-H), which exhibit increased mutational burdens that can be targeted by the immune system. These drugs work by blocking proteins that inhibit immune cell activity, thereby enhancing the immune system's ability to attack cancer cells (Guyot D'Asnières De Salins et al., 2021). Additionally, cancer vaccines, which stimulate the immune system to recognize and target cancer-specific antigens, and adoptive cell therapies, which involve modifying a patient's own immune cells to better target cancer, are under investigation (Boland & Goel, 2010). Combination therapies, merging immunotherapy with standard treatments like chemotherapy, targeted therapy, or radiation, are also being explored to improve efficacy. Moreover, ongoing research aims to refine biomarker-driven approaches to identify which patients are most likely to benefit from immunotherapy. While the potential of immunotherapy in adjuvant settings is promising, clinical trials are necessary to determine its optimal role and effectiveness in various stages of colorectal cancer. Table 7.1 summarizes landmark adjuvant chemotherapy trials in stage II and III CRC patients.

Table 7.1 Landmark adjuvant chemotherapy trials conducted in stage II and III CRC patients

Study title	Clinical trials.gov ID	Sponsor	Phase	Purpose	Intervention
MRD-guided deferred adjuvant therapy in resectable early-stage colon cancer	NCT06204484	Sun Yat-sen University	Phase II	To test if MRD status detected by ctDNA can guide precision therapy in resectable high-risk stage II and low-risk stage III colon cancer.	CAPEOX adjuvant therapy, deferred CAPEOX adjuvant therapy, mFOLFOXIRI intensive adjuvant therapy
Rofecoxib after surgery in treating patients with stage II or stage III colorectal cancer	NCT00031863	Cancer Research Campaign Clinical Trials Centre	Phase III	To determine the effectiveness of rofecoxib in improving survival and relapse-free survival in stage II/III colorectal cancer patients after surgery.	Rofecoxib, placebo
ePRO for adjuvant therapy of colorectal adenocarcinoma	NCT04069455	Zhejiang University	Not Applicable	To observe the improvement of QoL using ePRO to manage patients receiving oxaliplatin-based adjuvant chemotherapy for colorectal cancer.	Electronic Patient Reported Outcomes (ePRO)
mFOLFOXIRI Versus mFOLFOX6 as adjuvant chemotherapy for locally advanced colorectal cancer	NCT04338191	Sun Yat-sen University	Phase III	To compare the efficacy of mFOLFOXIRI vs. mFOLFOX6 as adjuvant chemotherapy after neoadjuvant oxaliplatin-based treatment in high-risk stage II and stage III colorectal cancer patients.	mFOLFOXIRI adjuvant chemotherapy, mFOLFOX6 adjuvant chemotherapy
Leucovorin and fluorouracil compared with observation in treating patients with colorectal cancer that has been surgically removed	NCT00005586	Institute of Cancer Research, United Kingdom	Phase III	To compare the efficacy of L-leucovorin and fluorouracil chemotherapy versus observation in improving survival and recurrence rates in patients with resected colorectal cancer.	L-leucovorin, fluorouracil, observation
PTC drug sensitivity detection to guide postoperative adjuvant of colorectal cancer	NCT05978349	Peking Union Medical College Hospital	Not Applicable	To evaluate the consistency between in vitro 3D drug sensitivity test results of micro tumors (PTC) and clinical outcomes, guiding adjuvant treatment strategy in colorectal cancer.	Adjuvant therapy based on 3D drug sensitivity test results, standard adjuvant therapy
Adjuvant PD-1 blockade for high-risk stage II dMMR/MSI-H colorectal cancer	NCT06520683	Sun Yat-sen University	Phase III	Colorectal cancer, microsatellite instability high	Drug: tislelizumab, adjuvant chemotherapy
Celecoxib, leucovorin, fluorouracil, and oxaliplatin in treating patients with metastatic colorectal cancer	NCT00072553	GERCOR—Multidisciplinary Oncology Cooperative Group	Phase II	Drug: celecoxib, leucovorin, fluorouracil, oxaliplatin	Primary: determine response rate. Secondary: assess toxicity, progression-free survival, salvage surgery rate, and quality of life.
Randomized phase III adjuvant study for stage III colorectal cancer	NCT00209742	Hokkaido Gastrointestinal Cancer Study Group	Phase III	Drug: UFT, Leucovorin, Krestin	Evaluate disease-free survival rate, overall survival rate, adverse events, and quality of life among different adjuvant therapies.

7.3 NEOADJUVANT THERAPY

Neoadjuvant therapy involves administering cancer medications before a potentially curative final local treatment, such as surgery or radiation (Figure 7.2). The choice of treatment modalities and the timing of their use are influenced by various factors, including the tumor's location, its biological characteristics, and the goals of the final treatment. This approach aims to shrink the tumor, making it easier to remove or treat effectively and potentially improving overall treatment outcomes. The impact of these many variables is aptly demonstrated in individuals suffering from colorectal cancer. There are significant differences between patient groups, with rectal cancer patients generally having a more established role for neoadjuvant treatment compared to colon cancer patients. Neoadjuvant therapy is used alongside primary treatment, usually surgery, to reduce the risk of tumor metastasis during the procedure or to shrink the tumor, making it easier to remove (Teshome & Hunt, 2014).

For locally advanced rectal cancer, neoadjuvant therapy, which involves radiation and chemotherapy before surgery, is often the primary approach. This treatment strategy aims to reduce the tumor size, improving surgical outcomes and decreasing the likelihood of local recurrence. The choice of adjuvant and neoadjuvant therapies is significantly influenced by the cancer's stage and its location within the colon or rectum. Additionally, there is a growing trend towards tailoring treatment strategies based on the tumor's genetic and molecular characteristics, such as MSI status and KRAS/NRAS mutations, to better address individual patient needs (Hsu et al., 2016). Patients at increased risk of recurrence who could benefit most from neoadjuvant therapy can be identified by several adverse factors. These include the tumor's height, its cT-stage, its proximity to the mesorectal fascia (MRF), its cN-stage, the presence of extramural vascular invasion (EMVI), and

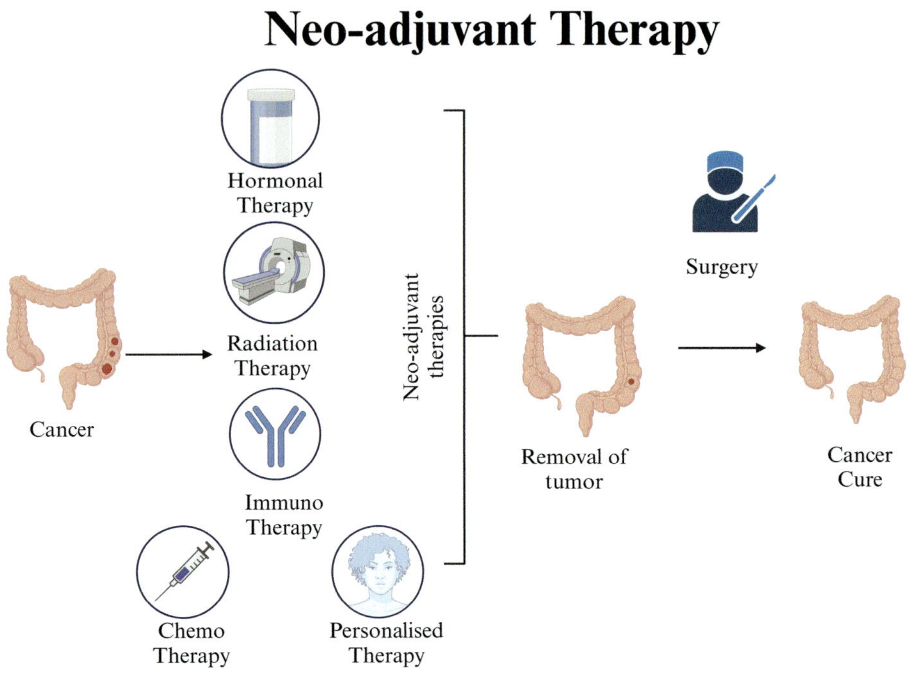

Figure 7.2 The neo-adjuvant therapy is administered before the surgery of the tumor to reduce the size of the tumor and ease the surgical procedure.

the presence of enlarged or suspicious lateral lymph nodes (Lambregts et al., 2022). North American patients with cT1-2 tumors are advised to get neoadjuvant treatment if there is a clinical suspicion of nodal involvement, indicating a more aggressive approach (Smith et al., 2024). Opponents of this approach claim that it might result in people with tumors that are not as dangerous being overtreated and argue that the cN stage is not as valuable as other unfavorable characteristics. There is no known head-to-head comparison study of these outcomes; however, both chemoradiotherapy (CRT) and short-course radiotherapy (SCRT) can lead to late toxicities and a diminished quality of life.

The Total Neoadjuvant Therapy (TNT) strategy is becoming more and more popular, especially for rectal cancer (Johnson et al., 2023; Ochiai et al., 2024). TNT involves administering both chemotherapy and chemoradiotherapy before surgery. Treating micrometastases sooner and raising the pathologic complete response rates could enhance results. Another recent development is the introduction of TNT as a substitute for SCRT or CRT. Patients at higher risk of recurrence who may gain the most from neoadjuvant therapy can be identified by several adverse factors, including the tumor's height, its cT-stage, its proximity to the mesorectal fascia (MRF), its cN-stage, the presence of extramural vascular invasion (EMVI), and the presence of enlarged or suspicious lateral lymph nodes (Capdevila et al., 2022; Daprà et al., 2023). Only CRT patients were administered postoperative chemotherapy, and only in certain centers. 32% of the patients in this research had cT4 malignancies, 62% had an MRF that was threatened, and 65% had cN2 cancers (Smith et al., 2024). The TNT arm demonstrated a significantly lower rate of disease-related treatment failure with a median follow-up of 4.6 years, particularly among patients with cT3 tumors that are MRF-positive, cT4 cancers, and those with positive lateral lymph nodes (Anker et al., 2021; Boublikova et al., 2023). Despite this, the current National Comprehensive Cancer Network (NCCN) guidelines advocate for total neoadjuvant therapy (TNT) as a treatment option even for less advanced tumors. The latest research on this topic is from the PROSPECT trial, which randomized 1,194 patients to receive either chemoradiotherapy (CRT) or neoadjuvant chemotherapy with fluorouracil, leucovorin, and oxaliplatin (FOLFOX) (Biniaz et al., 2024; Quezada-Diaz & Smith, 2022). Before surgery, patients in the FOLFOX group were restaged, and those whose tumors had shrunk by less than 20% or who were unable to finish more than five cycles were given CRT. The patients in the study were classified as intermediate risk and had mid- or upper-rectal cancers staged as cT2N1-2 or cT3N0-2 (Daprà et al., 2023).

These RAPIDO/ PROSPECT trials are escalating the existing treatment. Considering that the MERCURY research suggests that comparable local recurrence-free and disease-free survival can be achieved with total mesorectal excision (TME) alone in a patient cohort similar to that in the PROSPECT trial, the additional benefit of neoadjuvant chemotherapy in this context may be questioned (Romero Zoghbi et al., 2024; Smith et al., 2024). Nonetheless, a more sophisticated strategy is required in light of the growing recognition of the variability in tumor biology and the introduction of innovative treatment alternatives in colorectal carcinoma. Figure 7.3 schematically represents some adjuvant and neoadjuvant therapies in CRC treatment.

7.4 CONCLUSION AND FUTURE PROSPECTS

Despite advancements in adjuvant and neoadjuvant therapy for CRC, challenges remain, particularly in predicting and optimizing responses to treatment. Rising CRC cases, especially among younger individuals, and high recurrence rates highlight these difficulties. Selecting the right patients for therapy and identifying candidates for non-operative management is also complex. Circulating tumor DNA (ctDNA) has emerged as a promising biomarker, offering potential for predicting recurrence and treatment response. Global health efforts must focus on prevention, early detection, and more effective treatments. Personalized medicine and innovative therapies are expected to improve CRC management, especially for high-risk patients.

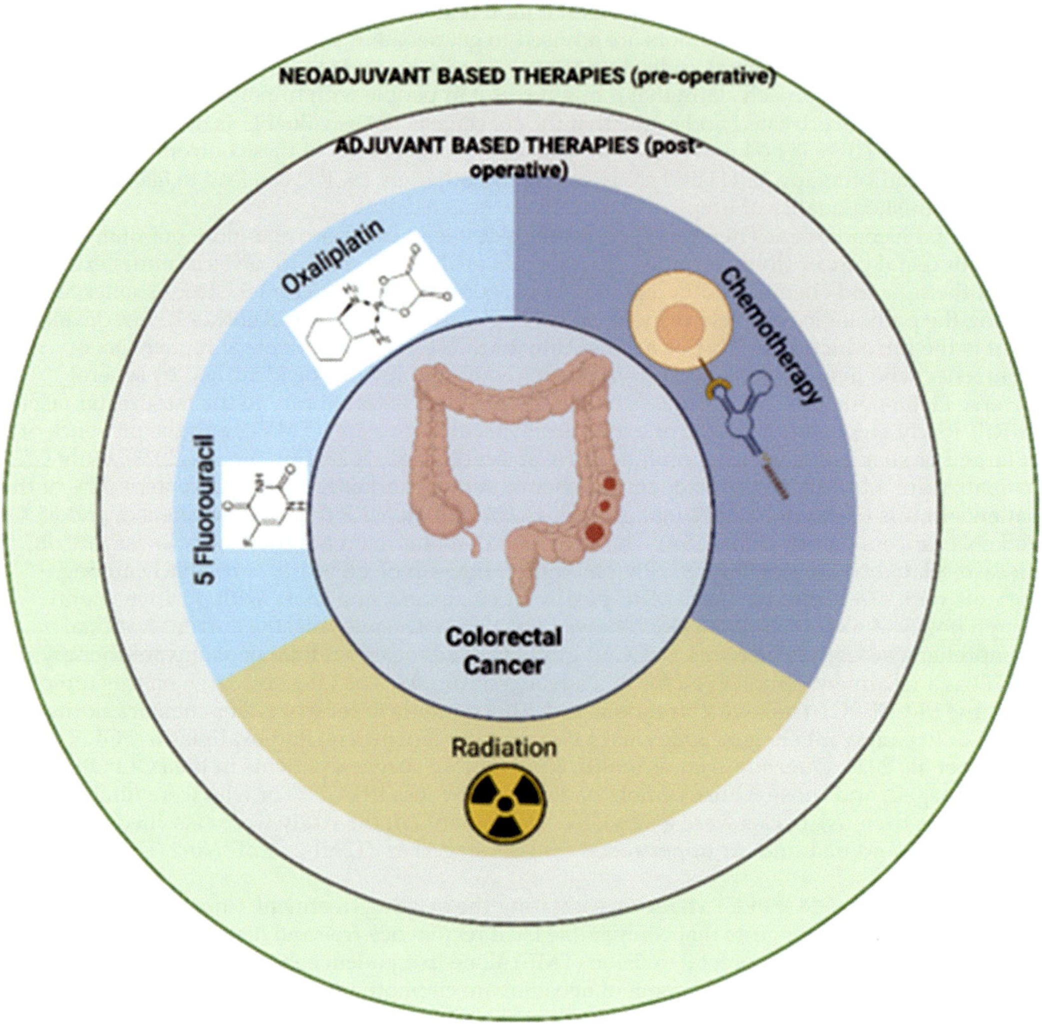

Figure 7.3 Adjuvant and neoadjuvant therapies in CRC treatment. This figure illustrates the various adjuvant therapies, including 5-fluorouracil (5-FU), oxaliplatin, radiation therapy, and chemotherapy, used in cancer treatment. Adjuvant therapy is typically administered post-operatively to eliminate residual cancer cells, reduce recurrence, and improve overall survival rates. The figure also depicts neoadjuvant therapy, which involves the same therapeutic agents administered pre-operatively, aiming to reduce tumor size, facilitate surgical resection, and improve surgical outcomes.

REFERENCES

Allegra, C. J., et al. (2015). Neoadjuvant 5-FU or capecitabine plus radiation with or without oxaliplatin in rectal cancer patients: a phase III randomized clinical trial. *J Nat Cancer Instit*, *107*(11):djv248.

André, T., et al. (2015). Adjuvant fluorouracil, leucovorin, and oxaliplatin in stage II to III colon cancer: updated 10-year survival and outcomes according to BRAF mutation and mismatch repair status of the MOSAIC study. *J Clinl Oncol*, *33*(35):4176–4187.

Anker, C. J., et al. (2021). The evolving personalized landscape of colorectal cancer therapies. *Int J Radiat Oncol Biol Phys*, *110*(5):1255–1262.

Biniaz, M., et al. (2024). Combination of neoadjuvant and adjuvant chemotherapy with FOLFOX compared with adjuvant chemotherapy in management of locally advanced rectal cancers: a randomized trial of a promising therapeutic approach. *BMC Cancer*, *24*(1):863.

Boland, C. R., Goel, A. (2010). *Microsatellite instability in colorectal cancer. Gastroenterol*, *138*(6):2073–2087. e3.

Boublikova, L., et al. (2023). Total neoadjuvant therapy in rectal cancer: the evidence and expectations. *Crit Rev Oncol Hematol*, *192*:104196.

Calvo, F. A., et al. (2013). Prognostic impact of external beam radiation therapy in patients treated with and without extended surgery and intraoperative electrons for locally recurrent rectal cancer: 16-year experience in a single institution. *Int J Radiat Oncol Biol Phys*, *86*(5):892–900.

Capdevila, J., et al. (2022). SEOM-GEMCAD-TTD clinical guidelines for localized rectal cancer (2021). *Clin Translat Oncol*, *24*(4):646–657.

Chakrabarti, S., et al. (2020). Early stage colon cancer: current treatment standards, evolving paradigms, and future directions. *World J Gastrointest Oncol*, *12*(8):808–832.

Chan, G. H. J., Chee, C. E. (2019). Making sense of adjuvant chemotherapy in colorectal cancer. *J Gastrointest Oncol*, *10*(6):1183–1192.

Chau, I., et al. (2005). A randomised comparison between 6 months of bolus fluorouracil/leucovorin and 12 weeks of protracted venous infusion fluorouracil as adjuvant treatment in colorectal cancer. *Ann Oncol*, *16*(4):549–557.

Colorectal cancer statistics | WCRF International. (n.d.). Available at: https://www.wcrf.org/cancer-trends/colorectal-cancer-statistics/ (accessed Aug 2024).

Daprà, V., et al. (2023). Total neoadjuvant treatment for locally advanced rectal cancer patients: where do we stand? *Int J Molec Sci*, *24*(15):12159.

Guyot D'Asnières De Salins, A., et al. (2021). Discordance between immunochemistry of mismatch repair proteins and molecular testing of microsatellite instability in colorectal cancer. *ESMO Open*, *6*(3):100120.

Hong, S. N., et al. (2012). The effect of the bowel preparation status on the risk of missing polyp and adenoma during screening colonoscopy: a tandem colonoscopic study. *Clin Endosc*, *45*(4):404.

Hsu, H.-C., et al. (2016). Mutations of *KRAS/NRAS/BRAF* predict cetuximab resistance in metastatic colorectal cancer patients. *Oncotarget*, *7*(16):22257–22270.

Ikematsu, H., et al. (2022). Detection of colorectal lesions during colonoscopy. *DEN Open*, *2*(1): e68. doi: 10.1002/deo2.68

Iveson, T. J., et al. (2018). 3 versus 6 months of adjuvant oxaliplatin-fluoropyrimidine combination therapy for colorectal cancer (SCOT): an international, randomised, phase 3, non-inferiority trial. *Lancet Oncol*, *19*(4):562–578.

Johnson, G. G. R. J., et al. (2023). Total neoadjuvant therapy for rectal cancer: a guide for surgeons. *Canad J Surg*, *66*(2):E196–E201.

Jones, E., et al (2023). Adjuvant 5-Fluorouracil/leucovorin, capecitabine, and oxaliplatin-related regimens for stage II/III colon cancer patients 66 years or older. *Cancer Med*, *12*(3):2389–2406.

Jonker, D., et al. (2006). Role of oxaliplatin combined with 5-fluorouracil and folinic acid in the first- and second-line treatment of advanced colorectal cancer. *Curr Oncol*, *13*(5):73–184.

Kasi, P. M., et al. (2023). Anti-EGFR antibodies in the management of advanced colorectal cancer. *Oncologist*, *28*(12):1034–1048.

Koka, K., et al. (2022). Technological advancements in external beam radiation therapy (EBRT): an indispensable tool for cancer treatment. *Cancer Managem Res*, *14*:1421–1429.

Kopetz, S., et al. (2019). Encorafenib, binimetinib, and cetuximab in *BRAF* V600E–mutated colorectal cancer. *New Engl J Med*, *381*(17):1632–1643.

Kurasaka, C., et al. (2022). Trapping of 5-fluorodeoxyuridine monophosphate by thymidylate synthase confers resistance to 5-fluorouracil. *ACS Omega*, *7*(7):6046–6052.

Lambregts, D. M. J., et al. (2022). Current controversies in TNM for the radiological staging of rectal cancer and how to deal with them: results of a global online survey and multidisciplinary expert consensus. *Eur Radiol*, *32*(7):4991–5003.

Li, Y., et al. (2021). Combination of anti-EGFR and anti-VEGF drugs for the treatment of previously treated metastatic colorectal cancer: a case report and literature review. *Frontiers Oncol*, 11:684309.

Lu, X., et al. (2023). Natural products targeting tumour angiogenesis. *Br J Pharmacol*. doi: 10.1111/bph.16232

Marshall, J. L., et al. (2007). Adjuvant therapy for stage II and III Colon Cancer: consensus report of the International Society of Gastrointestinal Oncology. *Gastrointest Cancer Res*, *1*(4):146–154.

Morse, M. A. (2005). Adjuvant therapy of colon cancer: current status and future developments. *Clin Colon Rectal Surg*, *18*(03):224–231.
Noordhuis, P., et al. (2004). 5-fluorouracil incorporation into RNA and DNA in relation to thymidylate synthase inhibition of human colorectal cancers. *Ann Oncol*, *15*(7):1025–1032.
Ochiai, K., et al. (2024). Total neoadjuvant therapy for rectal cancer: which regimens to use? *Cancers*, *16*(11):2093.
Osterman, E., et al. (2020). Recurrence risk after radical colorectal cancer surgery: less than before, but how high is it? *Cancers*, *12*(11):3308.
Patel, K., et al. (2004). Weekly 5-fluorouracil and leucovorin: achieving lower toxicity with higher dose-intensity in adjuvant chemotherapy after colorectal cancer resection. *Ann Oncol*, *15*(4):568–573.
PDQ Adult Treatment Editorial Board. (2002). Colon cancer treatment (PDQ®): health professional version. *PDQ Cancer Information Summaries*. Available at: http://www.ncbi.nlm.nih.gov/pubmed/11693338
Pectasides, D., et al. (2015). Randomized phase III clinical trial comparing the combination of capecitabine and oxaliplatin (CAPOX) with the combination of 5-fluorouracil, leucovorin and oxaliplatin (modified FOLFOX6) as adjuvant therapy in patients with operated high-risk stage II or stage III colorectal cancer. *BMC Cancer*, *15*(1):384.
Peng, H., et al. (2019). Tumor location as an indication for adjuvant radiotherapy in pT3N0 rectal cancer after surgery. *Radiat Oncol*, *14*(1):8.
Quezada-Diaz, F. F., Smith, J. J. (2022). Neoadjuvant therapy for rectal cancer. *Surg Oncol Clin N Am*, *31*(2):279–291.
Romero Zoghbi, S. E., et al. (2024). Total neoadjuvant therapy approach for the treatment of locally advanced rectal cancer: where do we stand? *Oncol*, *102*(7)646–658.
Sahin, I. H., et al. (2019). Immune checkpoint inhibitors for the treatment of MSI-H/MMR-D colorectal cancer and a perspective on resistance mechanisms. *Br J Cancer*, *121*(10):809–818.
Sargent, D., et al. (2011). Two or three year disease-free survival (DFS) as a primary end-point in stage III adjuvant colon cancer trials with fluoropyrimidines with or without oxaliplatin or irinotecan: data from 12,676 patients from MOSAIC, X-ACT, PETACC-3, C-06, C-07 and C89803. *Eur J Cancer*, *47*(7):990–996.
Sifaki-Pistolla, D., et al. (2022). Significant rise of colorectal cancer incidence in younger adults and strong determinants: 30 years longitudinal differences between under and over 50s. *Cancers*, *14*(19):4799.
Skóra, T., et al. (2015). Postoperative irradiation in patients with pT3-4N0 laryngeal cancer: results and prognostic factors. *Eur Arch Oto-Rhino-Laryngol*, *272*(3):673–679.
Smith, H. G., et al (2024). Neoadjuvant treatment of colorectal cancer: comprehensive review. *BJS Open*, *8*(3): zrae038. doi: 10.1093/bjsopen/zrae038
Taieb, J., Gallois, C. (2020). Adjuvant chemotherapy for stage III colon cancer. *Cancers*, *12*(9):2679.
Teshome, M., Hunt, K. K. (2014). Neoadjuvant therapy in the treatment of breast cancer. *Surg Oncol Clin N Am*, *23*(3):505–523.
Tomita, N., et al. (2019). Phase III randomised trial comparing 6 vs. 12-month of capecitabine as adjuvant chemotherapy for patients with stage III colon cancer: final results of the JFMC37-0801 study. *Br J Cancer*, *120*(7):689–696.
Triantafillidis, J. K., et al. (2015). Colonoscopy in colorectal cancer screening: current aspects. *Indian J Surg Oncol*, *6*(3):237–250.
Tsai, Y.-J., et al. (2016). Adjuvant FOLFOX treatment for stage III colon cancer: how many cycles are enough? *SpringerPlus*, *5*(1):1318.
Varghese, A. (2015). Chemotherapy for stage II colon cancer. *Clin Colon Rectal Surg*, *28*(04):256–261.
Vekic, B., et al. (2020). A correlation study of the colorectal cancer statistics and economic indicators in selected Balkan countries. *Frontiers Public Health*, *8*:29. doi: 10.3389/fpubh.2020.00029
Virag, P., et al. (2013). Oxaliplatin induces different cellular and molecular chemoresistance patterns in colorectal cancer cell lines of identical origins. *BMC Genomics*, *14*(1):480.
Xie, Y.-H. et al. (2020). Comprehensive review of targeted therapy for colorectal cancer. *Signal Transduct Targeted Ther*, *5*(1):22.
Yeom, S.-S., et al. (2019). The prognostic effect of adjuvant chemotherapy in the colon cancer patients with solitary lymph node metastasis. *Int J Colorect Dis*, *34*(8):1483–1490.
Zhan, Q.-H., et al. (2018). Survival and time to initiation of adjuvant chemotherapy among breast cancer patients: a systematic review and meta-analysis. *Oncotarget*, *9*(2):2739–2751.
Zhang, N., et al. (2008). 5-fluorouracil: mechanisms of resistance and reversal strategies. *Molecules*, *13*(8):1551–1569.

8 Targeted Therapies in Colorectal Cancer

Saxena Shivang, Sreekari Palakodeti, Thakral Mohit and Ashwin Kotnis

8.1 COLORECTAL CANCER: AN INTRODUCTION

Decades ago, colorectal cancer (CRC) was rare, but today it ranks as the second deadliest cancer globally, causing approximately one million deaths annually according to GLOBOCAN 2022. While genetic predisposition and aging remain key factors in its development, the increased incidence is also linked to modern lifestyle choices such as lack of physical activity, consumption of processed foods, obesity, alcohol consumption, and disruptions in gut microbiota balance (Thomas et al., 2020) (Table 8.1). Nearly half of the mortality due to colorectal cancers is attributable to preventable risk factors, underscoring the importance of screening, surveillance, and timely treatment. Posing a significant global health burden, colorectal cancer necessitates continued advancements in the strategies to effectively address its impact worldwide (Roshandel et al., 2024).

Conventionally, colorectal cancer was mainly treated using chemotherapy, radiation, and surgery. Although these approaches have been effective in improving patient survival, they have many limitations and side effects. Chemotherapy, for instance, unselectively targets rapidly dividing cells leading to ancillary healthy tissue damage and causing side effects like nausea, fatigue, and immunosuppression.

With emerging breakthroughs in cancer treatment and the limitations of traditional therapies becoming apparent, more selective therapies are warranted. This need has driven the development of targeted therapies—treatments specifically designed to attack cancer cells by disrupting key molecular pathways involved in tumor growth and progression (Roshandel et al., 2024; Asthana et al., 2021).

Along with many lifestyle risk factors, the pathogenesis of colorectal cancer involves several genetic and epigenetic aberrations that transform healthy colonic mucosal cells into cancer cells. Important mutations commonly observed in CRC include APC, KRAS, BRAF, and TP53 and disturbed WNT/β-catenin signaling pathway. These mutations drive uncontrolled cellular proliferation and inhibit apoptosis leading to cancer development. Targeted therapies signify a revolution in cancer treatment by focusing on specific molecular targets linked to cancer cell growth and survival. Unlike conventional chemotherapy, these treatments aim to enhance effectiveness while reducing harm to normal cells.

Several targeted agents like epidermal growth factor receptor (EGFR) inhibitors, vascular endothelial growth factor (VEGF) inhibitors, BRAF inhibitors, and immune checkpoint inhibitors are developed for worldwide use in colorectal cancer (Asthana et al., 2021).

Despite the significant progress, there are challenges in optimally utilizing targeted therapies. A primary concern is the development of resistance, which can result from tumor diversity and genetic adaptations. Resistance develops through secondary mutations or the ability of tumor cells to change their characteristics. Ongoing research is focusing on developing novel molecular

Table 8.1: Risk factors of colorectal cancer

Modifiable risk factors
Obesity and physical inactivity
Diet
Smoking
Alcohol
Diabetes and insulin resistance
Gut microbiota
Nonmodifiable risk factors
Race and ethnicity
Sex
Age
Hereditary mutations (HNPCC and FAP)
Inflammatory bowel disease
Abdominal radiations
Cystic fibrosis
Cholecystectomy

DOI: 10.1201/9781032704517-8

targets to address these resistance mechanisms. Furthermore, the high cost of targeted therapies can limit patient access.

Nevertheless, the treatment of colorectal cancer has witnessed notable advancements with the introduction of targeted therapies, ushering renewed optimism to both patients and healthcare providers. As our grasp of cancer biology is advancing, the prospects for even more accurate and effective treatments are expanding.

Hereditary CRCs representing about 5% of all CRCs include hereditary non-polyposis colorectal cancer (HNPCC) and familial adenomatous polyposis (FAP). The more common form HNPCC, also known as Lynch syndrome, accounts for 2% of cases of CRC and is attributed to germline mutations of mismatch repair (MMR) genes which play a key role in matching the DNA base-pair during the DNA replication and repair of damaged DNA. Genes of the MMR pathway are tumor-suppressor genes, and hence any defect in MMR causes the accumulation of mutations in critical genes resulting in tumorigenesis. The less common form familial adenomatous polyposis coli (FAPC) is inherited as an autosomal dominant disease wherein there is a mutation of the tumor suppressor adenomatous polyposis coli (APC) gene. The protein-coded APC is involved in the WNT **β**-catenin pathway which when aberrant results in uncontrolled proliferation of tumor cells (Hlubek et al., 2007).

8.2 PATHOGENESIS OF COLORECTAL CANCER

The primary function of the colon is to reabsorb water and remaining minerals and nutrients from the chyme, accomplished by columnar epithelial cells called enterocytes lining the intestinal mucosa. CRC development involves several pathophysiological mechanisms, including the loss of cell apoptosis and uncontrolled cell proliferation. CRC begins with non-cancerous lesions called polyps, which are mucosal proliferations into the intestinal lumen. These adenomas grow gradually over several years before becoming malignant. A heterogeneous combination of genetic and epigenetic abnormalities drives the adenoma-carcinoma sequence (Leslie et al., 2002) (Figure 8.1).

Nearly 70% of CRCs are of sporadic type and are attributed to chromosomal instability. About 5% of cases are hereditary/familial, such as Lynch syndrome and FAP, resulting from DNA mismatch repair genes and microsatellite instability. Additionally, 20% to 30% of CRC have a predisposition without association with any germline mutation (Ye et al., 2020). Approximately 50% of CRCs are located on the left side of the colon, while 25% occur on the right. Depending on their size, around 10% of adenomatous polyps progress into adenocarcinoma. The oncogenesis process involves the rapid accumulation of mutations in APC/WNT/**β**-catenin, EGFR, TP53, and transforming growth factor **β** (TGF-**β**) (Hlubek et al., 2007).

8.2.1 Role of Adenomatous Polyposis Coli (APC) Gene

Nearly 80% of sporadic cases of CRC are due to familial adenomatous polyposis syndrome which causes hundreds to thousands of benign polyps that develop along the inner mucosa of the colon and rectum. Notably, 100% of non-treated FAP develop CRC caused by over 300 types of frameshift mutations in both copies of the APC gene (Boland & Goel, 2010). These mutations result in the

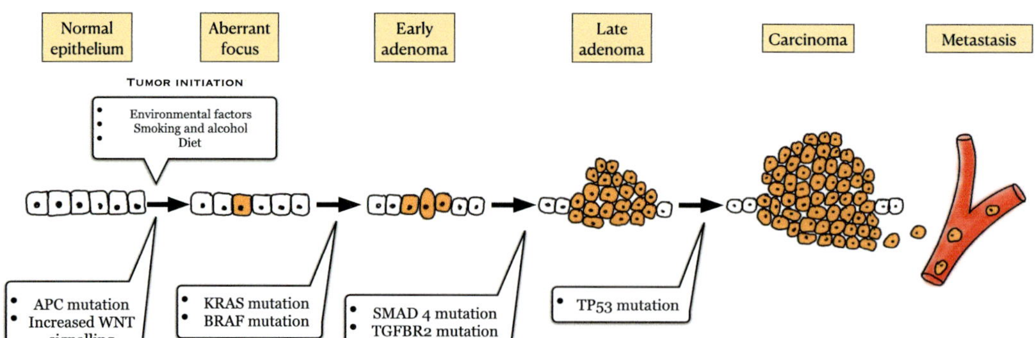

Figure 8.1 Schematic diagram showing mutations that lead to carcinoma development.

C-terminal truncations of the APC protein which leads to early progression of the disease (Bell & Findeis-Hosey, 2024). Normally, the APC protein binds to β-catenin and degrades it by proteosome-mediated degradation. β-catenin is a component of the WNT signaling pathway, acting as a transcription factor for genes coding for cell proliferation proteins. A loss-of-function mutation in the APC gene prevents APC complex formation and causes cytosolic accumulation of β-catenin, its nuclear translocation, forming a complex with a DNA-binding transcription factor of the TCF/LEF family, activating the transcription c-MycCyclin D1 and survivin thus promoting cancer progression by inhibiting apoptosis (Espada et al., 2009; Hlubek et al., 2007). Additional mutations, including gain-of-function in the KRAS oncogene and subsequent inactivation of tumor suppressor genes SMAD4 and TP53, further promote growth and inhibit apoptosis.

8.2.2 Role of Microsatellite Instability (MSI)

Microsatellite instability (MSI) is a crucial factor in the development of CRC, representing a distinct molecular pathway for tumor progression. This hypermutable condition arises from defects in the DNA mismatch repair (MMR) system, which repairs errors during the replication of DNA. When this repair system fails, mutations accumulate in microsatellite regions that are short, repetitive DNA sequences consisting of 1-6 base pairs. Around 15% of colorectal cancers show MSI, with a significant portion linked to Lynch syndrome, a hereditary disorder that increases the risk of various cancers, including colorectal cancer (Boland & Goel, 2010; Bell & Findeis-Hosey, 2024; Geiersbach & Samowitz, 2011).

MSI in CRC develops through two main pathways: hereditary and sporadic.

8.2.2.1 Hereditary Pathway

Lynch syndrome, which is also known as hereditary nonpolyposis colorectal cancer (HNPCC), is caused due to germline mutations in the genes coding for DNA mismatch repair pathways such as MSH2, MLH1, MSH6, and PMS2. These mutations result in a high frequency of MSI due to the ineffective repair of DNA replication errors. People with Lynch syndrome show a high risk of early onset of malignancies including colorectal cancer compared to the general population (Geiersbach & Samowitz, 2011; Li et al., 2020).

8.2.2.2 Sporadic Pathway

In sporadic cases, MSI often results from the hypermethylation of the MLH1 gene promoter, which silences the gene without changing its sequence. This epigenetic modification is common in tumors with the CpG island methylator phenotype (CIMP). About 12% of colorectal cancers arise through this sporadic mechanism, underscoring the complexity of MSI's role in cancer development (Bell & Findeis-Hosey, 2024; Geiersbach & Samowitz, 2011). (See Figure 8.1.)

8.2.3 MicroRNAs in Colorectal Cancer

MicroRNAs (miRNAs) are non-coding RNA molecules regulating gene expression. These small 18–25 nucleotides long miRNA bind to mRNAs post-transcriptionally leading to their destruction by preventing them from being translated. In CRC, miRNAs play pivotal roles in oncogenesis by variably acting as oncogenes or tumor suppressors and influencing fundamental cellular processes such as cell proliferation, cell death and oncogenic processes such as invasion and metastasis (Li et al., 2020).

8.2.3.1 Oncogenic miRNAs (OncomiRs)

- **miR-21** is one of the most important oncomiRs. miR-21 targets and downregulates tumor suppressor genes like PTEN, PDCD4, and TPM1. This leads to enhanced cell survival, proliferation, and invasion (Meng et al., 2007; Zhu et al., 2007).
- **miR-17-92 cluster** or oncomiR-1, promotes angiogenesis through the suppression of thrombospondin 1 (TSP1) and connective tissue growth factor (CTGF). OncomiR-1 is often upregulated in colorectal cancers (Martens-de Kemp et al., 2022).

8.2.3.2 Tumor Suppressor miRNAs

- **miR-143 and miR-145:** downregulation of these microRNAs increases expression of DNMT3A. upregulated DNMT3A promotes CRC oncogenesis (Ng et al., 2009).

8.2.3.3 Regulation of Metastasis

- **mi-200 family:** Its downregulation increases metastatic potential by facilitating epithelial-mesenchymal transition (EMT). miR200 targets transcription factors ZEB1 and ZEB2 which facilitate EMT (Klicka et al., 2022)).

8.3 CURRENT TREATMENT MODALITIES FOR COLORECTAL CANCER

CRC is a major health issue worldwide, necessitating a multifaceted approach to treatment. The primary modalities of treatment of colorectal cancer include surgery, chemotherapy, and radiation therapy. Each modality has subtypes and limitations that influence treatment decisions. Below is a comprehensive overview of each treatment modality, including its subtypes and limitations.

8.3.1 Surgery

Surgery is the primary treatment modality for colorectal cancer (irrespective of the stage) with various surgical options depending on the stage and location of the tumor. The types of surgical procedures are described below.

8.3.1.1 Polypectomy

Polypectomy is a surgical procedure aimed at removing polyps found in the colon or rectum. The primary purpose of this procedure is to prevent the progression of adenomatous polyps, which are precursors to colorectal cancer. By removing these polyps, healthcare providers can significantly reduce the risk of developing colorectal cancer. Indications for polypectomy include routine colonoscopy screening that often identifies polyps, which should be removed to prevent cancer, and symptomatic patients with symptoms such as rectal bleeding or changes in bowel habits may undergo polypectomy if polyps are detected. The major risk of polypectomy is incomplete removal, recurrence, and complications like bleeding, perforation, infection, etc. (Gmeiner, 2024; De Azevedo et al., 2019).

8.3.1.2 Local Excision

It is a surgical procedure used to remove tumors or abnormal tissue from the colon or rectum, primarily for the treatment of early-stage colorectal cancer. This approach is particularly relevant for small, localized tumors that have not invaded deeper layers of the bowel wall or spread to lymph nodes or distant sites. The major drawback of this method is that it has limited indications (only for the early stage) and a higher chance of incomplete resection (Pathak et al., 2024; De Azevedo et al., 2019).

8.3.1.3 Resection

This involves removing a portion of an organ or structure. In the context of colorectal cancer, resection is a common treatment option for various stages of the disease. Indications for resection include early-stage including stage I, II, and III tumors. Also, in advanced cases, resection may be performed to relieve symptoms such as obstruction or bleeding, even if complete removal of the tumor is not possible. It is highly invasive and carries complications such as bleeding, infection, and anastomotic leaks (where the two ends of the bowel do not heal properly after being reconnected). These complications can lead to prolonged hospitalization and additional surgeries (De Azevedo et al., 2019).

8.3.1.4 Laparoscopic Surgery

Laparoscopic surgery is a minimally invasive approach for treating colorectal cancer. It involves making several small incisions in the abdomen instead of one large incision used in traditional open surgery. It has less pain and discomfort after surgery, faster recovery time and shorter hospital stay, smaller scars, and a lower risk of infection also quicker return to normal activities is observed. Being a less invasive and relatively safer treatment option, it has some disadvantages like high cost, low availability, and greater skill required to handle the equipment and to carry out the procedure (Bokey et al., 1996; Weeks et al., 2002).

8.3.1.5 Trans Anal Endoscopic Microsurgery (TEM)

Total endoscopic mucosal resection (TEM) involves the complete removal of rectal tumors along with a margin of surrounding healthy tissue using endoscopic techniques. The procedure is typically performed through the anus, utilizing specialized instruments that provide access to the rectum while minimizing trauma to the surrounding tissues. TEM is less invasive than traditional surgical approaches, leading to reduced postoperative pain and quicker recovery times, and by removing a small portion of healthy tissue it also retains the normal bowel function. However, it has certain drawbacks; it is useful only for the removal of small tumors and not large tumors due to the lack of clear margins and also requires monitoring for a longer period of time (De Azevedo et al., 2019).

Surgery as a treatment option carries the following major risk factors:

- Complications: Risks include infection, bleeding, and damage to surrounding organs.
- Recurrence: Surgery alone does not guarantee a cure, especially in advanced stages.
- Postoperative quality of life: Patients may experience changes in bowel function, leading to long-term lifestyle adjustments.

In summary, surgery is the cornerstone of colorectal cancer treatment, with various options depending on the stage and location of the tumor. While it offers the best chance for a cure, especially in the early stages, it also carries risks and limitations that must be carefully considered.

8.3.2 Chemotherapy

Chemotherapy refers to the application of chemicals/drugs to destroy cancer cells or inhibit their growth. It can be administered systemically, affecting the whole body, or regionally, targeting specific areas. Chemotherapy is typically employed in three main contexts:

- Neoadjuvant chemotherapy is administered prior to surgery to shrink tumors, making them easier to remove.
- Adjuvant chemotherapy is administered post-surgery to eliminate any remaining cancer cells and reduce the risk of recurrence.
- Palliative chemotherapy is used for advanced cancer to alleviate pain and such symptoms to improve quality of life, even when curative treatment is not possible (Desai et al., 2007).

8.3.2.1 Types of Chemotherapy Drugs

Several chemotherapy drugs are commonly used to treat colorectal cancer, often in combination to enhance effectiveness. The most notable ones are tabulated in Table 8.2 along with the potential side effects of such regimens.

Table 8.2: Chemotherapeutic drugs: their types, actions, and adverse effects

Drug	Action	Route & combination	Adverse reactions
5-Fluorouracil	Antimetabolite- inhibition of maturation of pre-mRNA and post-transcriptional modification of tRNA	Given intravenously or orally via capecitabine, which converts to 5-FU in the body	1) Gastric effects- nausea, vomiting, diarrhea, particularly associated with Irinocetan, requiring careful monitoring and management 2) Fatigue is a common side effect that can persist even after the end of treatment 3) Neuropathy is often linked to oxaliplatin 4) Hand-foot syndrome (redness and blistering of palms and soles) particularly with capecitabine or 5-FU 5) Increased susceptibility to infections due to lowered immunity and hair loss are common to nearly all drugs
Oxaliplatin	Cytotoxic effects through DNA damage	Platinum based drug effective in combination with 5-FU and leucovorin (FOLFOX) regimen) or with capecitabine (XELOX regimen)	
Irinocetan	DNA topoisomerase I inhibitor (through complex formation); arrests the cell cycle at S-G2 phase	Often used in combination regimen (FOLFIRI); particularly effective in advanced cases	
Trifluridine/ Tipiracil (Lonsurf)	TFI is a nucleoside analogue thus inhibits DNA synthesis; TPI inhibits enzyme TP responsible for degradation of TFI; enhances anti-tumor action of TFI	This drug combination used in the molar ratio of 1:0.5 is used for metastatic CRC when other treatments have failed	

These drugs can be used in various combinations depending on the stage of cancer and individual patient factors. These drugs can be administered either by intravenous (IV) infusion, oral tablets, or as regional chemotherapy thereby decreasing the systemic adverse effects.

8.3.2.2 Treatment Cycles

Chemotherapy is typically given in cycles, which consist of a treatment period followed by a recovery period. A standard cycle lasts two to three weeks, allowing the body to recover from side effects. Depending on the specific regimen, patients may undergo multiple cycles over several months.

Thus, chemotherapy is a crucial component of colorectal cancer treatment, employed in various contexts to enhance surgical outcomes and manage advanced disease. With a range of drugs and administration methods available, oncologists can tailor treatment plans to individual patient needs, balancing efficacy with potential side effects (Labianca et al., 2007).

8.3.3 Radiation Therapy

Radiation therapy is an important treatment modality for colorectal cancer, particularly in cases of rectal cancer. It is often used in combination with other treatments like surgery and chemotherapy to improve outcomes. It is mainly used in:

- Neoadjuvant therapy: Radiation is given before surgery to shrink tumors and make them easier to remove.
- Adjuvant therapy: Radiation is used after surgery to eliminate any remaining cancer cells and reduce the risk of recurrence.
- Palliative therapy: Radiation can help relieve symptoms like bleeding, pain, or bowel dysfunction in advanced cases.
- Treating metastatic disease: If colorectal cancer spreads to other organs like the liver or bones, radiation can be used to target those areas (Tam & Wu, 2019).

The various types of radiotherapy and common side effects of radiation therapy are tabulated in Table 8.3.

To surmise, radiation therapy is a valuable tool in the treatment of colorectal cancer, particularly when combined with surgery and chemotherapy. While it carries some risks and limitations, it can improve outcomes and quality of life for many patients.

8.3.4 Immunotherapy

Immunotherapy is an emerging treatment modality for colorectal cancer (CRC) that utilizes the immune mechanisms to kill cancer cells. While not as widely used as surgery or chemotherapy, immunotherapy has shown promising results in specific subsets of CRC patients.

The most commonly used immunotherapies for CRC are immune checkpoint inhibitors that facilitate immune cells to better attack the cancer cells.

The commonly used FDA-approved checkpoint inhibitors are enlisted in Table 8.4 along with the salient side effects of immunotherapy.

Hence, immunotherapy represents a promising new approach for treating colorectal cancer, particularly in MSI-H or dMMR tumors. However, its use is currently limited to specific patient

Table 8.3: Radiotherapy and its types

Type of radiation therapy	Delivery	Adverse effects
EBRT—external beam radiation therapy	Most common type; a machine delivers high energy x-rays to the tumor from outside the body	Little effectiveness for colorectal cancer due to potential side effects of cancer therapy (fatigue, diarrhea, skin irritation and neuropathy) and long-term complications like bowel dysfunction and secondary cancers due to extended radiation exposure.
Brachytherapy	Involves placing radioactive material directly inside or near the tumor; less commonly used	
Radioembolization	Used when metastatic spread is to the liver; tiny radioactive beads are injected into the hepatic artery to deliver targeted radiation	

Table 8.4: Immunotherapeutic regimens and their side effects

Drug	Delivery	Adverse effects
Pembrolizumab (Keytruda)	Approved for MSI-H or dMMR mCRC that has progressed following chemotherapeutic treatment and VEGF-targeted therapy.	1) Although immunotherapy has shown efficacy in MSI-H or dMMR CRC, it is only effective in a small subset of patients as most CRCs are MSS. 2) There can be severe immune reactions including endocrine disorders.
Nivolumab (Opdivo)	Approved for MSI-H or dMMR mCRC that has progressed following chemotherapeutic treatment.	
Ipilimumab (Yervoy)	Used in combination with nivolumab for MSI-H or dMMR mCRC.	
Dostarlimab (Jemperli)	May be used off-label for early stage (II or III) rectal cancers to shrink tumors before surgery, although this indication is not yet FDA approved	

populations, and more research is needed to improve its efficacy and safety. As part of a multidisciplinary treatment approach, immunotherapy may help improve outcomes for some CRC patients.

8.4 CLINICAL TRIALS

Clinical trials are research studies that evaluate the effectiveness and safety of new treatments or interventions. They are essential for determining how well new therapies work compared to standard treatments and for discovering new ways to prevent, diagnose, or treat diseases.

The following are types of clinical trials for CRC.

1. Treatment trials: These trials focus on new treatments, including drugs, combinations of therapies, or novel approaches such as immunotherapy and targeted therapy.
2. Prevention trials: These studies investigate methods to prevent colorectal cancer in high-risk populations.
3. Screening trials: Focused on new ways to detect colorectal cancer early, improving outcomes through earlier intervention.
4. Observational studies: These studies monitor patients receiving standard treatments to gather data on outcomes and side effects without testing new interventions.

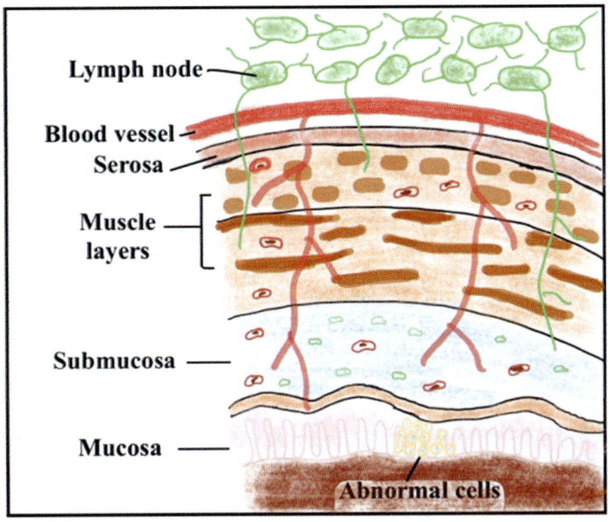

Figure 8.2 This diagram shows the affected layers of the colon wall in an individual suffering from stage 0 colorectal cancer (according to the TNM classification). The mainstay of treatment for this stage is surgery (polypectomy or local excision can be done).

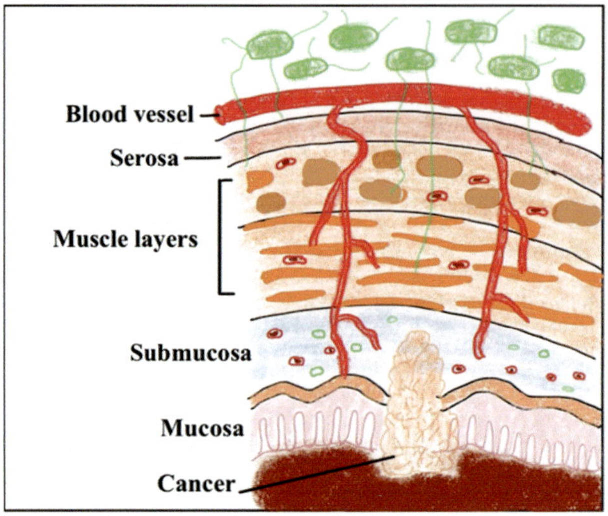

Figure 8.3 This figure shows the affected layers of the colon wall in an individual suffering from stage 1 colorectal cancer (according to the TNM classification). The mainstay of treatment for this stage is surgery same as stage 0.

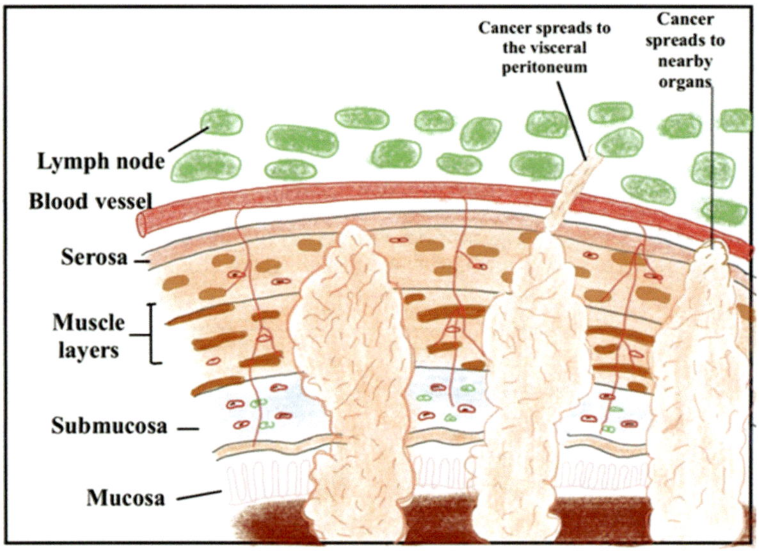

Figure 8.4 This diagram shows the affected layers of the colon wall in an individual suffering from stage-2 colorectal cancer (according to TNM classification). The mainstay of treatment for this stage is surgery (partial colectomy) and chemotherapy, neoadjuvant therapy and radiotherapy can be given before the treatment for shrinking the tumor.

Figures 8.2 to 8.4 illustrate how to identify the stage of colorectal cancer using the TNM staging system and how the treatment is tailored to each specific stage of cancer.

Figures 8.5A and 8.5B show the affected layers of the colon wall in an individual suffering from stage 3 colorectal cancer (according to the TNM classification). The mainstay of treatment for this stage is surgery(partial colectomy) and chemotherapy, neoadjuvant therapy, radiotherapy, and participating in some verified clinical trials.

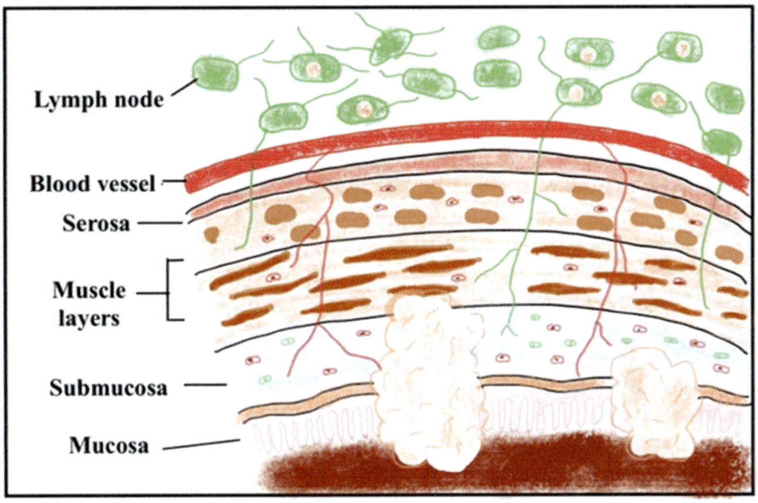

(a)

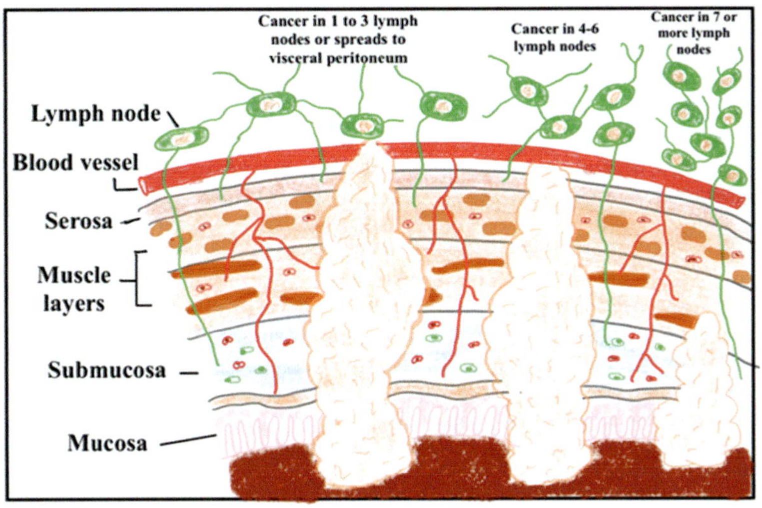

(b)

Figure 8.5 (a) Stage III-A (affecting 1–3 lymph nodes). (b) Stage III-B (affecting 4–6 lymph nodes.

8.4.1 Stage 4

In this stage colorectal cancer has spread to other parts of the body like the lung, liver, and abdominal wall involving distant lymph nodes. The mainstay of treatment in such patients includes surgery, chemotherapy, immunotherapy, and radiation therapy.

In advanced stages, palliative care becomes crucial to manage symptoms and improve the quality of life of the patient. This mainly includes pain management, nutritional support, and psychological support.

The treatment options mentioned above are the currently available and recommended options for colorectal cancer, but they can vary depending on the individual's specific symptoms and conditions. Colorectal cancer is a serious disease that can be cured and prevented if detected early. It is strongly advised to consult an oncologist if there is any suspicion of the disease. These treatment options should not be self-administered without medical supervision, as they can cause severe adverse reactions, potentially resulting in more harm than benefit.

8.5 EMERGING TREATMENT MODALITIES OF COLORECTAL CANCER

Having been diagnosed with colorectal cancer, it is crucial to understand that based on mutation, DNA repair, and stability, the cancer is classified in two. First, due to microsatellite instability (MSI-H), low beta 2 microglobulin (B2MG) expression, and aberrant DNA mismatch repair (dMMR). The other type shows stable microsatellites and is proficient in DNA mismatch repair (pMMR). The former, due to its DNA mismatch repair characteristic, has higher mutation rates, leading to the formation of new peptides on the tumor cell surface, called neoantigens- having better immunogenicity than the latter which shows a high level of DNA repair. This differentiation leads to their varied response to different immunotherapy treatments and hence, makes it essential to know about the type of colorectal cancer (He, X., et al. (2023)).

Based on etiology, CRC can be classified as sporadic, familial, and hereditary. Sporadic CRC accounts for three-fourths of the cases of colorectal cancer without the involvement of any germline mutations. It is influenced by lifestyle factors, including but not limited to diet, obesity, alcohol consumption, and smoking. Familial cases comprise nearly 20% of the total cases where positive family history is present though with an unknown underlying genetic cause, whereas high-risk mutations are seen in hereditary CRC contributing to the remaining 5–10% of the cases (Passardi et al., 2017).

DNA mismatch repair genes include MSH2, MLH1, MSH6, and PMS2 which when undergo mutations produce an entity called Lynch syndrome, an autosomal dominant disorder. DNA MMR pathway plays a central role in the development of this syndrome. Whereas in MMR-proficient cells, MMR can recognize and repair any glitch caused by DNA polymerase, the errors are more frequently observed in DNA microsatellite sequences. Hence, for reasons stated previously, microsatellite instability (MSI) ensues in cells that are MMR deficient, also developing in virtually all of the CRC cases that emanate from Lynch syndrome (Wolpin et al., 2007). Here, we have a look at some of the already discussed but crucial and unavoidable concepts of staging colorectal cancer along with the present barriers, and then we go on to discuss the emerging new trends in therapies for CRC including immune checkpoint inhibitors, c-ART T-cell therapy, etc.

8.5.1 Staging, Prognosis, and Challenges

Pathologic staging is a crucial prognostic factor for CRC. Based on the depth of invasion of the wall of the bowel, lymph node involvement, and metastatic sites of the disease, the TNM system of the American Joint Committee on Cancer has defined four stages, with a decreasing five-year survival rate from more than 90% in stage I to less than 10% in stage IV. This is a meticulously revised version of what Dukes classification had originally done in 1932.

In the early stages (stage 0), the pre-cancerous lesions are confined to the mucosa of the bowel, only invading the inner layers of the submucosa and muscular propria at a later stage (stage 1). It reaches the outermost serosa in stage 2, without penetrating it. Further penetration and progression to surrounding lymph nodes, blood vessels, and fat deposits occur in stage 3 of the disease. At its zenith in stage 4, the tumor metastasizes to distant organs, more commonly to the liver and lungs, and to parts of the peritoneal lining (Freitas et al., 2023).

The current treatment of CRC focuses on patient care through an array of different and advanced techniques including radiotherapy, immunotherapy, and surgery. As stated earlier, the staging of the tumor and its location guide us toward the treatment strategy (Freitas et al., 2023).

Surgical resection is employed for the early stage of CRC whereas for metastatic cancer, intensive chemotherapy is administered. Recurrent cancer is often treated with neoadjuvant therapy, i.e. radiotherapy or chemotherapy before surgical resection (De Azevedo et al., 2019). Several trials have been conducted to ascertain the efficacy of various neoadjuvant chemotherapeutic regimens (NACT), targeting mostly a cluster of populations. The efficacy of FOLFOXIRI in metastatic CRC has been well established and the triplet therapy of preoperative capecitabine with oxaliplatin (CAPOX) followed by adjuvant therapy (CAPOX) after resection has been administered as neoadjuvant therapy in stage IIIB CRC patients. The trial showed a plausible reduction in tumors in 91.3% of patients with only 26.1% left at the end of the trial with a two-year recurrence rate. (Sánchez-Gundín et al., 2018).

8.5.2 Emerging Standards: Target Specific Approaches—Immunotherapy as the Starting Point

Over the course of years of research and trials, immunotherapy has found a consistent place as a prominent therapeutic regimen for CRC and other solid tumors (Atkins et al., 1999). Table 8.5 helps us understand the currently used and emerging therapeutics in immunotherapy for CRC.

Table 8.5: Immunotherapies against CRC—current and emerging

Targeting EGF/EGFR & VEGF/VEGFR	
Cetuximab	EGFR
Panitumumab	
Nimotuzumab	
Necitumumab	VEGFR
Bevacizumab	
Ramucirumab	
Immune checkpoint inhibitors	
Anti-PDL1 antibodies	
Atezolizumab	
Avelumab	
Anti-PD1 antibodies	
Nivolumab	
Dostarlimab	
Pembrolizumab	
T-cell boosting therapy	
Chimeric antigen receptor T cell therapy (CAR-T)	
T cell receptor modification (TCR)	
Tumor infiltrating lymphocyte	

8.5.2.1 Targeting KRAS G12C Mutation

Kirsten rat sarcoma virus oncogene homolog (KRAS) mutation is present in more than half of the CRC patients. EGFR inhibitors are not recommended for the treatment of CRC patients with RAS mutations due to their limited therapeutic efficacy in this patient population (Cancer Genome Atlas Network, 2012). Beyond the second-line treatments, which include trifluridine/tipiracil (known as TAS-102—a combination of the cytotoxic nucleic acid analog trifluridine and the thymidine phosphorylase inhibitor tipiracil), regorafenib (an oral multikinase inhibitor targeting VEGFR1-3, KIT, RET, RAF1, BRAF, PDGFR, and FGFR), and the more recent combination of TAS-102 with bevacizumab, have shown clinical benefits and are now considered standard care. Despite these options, not all patients experience positive outcomes in the refractory setting, and overall survival rates continue to be low. The presence of a RAS mutation has a well-established negative prognostic impact, leading to poorer overall survival compared to tumors with a wild-type RAS/BRAF profile (Jones et al., 2017; Schirripa et al., 2020). Table 8.6 shows the trials that targeted KRAS G12C and EGRF KRAS G12C that have completed or have published their data along with the regimens or interventions that each of them investigates

Owing to high RTK activation, the colorectal cells have shown resistance to KRAS G12C inhibitors alone with evidence of EGFR signaling as the mechanism of resistance. This necessitated the need to target both EGFR and KRAS G12C, drastically improving patient outcomes (Amodio et al., 2020).

Table 8.6: Trials targeting KRAS G12C and EGFR-KRAS G12C in mCRC, either completed or with already published data

Trial name	Study population	Treatment regimen	Trial number
CodeBreaK 100	advanced *KRAS* G12C mutant solid tumors	Sotorasib	NCT03600883
CodeBreaK 101	advanced *KRAS* G12C mutant mCRC	Sotorasib + panitumumab	NCT04185883
KRYSTAL-1	advanced *KRAS* G12C mutant solid tumors	Adagrasib	NCT03785249
KRYSTAL-1 (combination arm)	advanced *KRAS* G12C mutant mCRC	Adagrasib 600 mg bid + cetuximab	NCT03785249

Although present in a small fraction of metastatic colorectal cancer patients, KRAS G12C mutations, when present, are associated with a poor overall prognosis and shorter overall survival than their non-G12C counterparts (Ros et al., 2024; Fakih et al., 2022). Whereas anti-EGFR has been added to overcome the EGFR-mediated adaptive feedback reactivation of the RAS-MAPK pathway following KRAS G12C inhibition, all patients will nevertheless progress.

8.5.2.2 PARP Inhibitors

An increasing body of evidence suggests that poly ADP-ribose polymerase (PARP) inhibitors may enhance treatment for CRC with aberrant DNA damage response mechanisms. The DNA damage response (DDR) is a highly conserved cellular mechanism that helps cells cope with DNA damage caused by both external and internal factors. This process involves detecting and repairing the damaged DNA. PAR is a signaling molecule for DDR, acts as a post-translator modifier, and promotes repair of defective DNA (Herceg & Wang, 2001). ARP inhibitors block this repair mechanism, leading to the accumulation of DNA damage. In cells with defective DNA repair pathways, such as those with BRCA1/2 mutations or other homologous recombination deficiencies, this results in synthetic lethality and tumor cell death.

8.5.2.3 Anti-HER2 Therapies

The evolution and exploration of the genetic roots of CRC have led to the development of specifically targeted therapies, aiming for higher precision and lesser "non-treatment" effects. Nearly 5% of CRC patients exhibit human epidermal growth factor receptor 2 (HER2) or HER2 protein overexpression. The evidence in support for HER2 targeted therapies has grown to the extent that the FDA has recently granted accelerated approval to tucatinib, a tyrosine kinase inhibitor of HER2 along with trastuzumab.

In the study by Nakamura et al. (2021) ctDNA genotyping was used to guide patient selection and assess their response to trastuzumab plus pertuzumab in KRAS wild-type HER2-amplified metastatic CRC. Of the 30 participants in the phase II TRIUMPH study, 22 had HER2 amplification detected in both tissue-based and ctDNA-based tests, 5 had HER2 amplification detected only in tissue testing, and 3 had it detected only in ctDNA testing.

The phase II HERACLES-B trial was the first to evaluate the efficacy of ADC in treating metastatic CRC with HER2-amplifications. Encouragingly, up to 67% of patients achieved stable disease with this regimen (Nakamura et al., 2021). For patients with chemorefractory HER2-positive, RAS wild-type metastatic CRC, current evidence indicates that a combination of two anti-HER2 therapies is recommended- tucatinib plus trastuzumab.

8.5.3 Emerging Evidence from Clinical Trials

The standard regimen for benign rectal cancer proceeds from chemotherapy, radiotherapy, surgery to adjuvant chemotherapy with a recovery period in between. A study in collaboration with the Ministry of Health of Brazil (BRAZIL-TNT) aims to find if a regimen with active chemotherapy (FOLFIRINOX) between radiation therapy and surgical intervention proves to improve and increase the complete response rate. It is one of the few trials for total neo-adjuvant therapy.

Recent clinical trial results have demonstrated that KRAS mutations, and to a lesser extent KRAS mutations combined with BRAF V600E mutations, significantly reduce the effectiveness of anti- EGFR monoclonal antibodies (mAbs) like panitumumab in treating metastatic colorectal cancer (mCRC). Although having wild-type KRAS (WT KRAS) is necessary for a favorable response to panitumumab, it is not sufficient by itself to ensure efficacy. The reasons for varying responses to EGFR-directed therapies in patients with WT RAS remain unclear. Possible factors influencing response include mutations in the EGFR receptor or activation through autocrine/ paracrine ligand binding. In contrast, tumors that do not respond might have activation of alternative pathways such as VEGF, PDGF, or insulin-like growth factor 1 receptor, or mutations in downstream signaling proteins like PI3K, Src, or RAF. Additionally, loss of function in genes such as phosphatase and tensin homolog (PTEN) could play a role.

Identifying prognostic and predictive biomarkers for EGFR-directed therapies is essential for selecting the most effective treatment combinations. Potential biomarkers include the number of EGFR gene copies and the levels of EGFR ligands expressed in tumors. This approach emphasizes the importance of KRAS status in determining patient eligibility for treatment. Patients with KRAS mutations, detected in pre-treatment liver biopsies, will be excluded from the trial. However, with patient consent, these biopsies will be analyzed by a CAP/CLIA-compliant independent laboratory to study pathway activation. This research aims to generate new hypotheses about how KRAS mutation status influences pathway activation and to further investigate predictive biomarkers for anti-EGFR therapies.

The NITMEC trial strives to assess the safety and efficacy of a tyrosine kinase inhibitor (imatinib mesylate) with or without a human monoclonal antibody to epidermal growth factor receptor (panitumumab).

A slightly different approach is followed in a study called CITOGENEX PA wherein the efficacy and safety of a dietary supplement called citogenix is tested. It is made up of insoluble polysaccharides of *Lactobacillus casei* and *Bifidobacterium Lactis*—both prebiotic bacteria. They have been proven to be immune enhancers. While they do not have any anti-tumor activity of their own, they are effective in lessening the side-effects of anti-tumor treatments.

Another trial, STRATEGIC-1, is a multi-line therapy study to carve out the best possible sequence of therapy in patients with metastatic colorectal cancer (mCRC). All the available treatments are used in this trial, in a different order, so as to monitor the response with the different orders used.

With many HER2 targeting agents in development and clinical trials of many underway, there are still lingering questions which warrant further investigation. Given that individuals with refractory CRC often deal with comorbid conditions and multiple medications, it will be critical to choose treatments wisely and manage the associated treatment-related adverse events (TRAEs) effectively to ensure sustained clinical benefits.

Finally, the integration of various treatment modalities, namely chemotherapy with targeted therapies or immunotherapies previously discussed, is gaining traction. They aim at mitigating the resistance that develops frequently in colorectal cancerous cells towards the therapeutic agents (Salva de Torres et al., 2024).

The landscape of colorectal cancer treatment is evolving with the introduction of new therapies that target specific genetic mutations and utilize innovative combinations of existing treatments. As clinical trials continue to validate these emerging options, they hold the promise of improving outcomes for patients with advanced CRC, particularly those with previously limited treatment options.

8.6 CONCLUSION

CRC arises from both genetic mutations—such as in APC, KRAS, and TP53—and environmental factors like diet and inflammation. Traditional treatments include surgery, chemotherapy, and radiation, with recent advances in targeted therapies (e.g., anti-EGFR and VEGF) and immunotherapy, particularly for MSI-H tumors. Personalized medicine, driven by molecular profiling, has improved treatment efficacy and patient outcomes. The future of CRC management lies in precision medicine, cutting-edge research, and multidisciplinary care, with ongoing trials expected to reveal new biomarkers and therapeutic targets. Collaboration is essential to improve survival and quality of life.

BIBLIOGRAPHY

Adebayo, A. S., et al. (2023). Colorectal cancer: Disease process, current treatment options, and future perspectives. *Pharmaceutics*, *15*(11):2620.

Amodio, V., et al. (2020). EGFR blockade reverts resistance to KRASG12C inhibition in colorectal cancer. *Cancer Disc*, *10*(8):1129–1139.

André, T., et al. (2009). Improved overall survival with oxaliplatin, fluorouracil, and leucovorin as adjuvant treatment in stage II or III colon cancer in the MOSAIC trial. *J Clin Oncol*, *27*(19):3109–3116.

André, T., et al. (2020). Pembrolizumab in microsatellite-instability-high advanced colorectal cancer. *New Engl J Med*, *383*(23): 2207–2218.

Asthana, S., et al. (2021). Incidence of colorectal cancers in India: A review from population-based cancer registries. *Curr Med Res Pract*, 11(2): 91–96.

Atkins, M. B., et al. (1999). High-dose recombinant interleukin 2 therapy for patients with metastatic melanoma: Analysis of 270 patients treated between 1985 and 1993. *J Clin Oncol*, *17*(7):2105.

Bagnardi, V., et al. (2015). Alcohol consumption and site-specific cancer risk: A comprehensive dose-response meta-analysis. *Br J Cancer*, 112(3):580–593.

Bell, P., Findeis-Hosey, J. (2024). Microsatellite instability pathway. *PathologyOutlines.com*. Available at: https://www.pathologyoutlines.com/topic/colontumormolecularmicrosatellite.html (accessed July 2024).

Bokey, E. L., et al. (1996). Morbidity and mortality following laparoscopic-assisted right hemicolectomy for cancer. *Dis Colon Rectum, 39*(10 Suppl):S24–S28.
Boland, C. R., Goel, A. (2010). Microsatellite instability in colorectal cancer. *Gastroenterol, 138*(6):2073–2087.e3.
Brosens, L. A. A., et al. (2007). Gastrointestinal polyposis syndromes. *Curr Molec Med*, 7(1):29–46.
Burkitt, D. P. (1971). Epidemiology of cancer of the colon and rectum. *Cancer*, 28(1):3–13.
Cancer Genome Atlas Network. (2012). Comprehensive molecular characterization of human colon and rectal cancer. *Nature, 487*(7407):330–337.
De Azevedo, J. M., et al. (2019). Current surgical strategies in the management of rectal cancer. *Curr Colorectal Cancer Rep, 15*(1):18–27.
Desai, M. J., et al. (2007). Optimizing quality of life through palliative care. *Journal of Osteopathic Medicine, 107*(s7):E9–E14.
Espada, J., et al. (2009). Wnt signaling and cancer stem cells. *Clin Translat Oncol, 11*(8):411–427.
Fakih, M., et al. (2022). Real-world study of characteristics and treatment outcomes among patients with KRAS p.G12C-mutated or other KRAS-mutated metastatic colorectal cancer. *Oncologist, 27*(8):663–674.
Fedirko, V., et al. (2011). Alcohol drinking and colorectal cancer risk: An overall and dose-response meta-analysis of published studies. *Ann Oncol, 22*(9):1958–1972.
Freitas, S. C., et al. (2023). New frontiers in colorectal cancer treatment combining nanotechnology with photo- and radiotherapy. *Cancers, 15*(2):383.
Geiersbach, K. B., Samowitz, W. S. (2011). Microsatellite instability and colorectal cancer. *Arch Pathol Laborat Med, 135*(10):1269–1277.
Gmeiner, W. H. (2024). Recent advances in therapeutic strategies to improve colorectal cancer treatment. *Cancers, 16*(5):1029.
Goto, A., et al. (2020). Diabetes and cancer risk: A Mendelian randomization study. *Int J Cancer*, 146(3):712–719.
He, X., et al. (2023). Can immunotherapy reinforce chemotherapy efficacy? A new perspective on colorectal cancer treatment. *Frontiers Immunol, 14*:1237764.
Herceg, Z., Wang, Z. Q. (2001). Functions of poly(ADP-ribose) polymerase (PARP) in DNA repair, genomic integrity, and cell death. *Mutation Res Fundamental Molec Mechanisms Mutagenesis, 477*(1–2):97–110.
Hlubek, F., et al. (2007). Wnt/FZD signaling and colorectal cancer morphogenesis. *Frontiers Biosci, 12*:458–470.
Jones, R. P., et al. (2017). Specific mutations in KRAS codon 12 are associated with worse overall survival in patients with advanced and recurrent colorectal cancer. *Br J Cancer, 116*(7):923–929.
Kim, D. S., Scherer, P. E. (2021). Obesity, diabetes, and increased cancer progression. *Diabet Metabol J, 45*(6):799–812.
Klicka, K., et al. (2022). The role of miR-200 family in the regulation of hallmarks of cancer. *Frontiers Oncol, 12*:965231.
Labianca, R., et al. (2007). The role of adjuvant chemotherapy in colon cancer. *Surgical oncology, 16*, 93–96.
Leslie, A., et al. (2002). The colorectal adenoma-carcinoma sequence. *Br J Surg, 89*(7):845–860.
Li, K., et al. (2020). Microsatellite instability: A review of what the oncologist should know. *Cancer Cell Int, 20*:16.
Lynch, H. T., de la Chapelle, A. (2003). Genomic medicine - Hereditary colorectal cancer. *New Engl J Med*, 348(10):919–932.
Martens-de Kemp, S. R., et al. (2022). Overexpression of the miR-17-92 cluster in colorectal adenoma organoids causes a carcinoma-like gene expression signature. *Neoplasia, 32*:100820.
Meng, F., et al. (2007). MicroRNA-21 regulates expression of the PTEN tumor suppressor gene in human hepatocellular cancer. *Gastroenterol, 133*(2):647–658.
Moertel, C. G., et al. (1995). Intergroup study of fluorouracil plus levamisole as adjuvant therapy for stage II/Dukes' B2 colon cancer. *J Clin Oncol, 13*(12):2936–2943.
Nakamura, Y., et al. (2021). Circulating tumor DNA-guided treatment with pertuzumab plus trastuzumab for HER2-amplified metastatic colorectal cancer: A phase 2 trial. *Nature Med, 27*(11):1899–1903.
Nelson, S., Nathke, I. S. (2013). Interactions and functions of the adenomatous polyposis coli (APC) protein at a glance. *J Cell Sci, 126*(4):873–877.
Ng, E. K. O., et al. (2009). MicroRNA-143 targets DNA methyltransferases 3A in colorectal cancer. *Br J Cancer, 101*(4):699–706.

Passardi, A., et al. (2017). Immune checkpoints as a target for colorectal cancer treatment. *Int J Molec Sci, 18*(6):1324.
Pathak, P. S., et al. (2024). State-of-the-art management of colorectal cancer: Treatment advances and innovation. *Am Soc Clin Oncol Educ Book, 44*(3):e438466.
Rawla, P., et al. (2019). Epidemiology of colorectal cancer: Incidence, mortality, survival, and risk factors. *Przegląd Gastroenterologiczny, 14*(2):89–103.
Rebersek, M. (2021). Gut microbiome and its role in colorectal cancer. *BMC Cancer, 21*(1):1325.
Ribic, C. M., et al. (2003). Tumor microsatellite-instability status as a predictor of benefit from fluorouracil-based adjuvant chemotherapy for colon cancer. *New Engl J Med, 349*(3):247–257.
Ros, J., et al. (2024). Targeting KRAS G12C mutation in colorectal cancer: A review of new arrows in the quiver. *Int J Molec Sci, 25*(6):3304.
Roshandel, G., et al. (2024). Colorectal cancer: Epidemiology, risk factors, and prevention. *Cancers,* 16(8):1530.
Salva de Torres, C., et al. (2024). Current and emerging treatment paradigms in colorectal cancer: Integrating hallmarks of cancer. *Int J Molec Sci, 25*(13):6967.
Sánchez-Gundín, J., et al. (2018). New trends in the therapeutic approach to metastatic colorectal cancer. *Int J Med Sci, 15*(7):659–665.
Schirripa, M., et al. (2020). KRAS G12C metastatic colorectal cancer: Specific features of a new emerging target population. *Clin Colorectal Cancer, 19*(3):219–225.
Tam, S. Y., Wu, V. W. C. (2019). A review on the special radiotherapy techniques of colorectal cancer. *Frontiers Oncol, 9*:208.
Thomas, V. M., et al. (2020). Trends in colorectal cancer incidence in India. *J Clin Oncol, 38*(15_suppl):e16084.
Weeks, J. C., et al. (2002). Short-term quality-of-life outcomes following laparoscopic-assisted colectomy vs open colectomy for colon cancer: A randomized trial. *JAMA, 287*(3):321–328.
Wolpin, B. M., et al. (2007). *Adjuvant treatment of colorectal cancer. CA, 57*(3):168–185.
Ye, P., et al. (2020). Linking obesity with colorectal cancer: Epidemiology and mechanistic insights. *Cancers, 12*(5):1408.
Zhu, S., et al. (2007). MicroRNA-21 targets the tumor suppressor gene tropomyosin 1 (TPM1). *J Biol Chem, 282*(19):14328–14336.

9 Immunotherapy and Immune Checkpoint Inhibitors

Yamna Khurshid and Kinza Rafi

9.1 INTRODUCTION

Initially celebrated for its remarkable success in melanoma treatment, immunotherapy has rapidly emerged as a leading therapeutic option for a variety of solid tumors including colorectal cancers (CRCs) (Golshani & Zhang, 2020). Immunotherapy addresses the critical issue of specificity, a major challenge in conventional radiotherapy and chemotherapy. It selectively aims antigens present on tumor cells, signaling the immune system to recognize and eliminate these abnormal cells. By orchestrating a coordinated immune response, cancer immunotherapy distinguishes between cancerous and normal cells, thus limiting the side effects (Li et al., 2023). While the outcomes of certain immunotherapies have yielded remarkable results, the efficacy is dependent on the individual patient's immune system. Those who respond favorably to cancer immunotherapy often experience improved prognosis and higher quality of life. Immunotherapies in CRCs include immune checkpoint inhibitors, monoclonal antibodies (mAbs), adoptive cell transfer, cytokine therapy, cancer vaccine, combination therapies, and microbiome modulation. Checkpoint inhibitors are the most promising approach out of all the existing therapies (Galluzzi et al., 2014). This chapter summarizes immunotherapy in CRC with a focus on important types, mechanisms, developments, shortfalls, and potential future treatments.

9.2 IMMUNE LANDSCAPE AND RATIONALE FOR IMMUNOTHERAPY IN CRC

Chemotherapy, surgery, and radiotherapy are the common treatments for CRC in practice and can be employed in combination depending on the anatomical location and stage of the disease (Schmoll et al., 2012; Van Cutsem et al., 2014; Van Cutsem et al., 2016; Yoshino et al., 2018). The most common strategies for localized and easily accessible CRC include mesorectal excision using laparoscopic and transanal surgery (Park et al., 2018). However, it is often challenging to eliminate all the cancer cells. In CRC stage II and stage III patients, chemotherapy, radiotherapy, or both are given as adjuvant therapy to about two-thirds of the patients (Miller et al., 2019). Many side effects are associated with these treatments because of their cytotoxicity and non-specificity towards rapidly dividing cells (Garg, et al., 2012; González-Perera et al., 2017). In spite of receiving neoadjuvant therapy, recurrence was reported in half of CRC patients (Ogura et al., 2019). Hence, moving to substitute and more effective methods is important in the treatment of CRC.

An approach that harnesses the potential of the immune system has modernized the treatment of cancer in recent years. Immunotherapy acts by enabling the patient's own immune system to address the limitations associated with chemotherapy and radiation therapy. Immunotherapy activates both innate and adaptive immune system and directs them to selectively combat tumor cells while causing no damage to the normal cells (Johdi, & Sukor, 2020). Immunotherapy can be either active or passive. Active immunotherapy involves dendritic cell-based immunotherapies, immunostimulatory cytokines, immunomodulatory mAbs, peptide- and DNA-based cancer vaccines inhibitors of immunosuppressive metabolism, pattern recognition receptors (PRR) agonists, and immunogenic cell death inducers while passive immunotherapy includes tumor-targeting mAbs, adoptive cell transfer (ACT) and oncolytic viruses. Some of these approaches are effective in the treatment of CRC (Galluzzi et al., 2014).

Immunotherapy for cancer treatment relies on the idea that cancer cells utilize regulatory T-cell-mediated immunosuppression as the main mechanism to evade the immune system. Cancer cells can evade attacks from the immune system through other mechanisms as well. Tumor cells can affect the cytokines that facilitate myeloid-derived suppressor cells and T regulatory cells to inhibit the function of cytotoxic T-cell. This consequently led to the suppression of CD4 and CD8+ T lymphocytes, rendering them incapable of recognizing foreign antigens. Additionally, tumor cells show loss of major histocompatibility complex (MHC) expression, and thus cannot be recognized by T-cells. Cancer cells also overexpress immune checkpoint proteins leading to the depletion of peripheral T-cells and preventing malignant cells from undergoing apoptosis (Tumeh et al., 2014; Golshani & Zhang, 2020). Initially, CRC was not considered an immunogenic cancer and would not respond to immunotherapy. However, extensive research has demonstrated that lymphocytic reaction serves as an important prognostic factor for CRC (Ogino et al., 2009).

 DOI: 10.1201/9781032704517-9

9.3 STRATEGIES IN IMMUNOTHERAPY FOR CRC

The poor five-year survival rate (Rawla et al., 2019), the side effects of the existing cytotoxic drugs, and resistance of the disease towards standard treatment options have driven efforts to explore and advance immunotherapy for CRC treatment. Immunotherapy appears as a successful treatment of metastatic CRC (mCRC) particularly for individuals with high microsatellite instability (MSI) tumors (Xie et al., 2020). This section details the molecular mechanisms of immunotherapies that have been studied for the management of CRC (Figure 9.1).

9.3.1 Immune Checkpoint Inhibitors (ICIs)

The body utilizes immunosuppressive methods to regulate immune balance and self-tolerance, including the activation of immunosuppressive cells, checkpoint proteins, and programmed cell death. Cancer cells exploit these methods to evade immune surveillance. Immune checkpoint inhibitors function by blocking the inhibitory receptor located on the surface of the immune cell, thereby triggering an anti-cancer immune response. Immunosuppressive checkpoint proteins that are frequently upregulated in cancers include programmed cell death 1 (PD-1), cytotoxic T-lymphocyte-associated protein 4 (CTLA-4), T-cell Immunoglobulin- and mucin domain-3-containing molecule 3 (LAG-3), and lymphocyte activation gene-3 (TIM-3). The ICIs are mostly immunomodulatory mAbs used to block immune checkpoint proteins PD-1 and CTLA-4 are extensively investigated immune checkpoint targets, primarily because of their upregulated expression in numerous solid tumors (He et al., 2021). There are three Food and Drug Administration (FDA) approved PD-1 and CTLA-4 targeting ICIs against CRC but the efficacy of these drugs is affected by the status of the MSI in each CRC patient (Rodriguez-Salas et al., 2017; Menter et al., 2019). According to the mutation patterns, CRC patients are categorized into three classes: microsatellite instability-high (MSI-H), microsatellite instability-low (MSI-L), and microsatellite stable (MSS) (Nojadeh et al., 2018). The immune checkpoint inhibitor drugs approved by the FDA and European Medicines Agency (EMA) for CRC treatment are presented in Table 9.1.

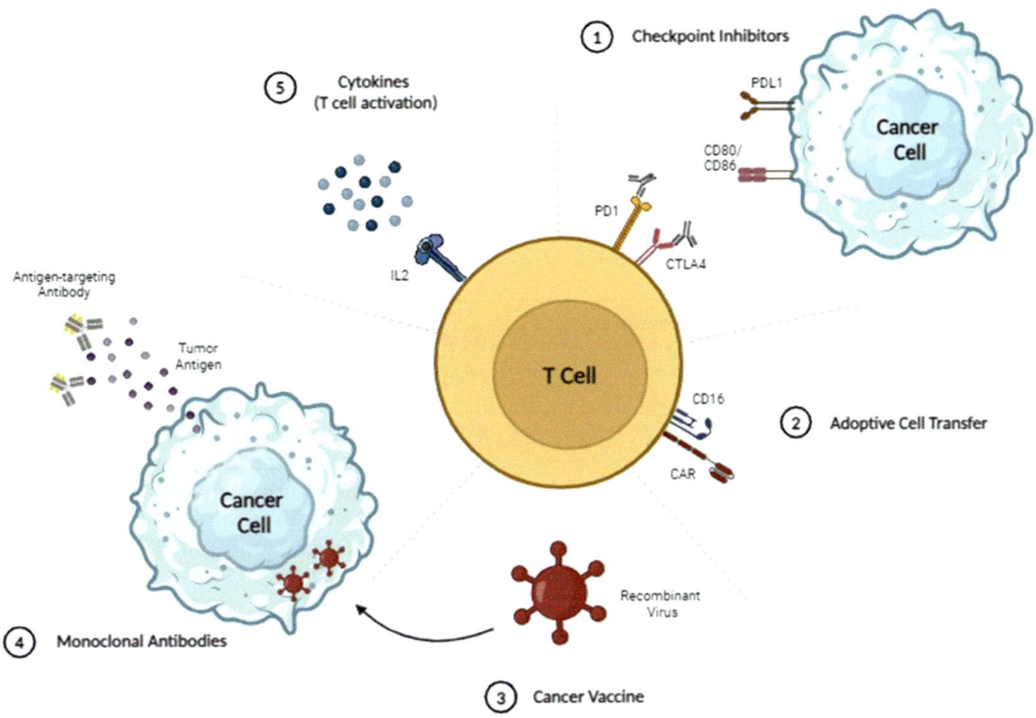

Figure 9.1 Molecular mechanisms of immunotherapies in colorectal cancer (CRC), including checkpoint inhibitors, adoptive cell transfer, cancer vaccines, monoclonal antibodies, and cytokines, illustrating key strategies to enhance the immune response against CRC.

Table 9.1: FDA and EMA approved immune check point inhibitors drugs for the treatment of colorectal cancer

Drug	Mechanism	Company	Agency	Key Trial	Treatment line
Pembrolizumab	Anti-PD-1	Merck & Co.	FDA	KEYNOTE-177	1st line
Nivolumab	Anti-PD-1	Bristol Myers Squibb Co	FDA, EMA	CheckMate-142	2nd/3rd line in combination with Ipilimumab
Ipilimumab	Anti-CTLA-4	Bristol Myers Squibb Co	FDA	NICHE-1/2	2nd in combination with Nivolumab

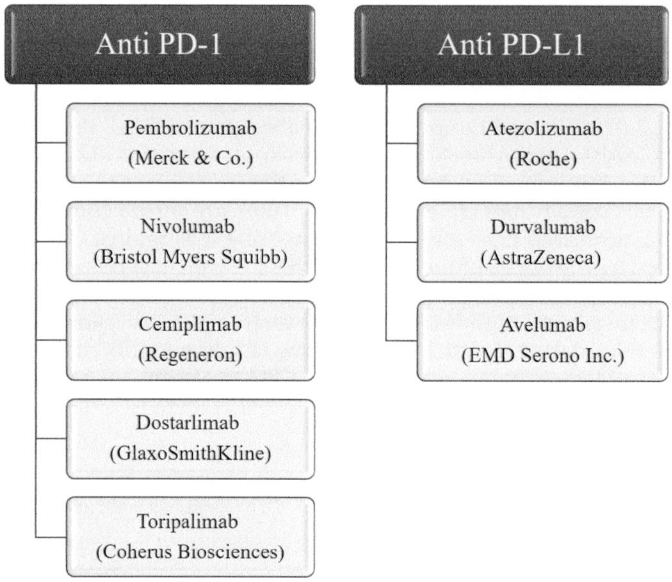

Figure 9.2 FDA-approved PD-1/PD-L1 inhibitors for cancer treatment, listing five anti-PD-1 and three anti-PD-L1 therapies currently in clinical use.

9.3.2 Programmed Cell Death Protein/Ligand 1 (PD-1/PD-L1)

The cell surface of T-cells, B-cells, and natural killer cells expresses the PD-1 receptor. A ligand for PD-1, PD-L1 is present on the outer membrane of certain cancer cells as well as healthy cells (Kitano et al., 2017). The engagement between PD-1/PD-L1 inhibits T-cell activity causing tumor cells to escape immune surveillance (Zhu et al., 2021). Pembrolizumab became the initial PD-1 inhibitor drug to receive FDA approval. To date, the FDA has approved eight PD-1/PD-L1 inhibitors for various cancers (Figure 9.2). In CRC patients with mismatch-repair deficient (dMMR) and microsatellite instability-high (MSI-H) treatment with Pembrolizumab, improved survival rates by up to 78% and 11% in comparison to mismatch-repair proficient and microsatellite stable (pMMR/MSS) respectively. Nivolumab, another PD-1 inhibitor proved effective in treating metastatic dMMR CRC. Furthermore, Nivolumab, in combination with anti-CTLA4 Ipilimumab exerted up to a 94% increase in response rate in metastatic dMMR CRC (Le et al., 2017).

9.3.2.1 Cytotoxic T-Lymphocyte-Associated Protein 4 (CTLA-4)

T-cell self-tolerance and homeostasis depend on co-inhibitory CTLA-4 and co-stimulatory CD28 signals. Upon encountering a foreign molecule, T-cell antigen receptors are stimulated by an increase in CD28 co-stimulatory signals, which in turn triggers downstream immunological signaling. Co-inhibitory signals by CTLA-4 are triggered to halt immunomodulation and avert unnecessary immunological reactions or autoimmunity once the foreign molecule has been eliminated (Sanchez-Lockhart et al., 2014). Nevertheless, in certain cases particularly in cancers, CTLA-4 co-inhibitory signals persistently remain elevated, causing the T-cell activation to stop and thus prevent immune response against tumor cells. The ligands CD80 and CD86 for CTLA-4 are expressed by cancer cells and antigen-presenting cells (APCs). These ligands can interact with either CTLA-4 or CD28, inducing either a co-inhibitory or costimulatory response, respectively. Furthermore, CTLA-4 is also present on regulatory T-cells (Tregs) in addition to CD4+ and CD8+ T-cells. The communication between CD80/86 on APCs and CTLA-4 on Tregs prevents APCs from activating naïve CD4+ T-cells (Egen et al., 2002; Buchbinder & Desai, 2016). This approach has paved the way for immunotherapy by aiming CTLA-4 to enhance the endogenous T-cell mediated immunity of the patient. Numerous CTLA-4 checkpoint inhibitors are available, facilitating the binding of CD28 to MHC, thus triggering T-cell immunity, initiating subsequent immune signaling, and ultimately eliminating tumor cells. For CRC patients with MSI-H mCRC, ipilimumab (anti-CTLA-4 mAbs) combined with nivolumab, is effective and approved. Remarkably, after just around four weeks of treatment, this combination resulted in 100% and 27% response rates in patients exhibiting MSI-H and MSS CRC, respectively (Overman et al., 2018a). Toxicity in patients undergoing immune checkpoint blockade is more prevalent with anti-CTLA-4 treatment compared to anti-PD-1/PD-L1 treatment.

9.3.2.2 T-cell Immunoglobulin- and Mucin Domain-3-Containing Molecule 3 (TIM-3)

TIM-3 inhibitory receptor is present on T-cells (CD8+, CD4+, Tregs), natural killer and APCs. TIM-3 induces apoptosis by interacting with its ligand galectin-9, expressed on or secreted by tumor cells, thereby maintaining immune homeostasis. The cell surface proteins such as phosphatidyl serine (PtdSer), CAECAM1, and soluble protein high mobility group box 1 protein (HMGB1) are also ligands for TIM-3 (Anderson et al., 2016). Two antibodies targeting TIM-3, known as MBG453 and TSR-022, are presently undergoing clinical trials, either alone or in parallel with anti-PD-1, for advanced-stage cancers, including CRC. The outcomes from phase I/II have shown the effectiveness of the combination, of MBG453 and spartalizumab (anti-PD1). The CD8+ T-cells with upregulated TIM-3 have the potential to restrict T-cell function in CRC patients (Acharya et al., 2020). Therefore, inhibiting TIM-3 may improve T-cell-mediated response in CRC patients. These studies suggest the exploration of anti-TIM-3 treatment in CRC patients.

9.3.2.3 Lymphocyte-Activation Gene 3 (LAG-3)

LAG-3 serves as an inhibitory receptor found on T-cells both CD8+ and CD4+, Tregs cells, and natural killer. LAG-3 expressed on CD4+ cells is expected to engage with MHC class II molecules on APC, whereas LAG-3 on natural killer cells and CD8+ T-cells possibly engage with L-SECtin on tumor cells, thereby inhibiting immunological reaction by reducing the cytokine formation, T-cell expansion, and Ca2+ flux. Similarly, T-regs cells with LAG-3 participate in immune suppression. However, the actual molecular mechanism exerted by LAG-3 still needs to be fully understood (Anderson et al., 2016). Many ongoing clinical phase II trials are exploring the combination of nivolumab or dostarlimab (anti-PD-1 mAbs) and anti-LAG-3 mAbs. Since anti-PD1/PD-L1 and anti-CTLA-4 treatments are not effective in some patients, the findings of anti-LAG-3 treatment are highly anticipated for such patients (Carlsen et al., 2022).

9.3.3 Adoptive Cell Transfer

Adoptive cell transfer involves the isolation and reinfusion of the patient's own immune cells, donors or stem cells-derived differentiated cells. Such cells are *in vitro* manipulated through gene editing to specifically target and eliminate cancer cells. Therefore, when administered to the patient enhances immune function and facilitates cancer removal (Fan et al., 2018). ACT involves tumor-infiltrating lymphocyte (TILs) therapy, chimeric antigen receptor T-cell (CAR T-cell) therapy, T-cell receptor (TCR)-engineered T-cell therapy, and natural killer cell therapy (Rosenberg & Restifo, 2015).

CRC represents a high level of carcinoembryonic antigen (CEA) compared to normal cells, which usually lack or have low levels of CEA (Hammarström, 1999). Considering this, several treatments have been developed to target CEA for adoptive transfer. Initially, in a phase I clinical trial, CAR T-cell therapy directed against CEA was examined in three mCRC patients (Parkhurst et al., 2011). All the patients exhibited significant reduction in the serum CEA level and one patient indicated objective regression to liver and lung metastasis. Furthermore, findings from a phase I trial (NCT02349724) demonstrated the efficacy of anti-CEA CAR T-cell therapy in CEA-positive mCRC patients. Out of ten patients, seven obtained stable disease with an acceptable toxicity profile (Katz et al., 2015).

Tumor cells release natural killer group 2D (NKG2D) ligands that downregulate the expression of NKG2D on natural killer cells and T-cells, thus escaping from immune surveillance (Lonez et al., 2017). An approach to amplify the immune response of the host against tumors is the genetic modification of CAR T-cells to express NKG2D and PD-L1 receptors. But, the effectiveness and safety of the method need further investigation.

CRC treatment through TILs also demonstrated promising outcomes. In a clinical phase II study (NCT01174121), assessing the efficiency of patient-derived TILs against various solid tumors including breast, ovarian, endometrial, digestive tract, and urothelial cancers, objective tumor response was noticed in one patient with mCRC. After receiving a single infusion of TILs targeting tumor-identified KRAS G12D mutation, the patient showed a response against all seven lung metastases. However, one lesion showed progression on evaluation after nine months of therapy and undergone resection. Following resection, the patient experienced no clinical disease and signs of recurrence for four months (Tran et al., 2016).

9.3.4 Monoclonal Antibodies (mAbs)

Monoclonal antibodies for cancer immunotherapy consist of cancer cell-targeting mAbs and immunoregulatory mAbs. Tumor-targeting mAbs are a type of passive immunotherapy and are widely employed in the treatment of hematologic malignancies and different types of solid tumors. The immunomodulatory mAbs usually function as ICIs which block the interaction between immune checkpoint proteins and cause the resurrection of T-cell mediated cytotoxic effect against cancer cells (Johdi & Sukor, 2020).

The most extensively studied immune checkpoints are PD-L1 and CTLA-4 as these are abundantly present in various types of solid and hematological malignancies (Seidel et al., 2018). However, other checkpoints, such as LAG-3, TIM-3, and T-cell immunoglobulin with ITIM domain (TIGIT), are currently being investigated for their possible contribution to the regulation of tumor immunity (Kong et al., 2016; Joller & Kuchroo, 2017).

The anti-CTLA-4 and anti-PD-L1 therapies are more effective against mCRC patients representing high MSI, attributed to higher tumor mutation burden (TMB). The high TMB increases tumor immunogenicity due to a positive correlation with the presence of high tumor antigen (Giannakis et al., 2016). Most of the CRC patients with MSI-H experienced significant outcomes from the combined use of Nivolumab and Ipilimumab immunotherapy with a disease control rate reached 80%, in contrast to patients with microsatellite-stable disease (Overman et al., 2018b).

9.3.5 Cancer Vaccine

Cancer vaccine consists of DNA vaccines, peptide vaccines, tumor antigen-expressing recombinant virus vaccines, patient-derived immune cell vaccines, and heterologous whole-cell vaccines obtained from established human cancer cell lines. Cancer cells trigger weaker immune responses compared to foreign antigens due to the altered expression of self-antigens. Therefore, adjuvants or immune stimulants are usually conjugated with the cancer vaccine to improve the efficacy (Thomas & Prendergast, 2016). Cancer vaccines can have both preventive and therapeutic effects. Preventive cancer vaccines aim to reduce cancer incidence as well as cancer-associated morbidity and mortality while therapeutic vaccines address existing malignancies and may limit further cancer development (Schlom et al., 2014; Lollini et al., 2015).

The CRC vaccine is directed against CEA, mucin 1 (MUC1), and melanoma-associated antigen (Shahnazari et al., 2020). During a clinical phase II study, the efficiency of a vaccine based on pox vectors encoding CEA and MUC1 (PANVAC) modified dendritic cells was evaluated on the survival rate of metastasectomy and perioperative chemotherapy experienced disease-free mCRC patients (Morse et al., 2013). However, the evidence to support a consistent survival benefit remains insufficient despite the ability of cancer vaccine tumor-associated antigens (TAAs) to augment the immune response and exhibit low toxicities (Jiang et al., 2019).

9.3.6 Cytokines Therapy

Cytokines are small cell-signaling molecules of less than 30 kDa (Berraondo et al., 2019). These signaling molecules are released by a variety of cells mainly endothelial cells, fibroblasts other stromal cells, and immune cells such as T-cells, neutrophils, and macrophages (Rusch et al., 2018). Because cytokines can effectively increase immune responses, they are one of the possible polypeptides to be used in immunotherapy. The FDA and European Medicines Agency (EMA) have approved three recombinant cytokines, including recombinant interferon-alpha 2a, interferon-alpha 2b, and interleukin-2 for the treatment of various types of cancers (García-Martínez et al., 2018). Due to an increased interest in the anticancer properties of cytokines, there is a rise in conducting clinical trials of cytokine-based drugs, exploring the efficacy and safety as separate regimes as well as in combination with other immunomodulatory drugs.

9.3.7 Combination Therapies

Combination therapy involves employing multiple drugs or treatment strategies to target cancer cells through diverse mechanisms to improve treatment outcomes. Combining various immunotherapy strategies or integrating immunotherapy with standard chemotherapy or radiotherapy treatments may increase the efficacy and response rates of immunotherapy in treating CRC (Ott et al., 2017). Several combination therapies for CRC have been studied. In CRC patients, combining ICIs, such as anti-PD-L1 and anti-CTLA-4 mAbs, has improved response rates and overall survival (Klein et al., 2021). CRC patients demonstrated better prognosis after checkpoint inhibitors were combined with chemotherapy and targeted therapy drugs, such as anti-vascular endothelial growth factor (VEGF) or anti-epidermal growth factor receptor (EGFR) inhibitors (Xu et al., 2023). In preclinical research, combining immunotherapy with radiation therapy improved the immune response and enhanced the prognosis in CRC patients (Parikh et al., 2021). Thus, combining immunotherapy with other treatments may appear as a promising alternative for patients who are unresponsive to standard cancer treatments.

9.3.8 Microbiome Modulation

The colon and rectum are exposed to over 100 trillion microorganisms from their proximal to distal ends as the luminal tract organs are exposed to the external environment (Fong et al., 2020). The presence of intestinal microbiome is greatly different in CRC patients with reduced diversity. For example, CRC tissues have large measures of *Fusobacterium nucleatum* and *Peptostreptococcus anaerobius* but low numbers of *Clostridium butyricum* and *Bifidobacterium animalis*. It is well known that the gut microbiota and its metabolic products play a role in immune responses and tumor development (Dai et al., 2018; Fong et al., 2020) (Figure 9.3). Opposing, probiotics that are the large

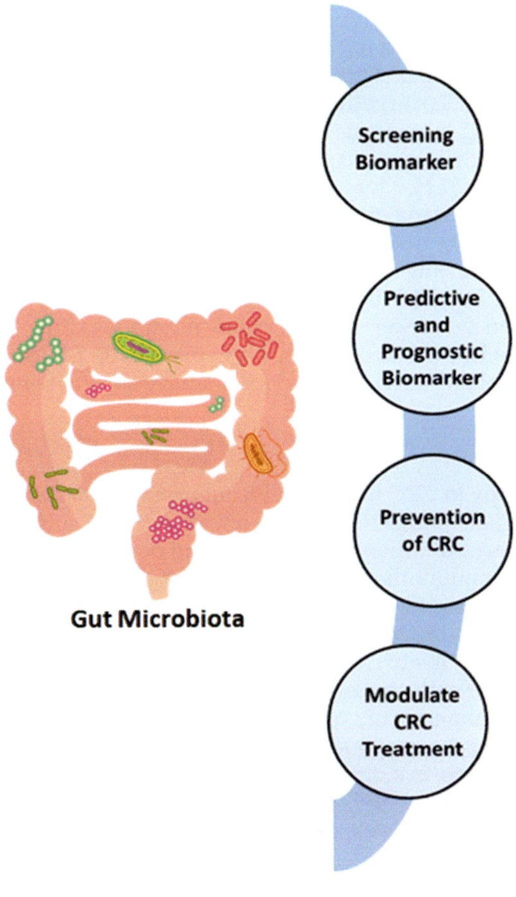

Figure 9.3 Role of gut microbiota in colorectal cancer (CRC): from screening and prognostic biomarkers to prevention and modulation of treatment.

commensal bacteria in the intestine have a profound effect on tumor regression and the efficacy of immunotherapies. *Clostridium butyricum* prevents histone deacetylase (HDAC) activity by releasing butyrate into the tumor microenvironment (TME) which prevents the growth of CRC (Stoeva et al., 2021). New research indicates that administration of *Lactobacillus rhamnosus* GG and *Lactobacillus* casei promote the infiltration of CD8+ TILs and increase secretion of cytotoxic cytokines, thus improving the anti-tumor efficacy of PD-L1 inhibitors (Aindelis et al., 2020; Owens et al., 2021). However, as a recently accepted factor of the TME, the impact of the microbiota on tumor cells and CD8+ T-cells in CRC is intricate and needs further investigation. Since CRC patients usually demonstrate an imbalance favoring the enrichment of pro-tumor microbiota, the prognosis resulting from CRC immunotherapy is strongly linked to the preservation of intestinal microbiota homeostasis or the supplement of exogenous anti-tumor microbiota and its metabolites (Shan et al., 2022).

9.4 CHALLENGES AND LIMITATIONS

Immunotherapy, particularly ICIs, has revolutionized oncology by offering a novel therapeutic alternative. But it also comes with some challenges and limitations. In the early stages of CRC, intra-tumor heterogeneity tends to be high, with diverse genetic and phenotypic characteristics within the tumor mass (Losi et al., 2005). This heterogeneity can impact treatment response and the development of resistance mechanisms over time (Jones et al., 2017).

CRC patients receiving mAb therapy experience hypersensitivity reactions triggered due to the immune response. Chimeric mAbs, which contain partial non-humanized sequences, may provoke an immunogenic response. Despite relatively low frequency, these types of mAbs can elicit immediate anaphylactic reactions. ICIs cause hypersensitivity pneumonitis as well. Additionally, both humanized and chimeric mAbs can lead to serum sickness-like reactions (Baldo, 2013; Baldo, 2022).

Antibody-drug resistance poses a significant limitation in CRC treatment. Treatment with mAbs, directed against growth factors like VEGF and epidermal growth factor (EGF) have demonstrated a limited survival rate. Anti-VEGF therapy has demonstrated short-term benefits in patients with mCRC patients. ICIs, targeting PD-1 ligand and receptor and CTLA-4 have presented clinical efficacy, including long-term remissions and improved overall survival in CRC patients. However, ICIs are effective only in a small subset of mCRC patients that represented DNA mismatch repair (dMMR) or MSI-H. Moreover, there remains a need for reliable biomarkers to predict ICIs therapy outcomes (Manz et al., 2021; Ganesh, 2022).

CAR T-cell therapy has demonstrated limited clinical response in CRC due to factors such as restricted molecular trafficking, low antigen expression, or less tumor infiltration (Ghazi et al., 2022). Cancer vaccines targeting CRC have been designed for patients with small lesion residues or as adjuvant therapy at advanced cancer stages. Despite numerous clinical trials, no CRC vaccine has been approved in phase III clinical trials. One possible reason for this outcome is the short-term immunity induced by the vaccine, which does not have clinical benefits on patients' survival rates (Jia et al., 2022).

9.5 FUTURE DIRECTIONS

The future of immunotherapy in CRC holds significant promise, marked by ongoing advancements and research endeavors. Future research on a wide range of subjects is necessary to enhance clinical response rates. An area of utmost importance is to develop and validate predictive biomarkers for response to immunotherapy in CRC patients. At present, the only well-characterized biomarker is MSI-H/dMMR to guide CRC immunotherapy. Nonetheless, the mechanism behind why MSI-H/dMMR serves as a biomarker for CRC immunotherapy remains unclear (Li & Martin, 2016). The likely explanation is the association between MSI-H/dMMR and elevated mutational load that results in the formation of neoantigens and consequent activation of the immune system (Timmermann et al., 2010; Viale et al., 2017). Only a few colon cancer patients show MSI-H and some among them may show inherent or acquired resistance to immunotherapy (Hou et al., 2022). Therefore, there is a dire need to look for predictive biomarkers to stratify colon cancer patients for immunotherapy. Other potential predictive biomarkers suggested for CRC immunotherapy include expression levels of PD-L1, tumor-infiltrating lymphocytes, tumor mutation burden, circulating tumor DNA, gut microbiota, and peripheral immune cells (Zeng et al., 2021; Temraz et al., 2019; Zou et al., 2021; Schrock et al., 2019; Li et al., 2016). The parameters reflecting the tumor microenvironment have also been anticipated as biomarkers for immunotherapy in CRC (Wang et al., 2022).

With the identification of novel targets such as TIM-3, LAG-3, VISTA, and TIGIT research interest in immune checkpoints has increased notably (Popovic et al., 2018). A study has shown a decrease in the quantity of IFN-γ-producing tumor-infiltrating CD8+ lymphocytes when TIM-3 and PD-1 are expressed simultaneously (Xu et al., 2015). It could be inferred from these findings that relying only on an anti-PD-1 may not be sufficient to recover the function of T-cells. Numerous clinical studies exploring next-generation checkpoint inhibitors are currently in clinical trials. A study reported improved T-cell response after treatment with anti-VISTA mAbs (Marin-Acevedo et al., 2018). An alternative approach for improving immunotherapy efficacy involves combining co-stimulatory checkpoint molecules, such as anti-ICOS, TNFRSF7, CD28, TNFRSF9, and glucocorticoid-induced TNFR-related protein with immune checkpoint inhibitors. OX40 antigen (CD134), belonging to the family of tumor necrosis factor receptors, along with the OX40L ligand, promotes the stimulation and expansion of CD4+ and CD8+ T-cells (Croft et al., 2009). Numerous trials are currently assessing the efficacy and toxicity of OX40 agonists either alone or in combination with immune checkpoint inhibitors (Fu et al., 2020; Poropatich et al., 2020).

To expand the therapeutic options of cancer immunotherapies advanced delivery systems are required for targeted delivery and to reduce side effects. Research is currently underway to explore a range of delivery platforms including nanoparticles, implants, scaffolds, and biomaterials (Wang et al., 2017).

The development of approaches that can transform cold tumors into hot tumors may extend the potential application of immunotherapy. This area of research also holds the potential to represent a breakthrough in the treatment of CRC (Weng et al., 2022).

9.6 CONCLUSION

Immunotherapy has become a key approach in CRC treatment, especially with immune checkpoint inhibitors targeting PD-1/PD-L1 and CTLA-4. Other strategies include adoptive cell transfer, monoclonal antibodies, cancer vaccines, and combination therapies. However, challenges like tumor heterogeneity and limited response in some patients persist. Future research aims to improve biomarkers, explore new checkpoints, and enhance targeted therapy delivery, offering promising potential for better patient outcomes.

REFERENCES

Acharya, N., et al. (2020). Tim-3 finds its place in the cancer immunotherapy landscape. *J Immunother Cancer*, *8*(1):e000911.

Aindelis, G., et al. (2020). Immune responses raised in an experimental colon carcinoma model following oral administration of Lactobacillus casei. *Cancers*, *12*(2):368.

Anderson, A. C., et al. (2016). Lag-3, Tim-3, and TIGIT: co-inhibitory receptors with specialized functions in immune regulation. *Immunity*, *44*(5):989–1004.

Baldo, B. A. (2013). Adverse events to monoclonal antibodies used for cancer therapy: focus on hypersensitivity responses. *Oncoimmunology*, *2*(10):e26333.

Baldo, B. A. (2022). Immune-and non-immune-mediated adverse effects of monoclonal antibody therapy: a survey of 110 approved antibodies. *Antibodies*, *11*(1):17.

Berraondo, P., et al. (2019). Cytokines in clinical cancer immunotherapy. *Br J Cancer*, *120*(1):6–15.

Buchbinder, E. I., Desai, A. (2016). CTLA-4 and PD-1 pathways: similarities, differences, and implications of their inhibition. *Am J Clin Oncol*, *39*(1):98.

Carlsen, L., et al. (2022). Immunotherapy for colorectal cancer: mechanisms and predictive biomarkers. *Cancers*, *14*(4):1028.

Croft, M., et al. (2009). The significance of OX40 and OX40L to T-cell biology and immune disease. *Immunolog Rev* *229*(1):173–191.

Dai, Z., et al. (2018). Multi-cohort analysis of colorectal cancer metagenome identified altered bacteria across populations and universal bacterial markers. *Microbiome*, *6*(1):1–12.

Egen, J. G., et al. (2002). CTLA-4: new insights into its biological function and use in tumor immunotherapy. *Nature Immunol*, *3*(7):611–618.

Fan, J., et al. (2018). Adoptive cell transfer: is it a promising immunotherapy for colorectal cancer? *Theranost*, *8*(20):5784.

Fong, W., et al. (2020). Gut microbiota modulation: a novel strategy for prevention and treatment of colorectal cancer. *Oncogene*, *39*(26), 4925–4943.

Fu, Y., Lin, Q., Zhang, Z., Zhang, L. Therapeutic strategies for the costimulatory molecule OX40 in T-cell-mediated immunity. *Acta Pharm Sin B*. 2020;*10*(3):414–433. doi:10.1016/j.apsb.2019.08.010

Galluzzi, L., et al. (2014). Classification of current anticancer immunotherapies. *Oncotarget*, *5*(24):12472.

Ganesh, K. (2022). Optimizing immunotherapy for colorectal cancer. *Nat Rev Gastroenterol Hepatol, 19*(2), 93–94.

García-Martínez, E., et al. (2018). Trial watch: immunostimulation with recombinant cytokines for cancer therapy. *Oncoimmunology, 7*(6):e1433982.

Garg, M. B., et al. (2012). Predicting 5-fluorouracil toxicity in colorectal cancer patients from peripheral blood cell telomere length: a multivariate analysis. *Br J Cancer, 107*(9):1525–1533.

Ghazi, B., El Ghanmi, A., Kandoussi, S., Ghouzlani, A., Badou, A. (2022). CAR T-cells for colorectal cancer immunotherapy: Ready to go? *Front. Immunol, 13*: 978195.

Giannakis, M., et al. (2016). Genomic correlates of immune-cell infiltrates in colorectal carcinoma. *Cell Rep, 15*(4):857–865.

Golshani, G., Zhang, Y. (2020). Advances in immunotherapy for colorectal cancer: a review. *Therapeut Adv Gastroenterol, 13*:1756284820917527.

González-Perera, I., et al. (2017). 5-fluorouracil toxicity in the treatment of colon cancer associated with the genetic polymorphism 2846 A> G (rs67376798). *J Oncol Pharm Pract, 23*(5):396–398.

Hammarström, S. (1999). The carcinoembryonic antigen (CEA) family: structures, suggested functions and expression in normal and malignant tissues. *Seminars Cancer Biol, 9*(2):67–81.

He, R., et al. (2021). Progress in the application of immune checkpoint inhibitor-based immunotherapy for targeting different types of colorectal cancer. *Frontiers Oncol, 11*:764618.

Hou, W., et al. (2022). Predictive biomarkers of colon cancer immunotherapy: Present and future. *Frontiers Immunol, 13*:1032314.

Jia, W., et al. (2022). Colorectal cancer vaccines: The current scenario and future prospects. *Frontiers Immunol, 13*:942235.

Jiang, S., et al. (2019). Vaccinations for colorectal cancer: progress, strategies, and novel adjuvants. *Int J Molec Sci, 20*(14):3403.

Johdi, N. A., Sukor, N. F. (2020). Colorectal cancer immunotherapy: options and strategies. *Frontiers Immunol, 11*:1624.

Joller, N., Kuchroo, V. K. (2017). Tim-3, lag-3, and TIGIT. *Curr Top Microbiol Immunol 410*:127–156.

Jones, H. G., et al.(2017). Genetic and epigenetic intra-tumour heterogeneity in colorectal cancer. *World J. Surg, 41*:1375–1383.

Katz, S. C., et al. (2015). Phase I hepatic immunotherapy for metastases study of intra-arterial chimeric antigen receptor–modified T-cell therapy for CEA+ liver metastases. *Clin Cancer Res, 21*(14):3149–3159.

Kitano, A., et al. (2017). Tumour-infiltrating lymphocytes are correlated with higher expression levels of PD-1 and PD-L1 in early breast cancer. *ESMO Open, 2*(2):e000150.

Klein, O., et al. (2021). Evaluation of TMB as a predictive biomarker in patients with solid cancers treated with anti-PD-1/CTLA-4 combination immunotherapy. *Cancer Cell, 39*(5):592–593.

Kong, Y., et al. (2016). T-cell immunoglobulin and ITIM domain (TIGIT) associates with CD8+ T-cell exhaustion and poor clinical outcome in AML patients. *Clinical Cancer Res, 22*(12):3057–3066.

Le, D. T., et al. (2017). Mismatch repair deficiency predicts response of solid tumors to PD-1 blockade. *Science, 357*(6349):409–413.

Li, Q., et al. (2023). Radiotherapy/chemotherapy-immunotherapy for cancer management: from mechanisms to clinical implications. *Oxid Med Cell Longev* 7530794.

Li, S. K., Martin, A. (2016). Mismatch repair and colon cancer: mechanisms and therapies explored. *Trends Molec Med, 22*(4):274–289.

Li, Y., et al. (2016). Prognostic impact of programed cell death-1 (PD-1) and PD-ligand 1 (PD-L1) expression in cancer cells and tumor infiltrating lymphocytes in colorectal cancer. *Molec Cancer, 15*(1):1–15.

Lollini, P. L., et al. (2015). The promise of preventive cancer vaccines. *Vaccines, 3*(2):467–489.

Lonez, C., et al. (2017). Study protocol for THINK: a multinational open-label phase I study to assess the safety and clinical activity of multiple administrations of NKR-2 in patients with different metastatic tumour types. *BMJ open, 7*(11):e017075.

Losi, L., et al. (2005). Evolution of intratumoral genetic heterogeneity during colorectal cancer progression. *Carcinogenesis, 26*(5):916–922.

Manz, S. M., et al. (2021). Efficacy and side effects of immune checkpoint inhibitors in the treatment of colorectal cancer. *Therap Adv Gastroenterol, 14*:17562848211002018.

Marin-Acevedo, J. A., et al. (2018). Next generation of immune checkpoint therapy in cancer: new developments and challenges. *J Hematol Oncol, 11*:1–20.

Menter, D. G., et al. (2019). Back to the colorectal cancer consensus molecular subtype future. *Curr Gastroenterol Rep, 21*:1–12.

Miller, K. D., et al. (2019). Cancer treatment and survivorship statistics, 2019. *CA*, *69*(5):363–385.
Morse, M. A., et al. (2013). A randomized phase II study of immunization with dendritic cells modified with poxvectors encoding CEA and MUC1 compared with the same poxvectors plus GM-CSF for resected metastatic colorectal cancer. *Ann Surg*, *258*(6):879–886.
Nojadeh, J. N., et al. (2018). Microsatellite instability in colorectal cancer. *EXCLI Journal*, *17*:159.
Ogino, S., et al. (2009). Lymphocytic reaction to colorectal cancer is associated with longer survival, independent of lymph node count, microsatellite instability, and CpG island methylator phenotype. *Clin Cancer Res*, *15*(20):6412–6420.
Ogura, A., et al. (2019). Neoadjuvant (chemo) radiotherapy with total mesorectal excision only is not sufficient to prevent lateral local recurrence in enlarged nodes: results of the multicenter lateral node study of patients with low cT3/4 rectal cancer. *J Clin Oncol*, *37*(1):33.
Ott, P. A., et al. (2017). Combination immunotherapy: a road map. *J Immunother Cancer*, *5*:1–15.
Overman, M. J., et al. (2018a). Where we stand with immunotherapy in colorectal cancer: deficient mismatch repair, proficient mismatch repair, and toxicity management. *Am Soc Clin Oncol Educational Book*, *38*:239–247.
Overman, M. J., et al. (2018b). Durable clinical benefit with nivolumab plus ipilimumab in DNA mismatch repair-deficient/microsatellite instability-high metastatic colorectal cancer. *J Clin Oncol*, *36*(8):773–779.
Owens, J. A., et al. (2021). Lactobacillus rhamnosus GG orchestrates an antitumor immune response. *Cellular Molec Gastroenterol Hepatol*, *12*(4):1311–1327.
Parikh, A. R., et al. (2021). Radiation therapy enhances immunotherapy response in microsatellite stable colorectal and pancreatic adenocarcinoma in a phase II trial. *Nature Cancer*, *2*(11):1124–1135.
Park, S. C., et al. (2018). Phase II clinical trial to evaluate the efficacy of transanal endoscopic total mesorectal excision for rectal cancer. *Dis Colon Rectum*, *61*(5):554–560.
Parkhurst, M. R., et al. (2011). T cells targeting carcinoembryonic antigen can mediate regression of metastatic colorectal cancer but induce severe transient colitis. *Molec Ther*, *19*(3):620–626.
Popovic, A., et al. (2018). Emerging strategies for combination checkpoint modulators in cancer immunotherapy. *J Clin Investig*, *128*(8):3209–3218.
Poropatich, K., et al. (2020). OX40+ plasmacytoid dendritic cells in the tumor microenvironment promote antitumor immunity. *J Clin Investig*, *130*(7):3528–3542.
Rawla, P., et al. (2019). Epidemiology of colorectal cancer: incidence, mortality, survival, and risk factors. *Gastroenterol Rev*, *14*(2):89–103.
Rodriguez-Salas, N., et al. (2017). Clinical relevance of colorectal cancer molecular subtypes. *Critl Rev Oncol Hematol*, *109*:9–19.
Rosenberg, S. A., Restifo, N. P. (2015). Adoptive cell transfer as personalized immunotherapy for human cancer. *Science*, *348*(6230):62–68.
Rusch, T., et al. (2018). Immunotherapy as an option for cancer treatment. *Archiv Immunolog Therap Experim*, *66*:89–96.
Sanchez-Lockhart, M., et al (2014). T cell receptor signaling can directly enhance the avidity of CD28 ligand binding. *PLoS One*, *9*(2):e89263.
Schlom, J., et al. (2014). Therapeutic cancer vaccines. *Adv Cancer Res*, *121*:67–124.
Schmoll, H. J., et al. (2012). ESMO Consensus Guidelines for management of patients with colon and rectal cancer. a personalized approach to clinical decision making. *Ann Oncol*, *23*(10):2479–2516.
Schrock, A. B., et al. (2019). Tumor mutational burden is predictive of response to immune checkpoint inhibitors in MSI-high metastatic colorectal cancer. *Ann Oncol*, *30*(7):1096–1103.
Seidel, J. A., et al. (2018). Anti-PD-1 and anti-CTLA-4 therapies in cancer: mechanisms of action, efficacy, and limitations. *Frontiers Oncol*, *8*:86.
Shahnazari, M., et al. (2020). Therapeutic vaccines for colorectal cancer: The progress and future prospect. *Int Immunopharmacol*, *88*:106944.
Shan, J., et al. (2022). Mechanism and strategies of immunotherapy resistance in colorectal cancer. *Frontiers Immunol*, *13*:1016646.
Stoeva, M. K., et al. (2021). Butyrate-producing human gut symbiont, Clostridium butyricum, and its role in health and disease. *Gut Microb*, *13*(1):1907272.
Temraz, S., et al. (2019). Gut microbiome: a promising biomarker for immunotherapy in colorectal cancer. *Int J Molec Sci*, *20*(17):4155.
Thomas, S., Prendergast, G. C. (2016). Cancer vaccines: a brief overview.: *Meth Molec Biol*, *1403*:755–761.

Timmermann, B., et al. (2010). Somatic mutation profiles of MSI and MSS colorectal cancer identified by whole exome next generation sequencing and bioinformatics analysis. *PloS one, 5*(12):e15661.
Tran, E., et al. (2016). T-cell transfer therapy targeting mutant KRAS in cancer. *New Engl J Med, 375*(23):2255–2262.
Tumeh, P. C., et al. (2014). PD-1 blockade induces responses by inhibiting adaptive immune resistance. *Nature, 515*(7528):568–571.
Van Cutsem, E., et al. (2014). Metastatic colorectal cancer: ESMO Clinical Practice Guidelines for diagnosis, treatment and follow-up. *Ann Oncol, 25*:iii1–iii9.
Van Cutsem, E., et al. (2016). ESMO consensus guidelines for the management of patients with metastatic colorectal cancer. *Ann Oncol, 27*(8):1386–1422.
Viale, G., et al. (2017). Mismatch repair deficiency as a predictive biomarker for immunotherapy efficacy. *BioMed Res Int.* 4719194.
Wang, C., et al. (2022). Identification of hypoxia-related subtypes, establishment of prognostic models, and characteristics of tumor microenvironment infiltration in colon cancer. *Frontiers Genet, 13*:919389.
Wang, C., et al. (2017). Tailoring biomaterials for cancer immunotherapy: emerging trends and future outlook. *Adv Mat, 29*(29):1606036.
Weng, J., et al. (2022). Exploring immunotherapy in colorectal cancer. *J Hematol Oncol, 15*(1):1–28.
Xie, Y. H., et al. (2020). Comprehensive review of targeted therapy for colorectal cancer. *Signal Transduct Targeted Ther, 5*(1):22.
Xu, B., et al. (2015). Circulating and tumor-infiltrating Tim-3 in patients with colorectal cancer. *Oncotarget, 6*(24):20592.
Xu, M., et al. (2023). Anti-VEGF and anti-EGFR antibody therapy on T-Cell infiltration and TCR variation in metastatic colorectal cancer. *Anticancer Res 43*(2):613–620.
Yoshino, T., et al. (2018). Pan-Asian adapted ESMO consensus guidelines for the management of patients with metastatic colorectal cancer: a JSMO–ESMO initiative endorsed by CSCO, KACO, MOS, SSO and TOS. *Ann Oncol, 29*(1):44–70.
Zeng, Z., et al. (2021). Biomarkers in immunotherapy-based precision treatments of digestive system tumors. *Frontiers Oncol, 11*:650481.
Zhu, D., et al. (2021). Deubiquitinating enzyme OTUB1 promotes cancer cell immunosuppression via preventing ER-associated degradation of immune checkpoint protein PD-L1. *Cell Death Differentiat, 28*(6):1773–1789.
Zou, Q., et al. (2021). DNA methylation-based signature of CD8+ tumor-infiltrating lymphocytes enables evaluation of immune response and prognosis in colorectal cancer. *J Immunother Cancer, 9*(9):e002671.

10 Personalized Medicine and Precision Oncology in Colorectal Cancer

Roy Rillera Marzo

10.1 INTRODUCTION

10.1.1 Overview of Colorectal Cancer (CRC)

10.1.1.1 Epidemiology and Public Health Impact

Colorectal cancer (CRC) is a significant global health concern, ranking as the third most commonly diagnosed cancer in men and the second in women, accounting for 10% of all cancer cases worldwide (Bray et al., 2018). Developed countries like the US, Canada, Australia, and Western Europe have higher CRC incidence, largely due to lifestyle factors such as high consumption of red and processed meats, low fiber intake, physical inactivity, and obesity (Arnold et al., 2017). While low- and middle-income countries have historically had lower CRC rates, these are rising due to urbanization and Western lifestyle adoption (Bishehsari et al., 2014).

CRC's mortality burden is also substantial. In 2020, CRC caused approximately 880,000 deaths, making it the second leading cause of cancer-related deaths globally (Sung et al., 2021). Despite advancements in screening and early detection, CRC remains a significant cause of morbidity and mortality, particularly in regions with limited access to screening programs. Early-stage CRC is often asymptomatic, leading to delayed diagnosis and treatment, especially in resource-poor settings (Haggar & Boushey, 2009).

Screening methods like colonoscopy and fecal occult blood testing have proven effective in reducing CRC incidence and mortality by enabling early detection of precancerous lesions (Brenner et al., 2014). However, the success of these programs relies on widespread accessibility, which remains a challenge in many low-resource regions (Zheng et al., 2023). Thus, public health strategies for CRC must include prevention (healthy diets and physical activity) and improved access to screening and early detection programs (Kolligs, 2016).

10.1.1.2 Current Treatment Paradigms

CRC treatment has evolved significantly in recent decades, with surgery remaining the primary curative option for early-stage disease. Surgical resection, often followed by adjuvant chemotherapy, remains the standard of care for localized CRC (Benson et al., 2018). Adjuvant chemotherapy regimens such as FOLFOX (5-fluorouracil, leucovorin, and oxaliplatin) aim to reduce the risk of recurrence by eradicating residual microscopic disease (André et al., 2004).

For metastatic CRC, systemic chemotherapy, such as FOLFIRI (5-fluorouracil, leucovorin, and irinotecan), remains a primary treatment option. However, the prognosis for metastatic CRC is poor, with a five-year survival rate of about 14% (Siegel et al., 2020). Recent developments in targeted therapies have improved outcomes for metastatic disease. Monoclonal antibodies targeting pathways involved in tumor growth, such as EGFR and VEGF, have shown significant clinical benefits. The addition of cetuximab or bevacizumab to chemotherapy has extended survival for patients with advanced CRC (Hurwitz et al., 2004; Van Cutsem et al., 2009).

Despite these advances, CRC is characterized by genetic and molecular heterogeneity, contributing to treatment resistance and variability in response. This highlights the need for more personalized approaches to treatment.

10.1.2 Evolution of Oncology: From Traditional to Precision Medicine

10.1.2.1 Historical Perspective on Cancer Treatment

Cancer treatment has progressed from generalized approaches, like surgery and radiation, to more targeted therapies. In the early 20th century, cancer treatment was largely limited to surgery, with radiation therapy emerging later as an adjunct for local control (Delaney et al., 2005). The mid-20th century saw the development of chemotherapy, offering a systemic approach to target rapidly dividing cells (Chabner & Roberts, 2005). Chemotherapeutic agents like 5-fluorouracil, still a cornerstone in CRC treatment, were developed during this period, although their nonspecific nature led to substantial toxicity and varying efficacy (Longley et al., 2003).

10.1.2.2 Shift Towards Personalized and Precision Approaches

With advances in molecular biology and genomics, personalized and precision medicine emerged. Personalized medicine tailors treatment to individual characteristics, including genetic,

DOI: 10.1201/9781032704517-10

environmental, and lifestyle factors (Garraway et al., 2013). Precision oncology, a subset of personalized medicine, targets specific genetic mutations and molecular profiles in tumors (Collins & Varmus, 2015). This approach has led to the development of targeted therapies, such as tyrosine kinase inhibitors and monoclonal antibodies, demonstrating significant efficacy in cancers with specific molecular alterations, including CRC (Reifenberger et al., 2017).

The introduction of precision oncology into clinical practice marks a paradigm shift in cancer treatment. For example, RAS mutation status now guides the use of EGFR-targeted therapies in CRC (Van Cutsem et al., 2015). Additionally, the discovery of biomarkers such as microsatellite instability (MSI) and mismatch repair (MMR) deficiency has expanded treatment options, including immunotherapy, for CRC patients (Le et al., 2017).

As molecular complexities of CRC continue to be uncovered, precision medicine promises to improve patient outcomes, reduce treatment toxicities, and transform the future of cancer care.

10.2 THE BIOLOGY OF COLORECTAL CANCER

CRC is driven by numerous genetic and molecular alterations. Understanding its biology is key to developing effective treatments, particularly in the realm of precision oncology. This section explores the main molecular pathways, key genetic mutations, the role of MSI and MMR status, and tumor heterogeneity.

10.2.1 Molecular Pathways Involved in CRC

10.2.1.1 Key Genetic Mutations

CRC development is a multistep process, progressing from normal epithelium to carcinoma, driven by genetic mutations.

- **KRAS:** Mutations in the KRAS gene, part of the RAS/MAPK signaling pathway, are found in 40% of CRC cases (Huang et al., 2021). These mutations cause continuous activation of KRAS, leading to uncontrolled cell growth and resistance to anti-EGFR therapies like cetuximab (Wu & Park, 2015).
- **NRAS:** NRAS mutations, found in 3–5% of CRC cases, also impact the RAS pathway and are associated with poor prognosis and treatment resistance (Yaeger et al., 2018).
- **BRAF:** The V600E mutation in BRAF occurs in about 10–15% of CRC cases and is linked to aggressive tumors. BRAF-mutant CRCs are resistant to standard therapies, but newer combinations of BRAF inhibitors show promise (Kopetz et al., 2019).
- **APC:** Mutations in APC, a tumor suppressor gene regulating the Wnt signaling pathway, are seen in around 80% of CRC cases and are an early event in CRC development (Hankey et al., 2018).

10.2.1.2 The Role of Microsatellite Instability (MSI) and Mismatch Repair (MMR) Status

MSI is caused by a loss of DNA mismatch repair (MMR) activity, leading to genetic hypermutability (Cortes-Ciriano et al., 2017).

- **MSI-H:** High MSI (MSI-H) occurs in 15% of CRC cases and is more common in right-sided colon cancers. MSI-H tumors tend to have a better prognosis in early stages but are less responsive to chemotherapy. However, MSI-H status is predictive of success with immune checkpoint inhibitors like pembrolizumab (André et al., 2020).
- **MMR deficiency:** Deficient MMR (dMMR) results from mutations in MMR genes such as MLH1 and MSH2. dMMR tumors can arise sporadically or as part of Lynch syndrome, an inherited cancer syndrome (Li et al., 2020). dMMR status is crucial for treatment planning, guiding the use of immunotherapy.

10.2.2 Tumor Heterogeneity

10.2.2.1 Inter- and Intra-Tumor Heterogeneity

CRC is highly heterogeneous, with genetic, epigenetic, and phenotypic variations both within and between tumors (Reifenberger et al., 2017).

- **Intra-tumor heterogeneity:** Different regions of a tumor may harbor distinct mutations and phenotypes, posing a challenge for treatment. Subclones within the tumor can evolve, leading to resistance to therapy and recurrence (Siravegna et al., 2015).

- **Inter-tumor heterogeneity:** CRC also displays substantial variability between patients' tumors, necessitating personalized treatment approaches. Molecular subtypes, such as those defined by KRAS or BRAF mutations, guide the use of targeted therapies (Yaeger et al., 2018).

10.2.2.2 Implications for Treatment Resistance and Disease Progression

Tumors with high heterogeneity are more likely to develop resistance to therapies, as resistant subclones emerge under treatment pressure (Saoudi González et al., 2023).

- Mutations in **KRAS**, **NRAS**, and **BRAF** confer resistance to anti-EGFR therapies, necessitating alternative treatment approaches, such as combining BRAF inhibitors, MEK inhibitors, and chemotherapy (Kopetz et al., 2019).

Understanding tumor heterogeneity is essential for improving CRC treatment outcomes. Advances in genomic profiling and liquid biopsy technologies help personalize treatment strategies, addressing the complexity of CRC.

10.3 PRECISION ONCOLOGY IN COLORECTAL CANCER

Precision oncology represents a transformative approach in treating CRC, integrating molecular biology, genomics, and advanced diagnostics to tailor treatments specifically to each patient's tumor. This section explores precision oncology concepts, the role of molecular profiling, and biomarkers guiding CRC treatment strategies.

10.3.1 Defining Precision Oncology

10.3.1.1 Concepts of Personalized and Precision Medicine

Personalized and precision medicine are often used interchangeably, but they are distinct concepts. Personalized medicine broadly refers to tailoring treatments to individual patient characteristics, including genetics, environment, and lifestyle. In oncology, this means customizing cancer treatments based on the patient's and the tumor's specific traits.

Precision oncology, a subset of personalized medicine, focuses on molecular and genetic tumor profiling to identify specific mutations, gene expressions, and biomarkers that can be targeted with therapies (Marquart et al., 2018). Recognizing that cancer is a heterogeneous disease, precision oncology aims to select the most effective and least toxic treatments for each patient (Tsimberidou et al., 2020).

CRC's genetic and molecular diversity makes precision oncology particularly impactful. CRC can be stratified into molecular subtypes, each with distinct biological traits and responses to therapy. This stratification allows for more accurate targeting of therapies, improving outcomes and minimizing unnecessary treatments (Reifenberger et al., 2017).

10.3.1.2 Integration of Genomic, Proteomic, and Other Omics Data

Precision oncology's success relies on integrating various "omics" technologies, including genomics, proteomics, transcriptomics, and metabolomics. Together, these technologies provide a comprehensive view of CRC's molecular landscape (Hasin et al., 2017). Genomic data, derived from next-generation sequencing (NGS), helps identify mutations and other genetic alterations driving tumor growth, critical for selecting targeted therapies (Ooki et al., 2021).

Proteomics, which studies protein behavior, complements genomics by offering insights into the tumor's functional state. Proteomic analyses identify aberrant protein expressions that serve as therapeutic targets or biomarkers for treatment response (Vasaikar et al., 2018). Similarly, transcriptomics examines gene expression profiles, identifying dysregulated pathways that can be therapeutically modulated.

Integrating multi-omics data provides a more nuanced understanding of tumor biology, enabling the identification of actionable targets and effective treatment strategies (Raufaste-Cazavieille et al., 2022). Advances in bioinformatics and computational biology further enhance oncology precision by analyzing the vast amounts of data generated from these technologies.

10.3.2 Molecular Profiling and Biomarkers

Molecular profiling matches patients with genotype-specific drug treatments, a complex process involving several stages. It begins with patient selection, where logistical and medical challenges can hinder participation. Improvements in patient consent, selection processes, and sample collection (e.g., using liquid biopsies) are suggested as solutions.

In the laboratory, technical issues with NGS and other assays can be addressed by advancing detection technologies. Decision-making is often complicated by limited access to molecular tumor boards (MTBs), though virtual MTBs offer a potential solution. Expanding access to precision medicine trials and simplifying trial matching can mitigate barriers to clinical trial participation.

10.3.2.1 The Role of Biomarkers in CRC

Biomarkers are biological molecules that reflect disease presence and progression, predict treatment response, and help guide CRC treatment decisions.

- **EGFR (epidermal growth factor receptor):** EGFR promotes cell proliferation and survival. Anti-EGFR therapies like cetuximab are effective in metastatic CRC patients who lack KRAS, NRAS, and BRAF mutations, which confer resistance to these therapies (Wu & Park, 2015).
- **VEGF (vascular endothelial growth factor):** VEGF drives angiogenesis, fueling tumor growth. Bevacizumab, an anti-VEGF monoclonal antibody, inhibits angiogenesis and slows CRC progression when combined with chemotherapy (Mazeda et al., 2020).
- **HER2 (human epidermal growth factor receptor 2):** Overexpressed in a subset of CRC cases, HER2-positive tumors, particularly in those with high microsatellite instability (MSI-H), can be targeted with therapies like trastuzumab (Horimoto et al., 2022).

10.3.2.2 Predictive and Prognostic Biomarkers

Biomarkers in CRC can be predictive (identifying which patients benefit from specific therapies) or prognostic (providing information on disease outcomes, independent of treatment).

- **Predictive biomarkers:** Mutations in KRAS, NRAS, and BRAF are well-established predictors of response to anti-EGFR therapies. For example, patients with wild-type KRAS/NRAS tumors are more likely to respond, while those with BRAF mutations may need alternative therapies (Di Nicolantonio et al., 2021). MSI-H status is another predictive biomarker, indicating a higher likelihood of response to immunotherapy (Overman et al., 2018).
- **Prognostic biomarkers:** MSI-H tumors generally have a better prognosis in early-stage disease but fare worse in advanced stages (Kawakami et al., 2015). Circulating tumor DNA (ctDNA) is also an emerging prognostic biomarker, with higher levels indicating a higher risk of recurrence and poorer survival (van Rees et al., 2023).

10.3.2.3 Techniques for Molecular Profiling

Advances in molecular profiling technologies have been instrumental in the implementation of precision oncology in CRC. These techniques allow for the comprehensive analysis of the genetic and molecular features of tumors, guiding the selection of targeted therapies.

- **Next-generation sequencing (NGS):** NGS is a powerful tool that enables the rapid and high-throughput sequencing of DNA and RNA, allowing for the identification of mutations, gene fusions, and other genetic alterations in CRC (Satam et al., 2023). NGS panels specific to CRC can detect a wide range of clinically relevant mutations, informing treatment decisions and identifying patients who may benefit from targeted therapies or clinical trials.
- **Liquid biopsy:** Liquid biopsy is a minimally invasive technique that analyzes ctDNA, circulating tumor cells (CTCs), and other tumor-derived materials in the blood (Siravegna et al., 2015). Liquid biopsies provide real-time insights into the molecular status of the tumor, allowing for the monitoring of treatment response and the detection of emerging resistance mechanisms. This technique is particularly valuable for patients with advanced CRC, where repeated tissue biopsies may not be feasible.
- **Immunohistochemistry (IHC) and in situ hybridization (ISH):** IHC and ISH are traditional techniques used to assess protein expression and gene amplification in tumor tissues. These methods are commonly used to evaluate biomarkers such as EGFR, HER2, and MMR proteins, providing critical information for the selection of targeted therapies (Vasaikar et al., 2018).

The integration of these molecular profiling techniques into clinical practice has revolutionized the treatment of CRC, enabling the delivery of personalized therapies that improve patient outcomes. As these technologies continue to evolve, they will further enhance the precision of oncology treatments, leading to more effective and tailored care for patients with CRC.

10.4 TAILORED THERAPEUTIC APPROACHES IN CRC

The treatment landscape for CRC has significantly evolved due to advances in molecular biology, genomics, and immunology. These developments have allowed therapeutic approaches to be tailored to the genetic and molecular characteristics of each patient's tumor, improving treatment efficacy and minimizing toxicity. This section explores tailored therapeutic approaches in CRC, including targeted therapies, immunotherapy, chemotherapy, radiotherapy, and combination therapies.

10.4.1 Targeted Therapies

Targeted therapies have revolutionized CRC treatment by precisely targeting specific molecular pathways responsible for tumor growth and progression.

10.4.1.1 Anti-EGFR Therapies

The epidermal growth factor receptor (EGFR) is a protein that triggers cell proliferation and survival when activated. Overexpression or aberrant activation of EGFR is common in CRC. Anti-EGFR therapies like cetuximab and panitumumab are monoclonal antibodies that prevent EGFR activation. These therapies are effective in metastatic CRC (mCRC), particularly in patients with wild-type KRAS and NRAS tumors, as mutations in these genes cause resistance to anti-EGFR treatments (Di Nicolantonio et al., 2021). In patients with wild-type KRAS and NRAS, anti-EGFR therapies improve progression-free survival (PFS) and overall survival (OS) (Wu & Park, 2015).

10.4.1.2 Anti-VEGF Therapies

Vascular endothelial growth factor (VEGF) promotes angiogenesis, which provides tumors with nutrients and oxygen. Bevacizumab, an anti-VEGF monoclonal antibody, prevents VEGF from binding to its receptors, inhibiting tumor growth (Mazeda et al., 2020). Bevacizumab, combined with chemotherapy, improves PFS and OS in mCRC (Bennouna et al., 2013). Other anti-angiogenic agents like aflibercept and ramucirumab have also shown efficacy by targeting different components of the VEGF signaling pathway.

10.4.1.3 Other Targeted Agents

Other targeted agents, such as BRAF inhibitors, have been developed for CRC patients with specific molecular alterations. BRAF V600E-mutant mCRC is associated with poor prognosis and resistance to conventional treatments. Combining BRAF inhibitors (e.g., encorafenib), MEK inhibitors (e.g., binimetinib), and anti-EGFR therapies (e.g., cetuximab) has shown improved response rates (Van Cutsem et al., 2019). HER2-targeted therapies like trastuzumab and pertuzumab are effective in HER2-positive mCRC, particularly in patients who have progressed after standard treatments (Horimoto et al., 2022).

Targeted therapies offer significant advancements in treating CRC, but their success depends on accurate molecular profiling to identify which patients will benefit.

10.4.2 Immunotherapy

Immunotherapy has emerged as a powerful approach in CRC, leveraging the body's immune system to eliminate cancer cells. Immune checkpoint inhibitors have transformed treatment, particularly for patients with microsatellite instability-high (MSI-H) or mismatch repair-deficient (dMMR) tumors.

10.4.2.1 Role of Immune Checkpoint Inhibitors in CRC

Immune checkpoint inhibitors, such as pembrolizumab and nivolumab, block molecules like PD-1 and CTLA-4, which tumors use to evade immune responses (Overman et al., 2018). These therapies have shown significant efficacy in patients with MSI-H or dMMR mCRC, which have a high mutational burden and increased neoantigen load, making them more susceptible to immune attack (André et al., 2020). Clinical trials have demonstrated durable responses and improved survival in MSI-H/dMMR mCRC patients treated with these inhibitors (Overman et al., 2018).

However, the benefits of immune checkpoint inhibitors are limited to MSI-H/dMMR tumors, which account for only 15% of CRC cases. Microsatellite stable (MSS) CRCs show limited response to immunotherapy, highlighting the need for further research on predictive biomarkers in MSS cases (Ooki et al., 2021).

10.4.2.2 MSI-High and dMMR as Biomarkers for Immunotherapy

MSI-H and dMMR status have become key biomarkers for predicting immunotherapy response in CRC. These tumors result from defects in the DNA mismatch repair system, which generates mutations and neoantigens recognized by the immune system (Li et al., 2020). Identifying MSI-H or dMMR status is crucial for guiding treatment decisions regarding immune checkpoint inhibitors. For MSI-H/dMMR mCRC patients, immunotherapy provides a promising alternative to conventional chemotherapy (André et al., 2020).

10.4.3 Chemotherapy and Radiotherapy

While targeted therapies and immunotherapy have broadened CRC treatment options, chemotherapy and radiotherapy remain essential, particularly for localized and advanced disease. Tailoring these conventional treatments based on molecular profiles enhances their effectiveness.

10.4.4 Personalization of Conventional Treatments

Chemotherapy, involving cytotoxic agents that kill rapidly dividing cells, has long been a cornerstone of CRC treatment. Standard chemotherapy regimens like FOLFOX and FOLFIRI are used in both adjuvant and metastatic settings. However, response to chemotherapy varies depending on molecular characteristics (Benson et al., 2021). For instance, patients with KRAS or NRAS mutations may respond better to chemotherapy combined with anti-VEGF agents like bevacizumab, while patients with BRAF V600E mutations may benefit from a combination of chemotherapy, targeted therapy, and anti-EGFR agents (Kopetz et al., 2019).

Radiotherapy, primarily used for rectal cancer, involves high-energy radiation to destroy cancer cells. Preoperative radiotherapy, combined with chemotherapy, is often used to shrink tumors before surgery. Advances in techniques like intensity-modulated radiotherapy (IMRT) and stereotactic body radiotherapy (SBRT) have improved precision, minimizing damage to healthy tissues (Raufaste-Cazavieille et al., 2022). Personalized radiotherapy plans based on tumor location and molecular profile further enhance outcomes.

10.4.5 Combination Therapies

Combination therapies, which use multiple treatment modalities concurrently or sequentially, are key to CRC management. Combining targeted therapy, immunotherapy, and chemotherapy has shown promise in overcoming resistance mechanisms and improving outcomes.

10.4.5.1 Synergistic Approaches

Combination therapies aim to target multiple pathways simultaneously, reducing resistance and improving treatment efficacy (Di Nicolantonio et al., 2021). For example, BRAF inhibitors, MEK inhibitors, and anti-EGFR therapies have shown improved response rates in patients with BRAF V600E-mutant mCRC (Kopetz et al., 2019). Similarly, combining chemotherapy with bevacizumab or cetuximab enhances the efficacy of these targeted therapies by disrupting angiogenesis and EGFR signaling.

Another promising approach is combining immunotherapy with targeted therapy or chemotherapy. For instance, combining immune checkpoint inhibitors with anti-angiogenic agents may improve immune responses in MSS CRC, which typically shows limited sensitivity to immunotherapy alone (Ooki et al., 2021). These strategies are being evaluated in clinical trials and may expand the use of immunotherapy to a broader CRC patient population.

The future of CRC treatment lies in the continued development and optimization of combination therapies. As we deepen our understanding of the molecular underpinnings of CRC and develop new therapeutic agents, designing personalized combination regimens will be critical for improving patient outcomes.

10.5 CHALLENGES AND OPPORTUNITIES IN PRECISION MEDICINE

Precision medicine in CRC represents a significant advancement in oncology, offering the potential for highly individualized treatments that improve patient outcomes. However, the implementation of precision medicine is not without its challenges. This section explores the barriers to the widespread adoption of precision oncology, including economic and accessibility issues, variability in patient response and toxicity, and ethical considerations related to genetic testing and data privacy. Additionally, we will discuss the opportunities presented by advances in technology and research, such as the role of artificial intelligence and machine learning, and future trends in molecular diagnostics and treatment strategies.

10.5.1 Barriers to Implementation

10.5.1.1 Economic and Accessibility Issues

One of the significant challenges in implementing precision medicine in CRC is the economic burden associated with advanced molecular diagnostics and targeted therapies. The costs of genetic testing, NGS, and molecular profiling can be prohibitively high, limiting access, particularly in low- and middle-income countries (Yabroff, et al., 2013). Moreover, the high cost of drug development is passed on to patients and healthcare systems through elevated drug prices.

In addition to direct costs, there are infrastructure challenges related to specialized laboratories, trained personnel, and data management systems required for precision medicine. The unequal distribution of these resources exacerbates disparities, with underserved areas less likely to benefit from cutting-edge cancer treatments.

Addressing these challenges requires collaboration between governments, healthcare providers, and pharmaceutical companies to lower the costs of diagnostics and therapies. Investment in healthcare infrastructure and workforce training is also crucial to ensure all patients, regardless of location or socioeconomic status, can access precision oncology.

10.5.1.2 Variability in Patient Response and Toxicity

Despite advances in molecular profiling, not all patients respond equally to precision treatments, and some may experience significant toxicities (Di Nicolantonio et al., 2021). This variability stems from tumor heterogeneity, genetic diversity, and the complexity of cancer biology.

Tumor heterogeneity means different regions of the same tumor may have distinct molecular characteristics, causing variations in treatment response. Additionally, genetic polymorphisms in drug-metabolizing enzymes influence how patients respond to therapies, further contributing to outcome variability.

Toxicity is another concern, with targeted therapies and immunotherapies sometimes triggering adverse events, particularly in patients with pre-existing conditions (Zugazagoitia et al., 2020). Biomarkers that predict both efficacy and toxicity are needed to minimize adverse effects and optimize treatment.

A more personalized approach to precision medicine is required, which considers the patient's genetic background, comorbidities, and other factors. Developing comprehensive biomarker panels that predict response and toxicity, along with adaptive clinical trial designs, will refine treatment strategies and improve patient outcomes.

10.5.1.3 Ethical Considerations in Genetic Testing and Data Privacy

Genetic testing and molecular profiling in precision medicine raise ethical concerns related to data privacy, informed consent, and the potential for genetic discrimination. Since precision oncology relies on collecting and analyzing genetic data, protecting this information is paramount (Ahmed, et al., 2023).

A key concern is the risk of unauthorized access to genetic data, which could be used for discriminatory purposes by employers or insurers. Robust legal frameworks are needed to protect individuals' data and prevent misuse (Tsamados et al., 2022).

Informed consent is also critical, ensuring patients fully understand the risks, benefits, and implications of genetic testing. Patients must be aware of the limitations of genetic tests and the potential psychological impact of discovering predispositions to diseases. Addressing these ethical challenges requires collaboration between healthcare providers, policymakers, and patient advocacy groups.

10.5.2 Advances in Technology and Research

10.5.2.1 Role of Artificial Intelligence and Machine Learning in Precision Oncology

Advances in artificial intelligence (AI) and machine learning (ML) are poised to transform precision oncology. These technologies can enhance molecular diagnostics, predict treatment outcomes, and identify therapeutic targets (Topol, 2019).

AI and ML algorithms can analyze vast amounts of genomic and clinical data, identifying patterns and correlations not easily apparent to humans. For instance, AI platforms can integrate multi-omics data to predict patient responses to therapies, enabling more personalized treatment plans (Esteva et al., 2019). AI can also assist in interpreting complex genomic data, improving molecular profiling accuracy.

Machine learning models are being developed to predict adverse events, allowing for proactive management of side effects and customized treatment regimens (Kourou et al., 2015). Additionally, AI can accelerate drug discovery by screening compounds and simulating interactions with molecular targets, reducing costs and speeding up the development process (Mak & Pichika, 2019).

While AI and ML offer significant opportunities, challenges include the need for large, high-quality datasets, the risk of algorithmic bias, and ethical concerns related to AI use in healthcare. Ensuring the transparency and fairness of AI-driven decisions is essential.

10.5.2.2 Future Trends in Molecular Diagnostics and Treatment Strategies

The future of precision oncology in CRC will be shaped by ongoing advances in molecular diagnostics and novel treatment strategies. Liquid biopsy is one such trend, offering a non-invasive method to monitor tumor dynamics in real-time. Liquid biopsy detects circulating tumor DNA (ctDNA) and circulating tumor cells (CTCs), allowing for early relapse detection and treatment adjustments (Siravegna et al., 2015).

Another emerging trend is the development of multi-targeted therapies that inhibit multiple oncogenic pathways simultaneously. This approach aims to overcome resistance from single-agent therapies by addressing tumor heterogeneity (Doostmohammadi et al., 2024).

Immunotherapy research is also expanding the applicability of precision medicine in CRC. Efforts are underway to identify biomarkers that predict immunotherapy response in microsatellite stable (MSS) tumors, which currently show limited sensitivity to immune checkpoint inhibitors (Overman et al., 2018). Combination therapies that enhance immune responses, such as using immune checkpoint inhibitors with anti-angiogenic agents, are being explored.

The integration of multi-omics data into clinical practice will further drive the evolution of precision oncology. By combining genomics, transcriptomics, proteomics, and metabolomics, clinicians can gain a more comprehensive understanding of tumor biology, leading to more precise and personalized treatments (Raufaste-Cazavieille et al., 2022).

10.6 THE PUBLIC HEALTH PERSPECTIVE

The integration of precision medicine into the treatment of CRC has significant implications at the population level. While personalized treatments have the potential to improve outcomes for individual patients, their broader impact on public health, healthcare access, and policy requires careful consideration. This section explores the population-level implications of precision medicine in CRC, the strategies to address disparities in access, and the role of health policy and regulation in facilitating the adoption of personalized medicine approaches.

10.6.1 Population-Level Implications

10.6.1.1 Impact of Precision Medicine on CRC Screening and Early Detection

Precision medicine has the potential to revolutionize CRC screening and early detection by identifying individuals at higher risk based on their genetic and molecular profiles. Traditional screening methods, such as colonoscopy and fecal occult blood tests, are generally applied uniformly across populations. However, precision medicine enables a more tailored approach, where screening strategies can be adjusted based on an individual's genetic predisposition, family history, and other risk factors (Hunter et al., 2017).

For example, individuals with a known hereditary cancer syndrome, such as Lynch syndrome, can benefit from more frequent and earlier screening, leading to the detection of CRC at a much earlier and more treatable stage (Vasen et al., 2013). Moreover, the identification of specific biomarkers associated with early tumor development could lead to the development of new non-invasive screening tests, such as blood-based assays for circulating tumor DNA (ctDNA), which could be used to detect CRC at an early stage in at-risk populations (Siravegna et al., 2015).

However, the implementation of precision medicine in screening programs also presents challenges. The widespread adoption of genetic testing and molecular profiling requires substantial resources, including the availability of advanced laboratory facilities and trained personnel. Additionally, the cost of these technologies may limit their accessibility, particularly in low-resource settings. To maximize the public health benefits of precision medicine, it is essential to develop strategies that ensure equitable access to these advanced screening and early detection tools (Yabroff, et al., 2013).

10.6.1.2 Public Health Strategies to Reduce Disparities in Access to Personalized Treatments

One of the primary concerns with the adoption of precision medicine is the potential to exacerbate existing disparities in healthcare access. Personalized treatments, including targeted therapies and immunotherapies, are often expensive and may not be readily available to all populations, particularly those in low- and middle-income countries. Disparities in access to precision medicine can lead to unequal outcomes, where only those with the financial means or access to specialized care can benefit from the latest advancements in oncology.

To address these disparities, public health strategies must focus on ensuring that all patients, regardless of socioeconomic status or geographic location, have access to personalized treatments. This can be achieved through several approaches:

i. **Government and non-governmental initiatives:** Governments and non-governmental organizations (NGOs) can play a crucial role in subsidizing the cost of genetic testing and targeted therapies, making them more affordable for underserved populations. Public health campaigns can also raise awareness about the availability of precision medicine and its benefits, encouraging at-risk individuals to seek out these treatments (Goldstein et al., 2017).

ii. **Telemedicine and digital health solutions:** The use of telemedicine and digital health platforms can help bridge the gap in access to precision medicine by connecting patients in remote or underserved areas with specialists in oncology and genetics. These platforms can facilitate remote consultations, genetic counseling, and the delivery of personalized treatment plans, ensuring that patients receive the care they need regardless of their location (Keesara et al., 2020).

iii. **Integration of precision medicine into national health systems:** Integrating precision medicine into national health systems can ensure that personalized treatments are part of standard care, rather than an option available only to a select few. This integration requires policy support, investment in healthcare infrastructure, and the training of healthcare professionals to incorporate precision medicine into routine clinical practice (Ginsburg & Phillips, 2018).

By implementing these strategies, public health systems can work towards reducing disparities in access to personalized treatments, ensuring that the benefits of precision medicine are shared equitably across populations.

10.6.2 Policy and Regulation

10.6.2.1 The Role of Health Policy in Facilitating Access to Precision Oncology

Health policy plays a critical role in facilitating access to precision oncology by setting the framework for the integration of personalized medicine into healthcare systems. Policymakers must address several key areas to ensure that precision medicine is accessible, affordable, and effectively implemented at the population level.

i. **Funding and reimbursement:** One of the primary barriers to the widespread adoption of precision medicine is the cost. Health policies must address the funding and reimbursement of genetic testing, molecular diagnostics, and targeted therapies to make these interventions affordable for all patients. This may involve expanding insurance coverage, providing subsidies for low-income patients, and ensuring that public healthcare systems include precision medicine in their standard care protocols (Johnson et al., 2021).

ii. **Regulatory frameworks:** The development and approval of precision medicine treatments require robust regulatory frameworks that ensure the safety, efficacy, and quality of these interventions. Regulatory bodies must establish clear guidelines for the approval of new genetic tests, biomarkers, and targeted therapies, balancing the need for innovation with patient safety. Additionally, regulations must address the ethical implications of genetic testing, including informed consent, data privacy, and the potential for genetic discrimination (Tsamados, et al., 2022).

iii. **Health technology assessment (HTA):** HTA is a systematic evaluation of the social, economic, organizational, and ethical issues of health technology. It is critical in determining the value of precision medicine interventions and guiding decisions about their inclusion in public health programs. Policymakers should support the integration of HTA in the decision-making process to ensure that precision medicine approaches are cost-effective and deliver tangible benefits to patients (EUnetHTA, 2020).

10.6.2.2 Regulatory Considerations for Personalized Medicine Approaches

Personalized medicine presents unique regulatory challenges that differ from those of traditional medical approaches. As personalized treatments are often based on individual genetic or molecular profiles, the regulation of these therapies must consider the variability in patient responses and the potential for off-label use.

i. **Adaptive regulatory pathways:** Traditional regulatory pathways may not be well-suited to the rapidly evolving field of precision medicine. Adaptive regulatory pathways, which allow for the iterative development and approval of therapies based on emerging data, can accelerate the availability of new treatments while ensuring ongoing monitoring of their safety and efficacy. These pathways can also facilitate the use of real-world evidence in regulatory decision-making, allowing regulators to respond more quickly to new scientific discoveries (Lievevrouw, et al., 2022).

ii. **Companion diagnostics:** Many precision medicine therapies are paired with companion diagnostics that identify patients most likely to benefit from the treatment. Regulatory bodies must establish clear guidelines for the co-development and approval of these diagnostics alongside their corresponding therapies. This includes ensuring that companion diagnostics are accurate, reliable, and validated for their intended use (Cooper & Chen, 2022).

iii. **Ethical and legal considerations:** The use of genetic information in personalized medicine raises significant ethical and legal concerns, particularly related to data privacy, informed consent, and the potential for genetic discrimination. Regulatory frameworks must protect patients' genetic information, ensuring that it is used only for its intended purposes and that patients are fully informed about the implications of genetic testing. Additionally, laws must be in place to prevent discrimination based on genetic information in employment, insurance, and other areas (Ahmed, et al., 2023).

In conclusion, while precision medicine holds great promise for improving CRC outcomes, its successful implementation at the population level requires careful consideration of public health strategies, health policy, and regulatory frameworks. By addressing the challenges and opportunities associated with precision medicine, policymakers can help ensure that these advancements benefit all patients, leading to more equitable and effective healthcare.

10.7 CONCLUSION

Precision oncology has transformed CRC treatment by enabling personalized therapies through advances in molecular biology, genomics, and immunotherapy. Targeted therapies like anti-EGFR and anti-VEGF, along with immunotherapy for MSI-H/dMMR CRC, have shown promise. However, challenges such as economic barriers, variability in response, and ethical issues with genetic testing must be addressed. Future directions include the integration of AI, machine learning, liquid biopsies, and combination therapies for adaptive treatment. Multidisciplinary collaboration is key to overcoming these hurdles and ensuring equitable access to precision medicine for CRC patients.

REFERENCES

Ahmed, L., et al. (2023). Patients' perspectives related to ethical issues and risks in precision medicine: a systematic review. *Frontiers Med, 10*:1215663.

André, T., et al. (2004). Oxaliplatin, fluorouracil, and leucovorin as adjuvant treatment for colon cancer. *New Engl J Med, 350*(23):2343–2351.

André, T., et al. (2020). Pembrolizumab in microsatellite-instability-high advanced colorectal cancer. *New Engl J Med, 383*(23):2207–2218.

Arnold, M., et al. (2017). Global patterns and trends in colorectal cancer incidence and mortality. *Gut, 66*(4):683–691.

Bennouna, J., et al. (2013). Continuation of bevacizumab after first progression in metastatic colorectal cancer (ML18147): A randomised phase 3 trial. *Lancet Oncol, 14*(1):29–37.

Benson, A. B., et al. (2018). Colon cancer, version 2.2018, NCCN clinical practice guidelines in oncology. *J Nat Comp Cancer Netw, 16*(4):359–369.

Benson, A. B., et al. (2021). Colon cancer, version 2.2021, NCCN clinical practice guidelines in oncology. *J Nat Comp Cancer Netw*, *19*(3):329–359.
Bishehsari, F., et al. (2014). Epidemiological transition of colorectal cancer in developing countries: environmental factors, molecular pathways, and opportunities for prevention. *World J Gastroenterol*, 20(20):6055–6072.
Bray, F., et al. (2018). Global cancer statistics 2018: GLOBOCAN estimates of incidence and mortality worldwide for 36 cancers in 185 countries. *CA*, *68*(6):394–424.
Brenner, H., et al. (2014). Effect of screening sigmoidoscopy and screening colonoscopy on colorectal cancer incidence and mortality: Systematic review and meta-analysis of randomised controlled trials and observational studies. *Br Med J*, *348*:g2467.
Chabner, B., Roberts, T. (2005). Chemotherapy and the war on cancer. *Nature Rev Cancer*, 5(1):65–72.
Collins, F. S., Varmus, H. (2015). A new initiative on precision medicine. *New Engl J Med*, *372*(9):793–795.
Cooper, L., Chen, J. (2022). Changes in companion diagnostic labelling: implementation of FDA's April 2020 guidance for industry for in vitro CDx labeling for specific oncology therapeutic groups. *Therapeut Innov Regulat Sci*, *56*(5):689–697.
Cortes-Ciriano, I., et al. (2017). A molecular portrait of microsatellite instability across multiple cancers. *Nature Communicat*, *8*:15180.
Delaney, G., et al. (2005). The role of radiotherapy in cancer treatment: estimating optimal utilization from a review of evidence-based clinical guidelines. *Cancer*, *104*(6):1129–1137.
Di Nicolantonio, F., et al. (2021). Precision oncology in metastatic colorectal cancer: from biology to medicine. *Nature Rev Clin Oncol*, *18*(8):506–525.
Doostmohammadi, A., et al. (2024). Potentials and future perspectives of multi-target drugs in cancer treatment: the next generation anti-cancer agents. *Cell Communicat Signal*, 22(1), 228.
EUnetHTA Network. (2020). Advancing cooperation on health technology assessment (HTA) and supporting the implementation of the new EU legal framework on HTA; available at: https://www.eunethta.eu/
Esteva, A., et al. (2019). A guide to deep learning in healthcare. *Nature Med*, *25*(1):24–29.
Garraway, L. A., et al. (2013). Precision oncology: An overview. *J Clin Oncol*, *31*(15):1803–1805.
Ginsburg, G. S., Phillips, K. A. (2018). Precision Medicine: From Science To Value. *Health affairs (Project Hope)*, *37*(5):694–701.
Goldstein, D. A., et al. (2017). A global comparison of the cost of patented cancer drugs in relation to global differences in wealth. *Oncotarget*, *8*(42):71548–71555.
Haggar, F. A., Boushey, R. P. (2009). Colorectal cancer epidemiology: Incidence, mortality, survival, and risk factors. *Clin Colon Rectal Surg*, *22*(4):191–197.
Hankey, W., et al. (2018). Functions of the APC tumor suppressor protein dependent and independent of canonical WNT signaling: implications for therapeutic targeting. *Cancer Metastasis Rev*, *37*(1):159–172.
Hasin, Y., et al. (2017). Multi-omics approaches to disease. *Genome Biology*, *18*(1):83.
Horimoto, Y., et al. (2022). Comparison of tumors with HER2 overexpression versus HER2 amplification in HER2-positive breast cancer patients. *BMC Cancer*, *22*(1):242.
Huang, L., et al. (2021). KRAS mutation: from undruggable to druggable in cancer. *Signal Transduct Targeted Ther*, *6*(1):386.
Hunter, J. E., et al. (2017). Universal screening for Lynch syndrome among patients with colorectal cancer: patient perspectives on screening and sharing results with at-risk relatives. *Fam Cancer*, *16*(3):377–387.
Hurwitz, H., et al. (2004). Bevacizumab plus irinotecan, fluorouracil, and leucovorin for metastatic colorectal cancer. *N Engl J Med*, *350*(23):2335–2342.
Johnson, K. B., et al. (2021). Precision medicine, AI, and the future of personalized health care. *Clin Translat Sci*, *14*(1):86–93.
Kawakami, H., et al. (2015). Implications of mismatch repair-deficient status on management of early stage colorectal cancer. *J Gastrointest Oncol*, *6*(6):676–684.
Keesara, S., Jonas, A., & Schulman, K. (2020). Covid-19 and health care's digital revolution. *N Engl J Med*, *382*(23), e82.
Kolligs, F. T. (2016). Diagnostics and epidemiology of colorectal cancer. *Visc. Med*, *32*(3), 158–164.
Kopetz, S., et al (2019). Encorafenib, binimetinib, and cetuximab in BRAF V600E–mutated colorectal cancer. *New Engl J Med*, *381*(17):1632–1643.
Kourou, K., et al. (2015). Machine learning applications in cancer prognosis and prediction. *Comput Struct Biotechnol J*, *13*, 8–17.

Le, D. T., et al. (2017). Mismatch repair deficiency predicts response of solid tumors to PD-1 blockade. *Sci, 357*(6349):409–413.
Li, C., et al. (2020). Integrated omics of metastatic colorectal cancer. *Cancer Cell, 38*(5):734–747.
Lievevrouw, E., et al. (2022). The FDA's standard-making process for medical digital health technologies: co-producing technological and organizational innovation. *BioSocieties, 17*(3):549–576.
Longley, D. B., et al. (2003). 5-fluorouracil: Mechanisms of action and clinical strategies. *Nature Rev Cancer, 3*(5):330–338.
Mak, K. K., & Pichika, M. R. (2019). Artificial intelligence in drug development: present status and future prospects. *Drug Discov. Today, 24*(3):773–780.
Marquart, J., et al. (2018). Estimation of the percentage of US patients with cancer who benefit from genome-driven oncology. *JAMA Oncol, 4*(8):1093–1098.
Mazeda, I., et al. (2020). VEGF expression in colorectal cancer metastatic lymph nodes: Clinicopathological correlation and prognostic significance. *Gastrointest Dis,* 2(3):267–280.
Ooki, A., et al. (2021). Immunotherapy in colorectal cancer: current and future strategies. *J Anus Rectum Colon, 5*(1):11–24.
Overman, M. J., et al. (2018). Durable clinical benefit with nivolumab plus ipilimumab in DNA mismatch repair-deficient/microsatellite instability-high metastatic colorectal cancer. *J Clin Oncol, 36*(8), 773–779.
Raufaste-Cazavieille, V., et al. (2022). Multi-omics analysis: Paving the path toward achieving precision medicine in cancer treatment and immuno-oncology. *Frontiers Molec Biosci, 9*:962743.
Reifenberger, G., et al. (2017). Advances in the molecular genetics of gliomas: Implications for classification and therapy. *Nature Rev Clin Oncol,* 14(7):434–452.
Saoudi González, N., et al. (2023). Unravelling the complexity of colorectal cancer: heterogeneity, clonal evolution, and clinical implications. *Cancers,* 15(16):4020.
Satam, H., et al. (2023). Next-generation sequencing technology: current trends and advancements. *Biology,* 12(7):997.
Siegel, R. L., et al. (2020). Cancer statistics, 2020. *CA, 70*(1):7–30.
Siravegna, G., et al. (2015). Clonal evolution and resistance to EGFR blockade in the blood of colorectal cancer patients. *Nature Med, 21*(7):795–801.
Sung, H., et al. (2021). Global cancer statistics 2020: GLOBOCAN estimates of incidence and mortality worldwide for 36 cancers in 185 countries. *CA, 71*(3), 209–249.
Topol, E. J. (2019). High-performance medicine: the convergence of human and artificial intelligence. *Nature Med, 25*(1):44–56.
Tsamados, A., et al. (2022). The ethics of algorithms: Key problems and solutions. *AI & Society,* 37(1):215–230.
Tsimberidou, A. M., et al. (2020). Review of precision cancer medicine: Evolution of the treatment paradigm. *Cancer Treatm Rev, 86*:102019.
Van Cutsem, E., et al. (2009). Cetuximab and chemotherapy as initial treatment for metastatic colorectal cancer. *New Engl J Med, 360*(14):1408–1417.
Van Cutsem, E., et al. (2019). Binimetinib, encorafenib, and cetuximab triplet therapy for patients with BRAF V600E-mutant metastatic colorectal cancer: Safety lead-in results from the phase III BEACON colorectal cancer study. *J Clin Oncol, 37*(17):1460–1469.
Van Cutsem, E., et al. (2015). Fluorouracil, leucovorin, and irinotecan plus cetuximab treatment and RAS mutations in colorectal cancer. *J Clin Oncol, 33*(7):692–700.
van Rees, J. M., et al. (2023). Circulating tumour DNA as biomarker for rectal cancer: A systematic review and meta-analyses. *Frontiers Oncol, 13*:1083285.
Vasaikar, S. V., et al. (2018). LinkedOmics: analyzing multi-omics data within and across 32 cancer types. *Nucleic Acids Res, 46*(D1):D956–D963.
Vasen, H. F., et al. (2013). Revised guidelines for the clinical management of Lynch syndrome (HNPCC): recommendations by a group of European experts. *Gut, 62*(6):812–823.
Wu, P. K., Park, J. I. (2015). MEK1/2 Inhibitors: Molecular Activity and Resistance Mechanisms. *Sem Oncol, 42*(6):849–862.
Yabroff, K. R., et al. (2013). Economic studies in colorectal cancer: challenges in measuring and comparing costs. *J Nat Cancer Instit Monographs, 2013*(46):62–78.
Yaeger, R., et al. (2018). Clinical sequencing defines the genomic landscape of metastatic colorectal cancer. *Cancer cell, 33*(1):125–136.
Zheng, S., et al. (2023). Effectiveness of colorectal cancer (CRC) screening on all-cause and CRC-specific mortality reduction: a systematic review and meta-analysis. *Cancers, 15*(7):1948.

11 Biomarkers for Diagnosis, Prognosis, and Treatment Response of Colorectal Cancer

Wasan A. M. Al Taie, May K. Ismail, Arwa Alnahdi, Eslam ElNebrisi, Eman Hassan Ibrahim and Saba Z. Al-Abachi

11.1 INTRODUCTION TO BIOMARKERS

Colorectal cancer (CRC) is one of the most prevalent and incident cancers globally and an important cause of mortality. Early detection is crucial for improving survival rates and reducing the mortality associated with CRC. Emphasising early detection strategies is imperative in the management of this disease (Zygulska & Pierzchalski, 2022) CRC ranks as the second most common cause of cancer-related fatalities in developed nations, accounting for 10% of the overall cancer burden and carrying an individual lifetime risk of approximately 6% in Western countries (Cronin et al., 2018; Siegel et al., 2020).

Despite the availability of multiple screening options that markedly reduce mortality, 40% of eligible patients do not adhere to screening recommendations in the United States. On a global scale, there are 1.4 million new CRC cases and 700,000 related deaths annually, and compliance with screening may be even lower than in the USA. Implementing universal colonoscopy for CRC screening across North America and Europe would necessitate screening 250 million individuals over a decade.

11.2 DIAGNOSTIC BIOMARKERS OF COLORECTAL CANCER

Early screening and diagnosis of CRC are crucial for reducing its incidence and mortality rates (Liang & Dominitz, 2019). There are various CRC screening and diagnosis methods, which can be classified into invasive and non-invasive strategies. Both play an essential role in CRC screening and diagnosis (Beniwal et al., 2023; Ferrari et al., 2021).

Invasive methods include optical colonoscopy (OC), computed tomographic colonography (CTC), flexible sigmoidoscopy, colonoscopy, and double-contrast barium enema. On the other hand, non-invasive methods can use biological samples like stool, blood, urine, or others to detect the presence of genetic markers and biomarkers or to identify DNA alterations indicating CRC. Blood-based biomarkers include circulating tumour DNA (ctDNA) and carcinoembryonic antigen (CEA), whereas stool-based assays include the faecal immunochemical test (FIT), faecal occult blood test (FOBT), guaiac-based faecal occult blood test (gFOBT), and stool DNA test as illustrated in Figure 11.1 (Beniwal et al., 2023; Chan & Liang, 2022; Gude et al., 2023; Pickhardt, 2016; Yao et al., 2023).

Diagnostic biomarkers of CRC play a crucial role in the early identification, diagnosis, and stratification of risk. They are essential for detecting CRC in their preliminary stages, allowing for prompt treatment and better health outcomes.

11.2.1 Blood-Based Biomarkers (CTCs, cfDNA)

Blood-based biomarkers hold great potential as diagnostic tools for CRC. Innovative techniques for CRC detection and diagnosis in liquid biopsies include examining circulating tumour cells (CTCs) and cell-free DNA (cfDNA). In this section, we discuss the pros and cons of both biomarkers to better understand their effectiveness.

11.2.1.1 Circulating Tumour Cells (CTCs)

Circulating tumour cells (CTCs) are cancer cells that break away from the primary tumour and enter the bloodstream. This happens via either passive or active intravasation mechanisms or both. Once in the bloodstream, CTCs can adapt, survive, and metastasise to other parts of the body, potentially creating new tumours. CTCs are intact epithelial cancer cells and can appear in the blood as single cells (Christou et al., 2019; Joosse et al., 2015; Ju et al., 2022; Marcuello et al., 2019).

Clusters of CTCs consisting of 2–50 cells including leukocytes, endothelial cells, fibroblasts, and platelets are present in the blood of CRC patients alongside single cells. CTC clusters are groups of circulating tumour cells that have a short half-life but high metastatic ability and survival capacity. These clusters can be found in the blood for only a few hours as they can be easily captured by

DOI: 10.1201/9781032704517-11

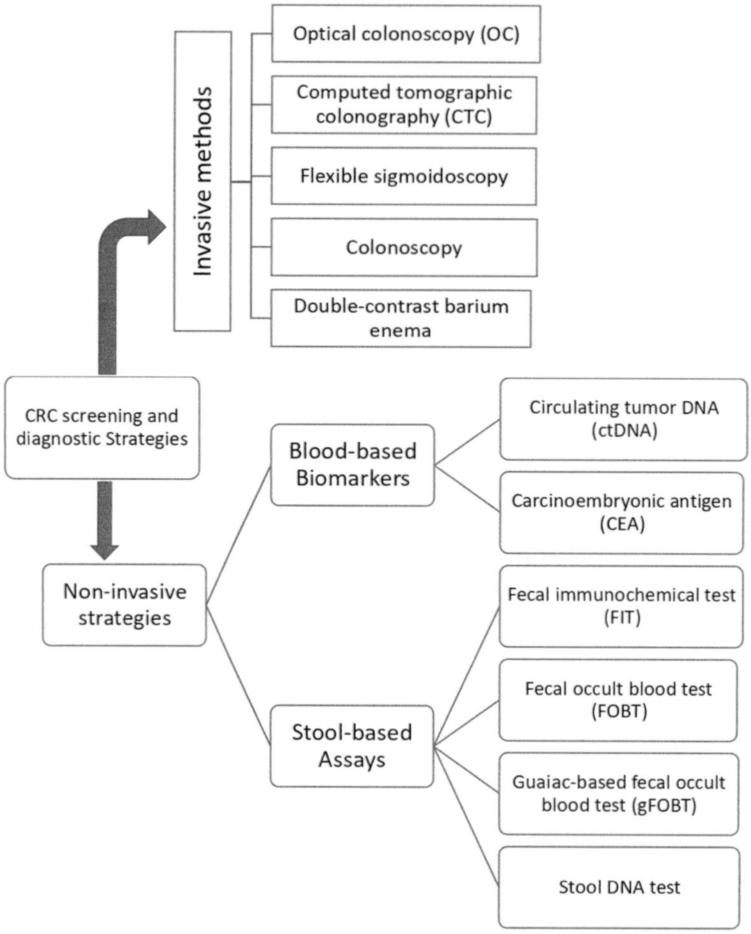

Figure 11.1 Various strategies for screening and diagnosis of colorectal cancer.

small vessels. The two main types of CTC clusters are homotypic and heterotypic clusters (Aceto et al., 2014; Hong et al., 2016; Petrik et al., 2022). Although multiple CTCs enter the bloodstream, only a subset of these cells can form metastases and travel to distant organs. Circulating tumour cells (CTCs) are present in three phenotypes: epithelial, mesenchymal, and mixed epithelial and mesenchymal cell phenotypes (Capuozzo et al., 2023; Christou et al., 2019; Zhong et al., 2021). According to Satelli et al. (2015), the epithelial-mesenchymal transformation (EMT) phenotype of CTCs may be associated with early metastasis. Additionally, Hou et al. (2020) found that the mesenchymal phenotype of CTCs, also known as M-CTCs, is related to the severity and metastasis of CRC. Therefore, patients diagnosed with the mesenchymal cell phenotype have a higher risk of recurrence in the preliminary stages of CRC (Zhong et al., 2021).

Zhong et al. (2021) found that the mesenchymal phenotype of CTCs exhibited a positive value when its quantity exceeded 1/7.5 mL of peripheral blood. The researchers also noted that the number of mesenchymal CTCs increased as the tumour developed, and these cells may play a more important role in the risk of metastasis than other CTC phenotypes. Surgery was found to be effective in decreasing the proportion of mesenchymal CTCs and number of CTCs.

During the epithelial-mesenchymal transition (EMT), the expression of epithelial markers such as keratins, E-cadherin, and EpCAM decreases. By contrast, both the expression of mesenchymal markers such as vimentin (VIM) and matrix metalloproteinase (MMP) activity increase to aid in the metastatic process. This helps the cancer cells navigate through the extracellular matrix (ECM), access the microvasculature, and gain migratory capacity (Ju et al., 2022; Saadi et al., 2023). CTC analysis offers a precise, dynamic, and treatment-related diagnostic and management approach for CRC.

11.2.1.2 *Cell-free Circulating DNA (cfDNA)*

Laboratory analysis of cell-free DNA in plasma or serum has become a potentially non-invasive technique for diagnosing and predicting the prognosis of CRC patients (Salvi et al., 2016). Cell-free DNA (cfDNA) and small and long non-coding RNA, also known as extracellular nucleic acids (ecNAs), are biomarkers that represent degraded and unencapsulated DNA fragments freely circulating in the blood, bile, urine, saliva, or stool (Petit et al., 2019; Pös et al., 2020; Vymetalkova et al., 2018).

The concentrations of cfDNA increase more in cancer patients than in those with benign diseases or healthy individuals. CRC patients have five times higher serum cfDNA concentrations compared to healthy control individuals, whereas plasma cfDNA concentrations are 25–50 times greater in CRC patients. Furthermore, the levels of cfDNA are higher in metastatic CRC cancer patients than in non-metastatic patients. cfDNA levels decrease following radiotherapy; however, they increase during the development of CRC disease or relapse (Vymetalkova et al., 2018).

cfDNA is found in the bloodstream, is usually around 200 base pairs in length (Arshad et al., 2022), and has a half-life from 30 minutes to 2 hours (Oellerich et al., 2021). The cfDNA biomarker can include distinct types of DNA, such as circulating tumour DNA (ctDNA), cell-free mitochondrial DNA (cf mtDNA), cell-free foetal DNA (cffDNA), and donor-derived cell-free DNA (dd-cfDNA).

The tumour-derived portion of cfDNA is called circulating tumour DNA (ctDNA). Plasma ctDNA is a type of cell-free DNA found in the blood; it represents a fraction ranging from 0.01% to 90% of the total cfDNA. Although ctDNA has several benefits in diagnosing CRC, some limitations have been identified. False-negative results may occur because of the small concentrations of ctDNA in the blood due to its fragmentation, making it hard to detect, especially in the early stage of the disease, compared to the analysis of CTCs (Tan et al., 2016).

cfDNA fragmentation affects the accuracy of the analysis techniques. It was reported that simple fluorometric and qPCR-based assays were influenced by cfDNA fragmentation, which lowers the measured concentrations of cfDNA. On the other hand, the sensitivity and accuracy of spectrophotometric measurements are unaffected by cfDNA fragmentation. Therefore, the cfDNA quantification test should be performed with equal fragmented control DNAs (Sedlackova et al., 2013). The analysis of cfNAs in plasma or serum has been less studied than in other body fluids. Therefore, analysis of cfNAs is not yet used in clinical practice (Pös et al., 2020).

11.2.2 Immunohistochemistry Biomarkers

Immunohistochemistry biomarkers are molecules identified within cells or tissues via antibody-based techniques and play a crucial role in diagnosing diseases, tracking disease progression, and guiding treatment decisions. Immunohistochemistry biomarkers are classified into primary IHC markers, cell-type specific markers, and metastatic markers for diagnosing CRC. Three immunohistochemistry biomarkers are discussed in this chapter.

11.2.2.1 *GPA33 (A33)*

The A33 antigen is a transmembrane glycoprotein found in human cells. The protein contains a single hydrophobic transmembrane domain, which has 23 amino acids, and a highly polar intracellular tail consisting of 62 amino acids, including four consecutive cysteine residues as illustrated in Figure 11.2 (NIH, 2024; Tabone & Heath, 2008).

The exact functions of the membranous protein A33 are yet to be discovered, but it is thought to play a role in intercellular adhesion, cell-cell recognition and signalling, and transmitting information between intestinal epithelial cells and the gut immune system. Additionally, it is associated with immune dysregulation and disorders (Opstelten et al., 2021; SinoBiology, n.d.; Tabone & Heath, 2008)

The A33 glycoprotein is a validated tumour-associated antigen and a single-pass type 1 membrane protein. It is expressed in the normal gastrointestinal epithelium and overexpressed in more than 95% of human primary and metastatic CRCs (Baptistella et al., 2016; Murer et al., 2020; SinoBiology, n.d.; Tabone & Heath, 2008).

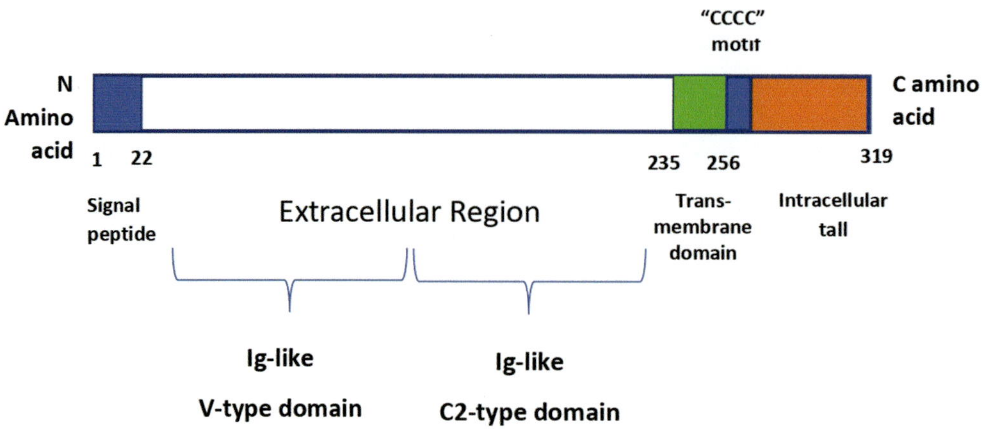

Figure 11.2 GPA33 structure (Griffin, 2023; Tabone & Heath, 2008).

11.2.2.2 Cytokeratins (CKs)

Cytokeratins (CKs) are proteins found in the cytoskeleton of epithelial cells. They are expressed by these cells and are commonly used to show the origin of metastatic CRC and distinguish it from other tumours. CK20 is expressed in a higher percentage in primary CRC and nodal metastasis, compared to CK7 (Al-Maghrabi et al., 2018; Nikolouzakis et al., 2018; Oh & Joo, 2020).

11.2.2.3 Caudal type homeobox 2 (CDX2)

The CDX2 protein is a homeotic transcription factor encoded by the CDX2 gene. This gene belongs to the caudal subgroup of homeobox genes. The CDX2 protein is expressed within the nucleus of intestinal epithelial cells and plays a critical role in cell differentiation within the gastrointestinal (GI) tract. Additionally, CDX2 suppresses colon tumours and prevents metastasis (Nikolouzakis et al., 2018; Oh & Joo, 2020; Platet et al., 2017; Zygulska & Pierzchalski, 2022).

The nuclear transcription factor CDX2 is an immune biomarker commonly used for diagnosing CRC. CDX2 is a specific and sensitive biomarker, with highly increased expression levels in CRC, and somewhat increased levels in other adenocarcinomas of the GI tract. CDX2 can be used alone to differentiate between adenocarcinomas of the GI tract. However, to ensure accuracy, the CDX2 test should be associated with the CK staining test (Oh & Joo, 2020).

11.3 PROGNOSTIC BIOMARKERS OF COLORECTAL CANCER

Recent advancements in molecular technologies and proteomics have allowed researchers to detect variations in DNA, RNA, protein, and small molecules from limited amounts of tissue. The gene signature prognostic model exhibits high predictive ability for the prognosis of colon cancer (Fang et al., 2021).

Cytokeratin 20 (CK 20) is a low-molecular-weight cytokeratin useful for detecting primary and metastatic cancer cells by immunohistochemistry and PCR analysis. Advanced cancers are more likely to show CK20+/cytokeratin 7+ (CK7+) profiles than early-stage cancers, which are predominantly CK20+/cytokeratin 7- (CK7-) (Hernandez et al., 2005). The number of pre-operative cytokeratin (CK) 20-positive circulating tumour cells (pCTCs) correlates with the tumour-node-metastasis stage and lymph node status in CRC patients. Thus, CK20 pCTC levels may be used to stratify CRC patients into poor and good prognosis groups, which may in turn affect the patients' treatment strategies (Wong et al., 2009).

CDX-2 could be easily detected in bioptic and cytologic samples of CRC metastases. CDX-2 is a reliable, specific, and sensitive immunohistochemical marker of the normal and neoplastic intestinal epithelium. CDX-2 can be easily applied to routine histologic and cytologic material and is, therefore, a useful marker in the differential diagnosis of primary versus metastatic adenocarcinomas, and supports their intestinal origin (Barbareschi et al., 2003).

CD10: The stromal expression of CD10 was associated with the accumulation of p53 and a larger tumour size. These results show that CD10 expression is an integral part of colorectal carcinogenesis. CD10 expression seems to contribute to the invasion and thus facilitates metastasis (Ogawa et al., 2002).

Carcinoembryonic antigen (CEA) is a glycoprotein commonly used as a clinical marker. As a tumour marker, CEA plays a pivotal role in determining the presence, therapeutic assessment, progression, surveillance, and prognosis of various cancer types. Currently, CRC patients undergo pre- and post-treatment monitoring to investigate potential metastasis (Chu et al., 2020).

The BRAF oncogene encodes the BRAF protein (also known as serine-threonine kinase). This oncogene serves as a prognostic biomarker and a potential target for therapies in patients with CRC (Chu et al., 2020). A significant correlation exists between BRAF mutation and MSI. Mutated BRAF tends to be more prevalent in females and individuals over 70 years old, predominantly affecting the right colon. Thus, mutation testing for BRAF in stage IV patients is recommended to enhance treatment strategies (Sahin et al., 2021).

The KRAS oncogene encodes small proteins that bind to guanine triphosphate and function as GTPase transducers. These KRAS proteins, also known as p21, are situated on the cell membrane and undergo temporary activation during signal transduction. Mutations in the KRAS gene primarily occur at codons 12 (82%–87%), associated with mucinous CRC, and 13 (13%–18%), linked to non-mucinous CRC, which tends to be more aggressive with higher rates of metastasis (Garcia-Carbonero et al., 2020).

Microsatellite instability (MSI) arises from a deficiency in the DNA mismatch repair system (MMR), resulting in an inability to rectify insertions or deletions of repetitive sequences during DNA replication. MSI is detected in approximately 15% of colorectal tumours. CRC with MSI typically manifests as mucinous with poor cellular differentiation and marked lymphocyte infiltration, and is predominantly observed in the right colon. MSI can serve as a potential prognostic indicator for CRC patients (Kang et al., 2018).

Adenomatous polyposis coli (APC) is a tumour suppressor gene identified in familial adenomatous polyposis (FAP). An epigenetic alteration in mutated APC accounts for most sporadic CRC cases, with 70%–80% of patients harbouring this mutation. APC functions as an antagonist of the WNT signalling pathway and regulates various cellular processes including migration, adhesion, transcriptional activation, and apoptosis (Eshghifar et al., 2017). Several researchers have highlighted the role of hypermethylated APC as an important biomarker for early CRC diagnosis and a potential target for personalised treatment directed at the specific mutation involved (Vacante et al., 2018).

The phosphatase and tensin homolog (PTEN) serves as a critical tumour suppressor gene regulating the cell proliferation pathway initiated by PI3K. This gene encodes an enzyme pivotal in governing cell division, thereby curbing uncontrolled cell growth and prompting apoptosis. Moreover, PTEN participates in cell migration, adhesion, angiogenesis, and DNA repair mechanisms. Mutations in PTEN are linked with advanced and metastatic tumours. Hypermethylation of the PTEN promoter is highly prevalent in high-grade CRC (Eissing et al., 2018).

The insulin-like growth factor receptor (IGF1R)) is a transmembrane glycoprotein whose activation plays a crucial role in cell proliferation, differentiation, angiogenesis, and apoptosis. Elevated IGF-1 expression serves as a crucial marker of CRC development and holds prognostic value because its overexpression correlates with an increased risk of CRC. Notably, metastatic CRC exhibits heightened levels of IGF-1R compared to primary cancer, leading to the targeting of IGF1R in treatments, particularly through monoclonal antibodies or tyrosine kinase inhibitors (Chen & Sharon 2013).

The transforming growth factor-β (TGF-β) signalling pathway exerts a vital role in inhibiting cell proliferation while influencing processes like cell invasion, immune modulation, and modifications to the microenvironment. Mutations in the TGF-β type II receptor (TGF-BR2) occur in around 30% of CRC cases. Numerous studies have illustrated TGF-β's involvement in CRC development and metastatic progression (Papavassiliou et al., 2024).

Stromal cell-derived factor-1 (SDF-1) is a small protein (8–14 kDa) that exists in six isoforms. Among these isoforms, SDF-1β was linked to metastasis, whereas SDF-1γ was associated with tumour size. All six isoforms of SDF-1 were detected in CRC tissues and their presence correlated with metastasis, indicating their potential as tumour markers for local tumour progression. Elevated SDF-1 expression in CRC was also associated with advanced clinical stage and lymphatic invasion, suggesting its indirect prognostic significance and demonstrating its role in promoting metastasis (Lalos et al., 2021).

CD26 was recently identified as a marker for cancer stem cells. CD26 interacts with type I and III collagens and fibronectin, facilitating metastasis. Studies have demonstrated elevated levels of this enzyme in the serum of patients with metastatic CRC compared to healthy individuals (De la Haba-Rodríguez et al., 2002).

β-catenin is a crucial protein involved in regulating cell-cell adhesion and gene transcription. This protein plays a crucial role in maintaining the stemness of normal intestinal cells. However, elevated levels of β-catenin expression in the cytoplasm, along with nuclear localisation, consistently induce cancer cell proliferation and survival (Gekas et al., 2016). Nuclear accumulation of β-catenin has been observed in approximately 80% of CRC cases. On the other hand, the reduction or loss of β-catenin expression in CRC has been linked to liver cancer metastasis (He & Gan, 2022).

11.4 TREATMENT RESPONSE PREDICTION OF COLORECTAL CANCER

CRC is recognised as a heterogeneous group of disorders characterised by various genetic mutations and epigenetic alterations that contribute to the onset and progression from colorectal adenoma to invasive carcinoma. Epigenetics are inheritable alterations in genetic expression that occur independently of modifications in the DNA sequence, involving processes like DNA methylation, chromatin remodelling, histone modification, noncoding RNA, and microRNAs. The disruption of epigenetics mechanisms is prevalent in CRC.

CRC presents a remarkable challenge in oncology because of its widespread occurrence and formidable nature. CRC continues to pose a substantial global health burden, urging the need for innovative strategies in personalised treatment. According to the World Health Organization (WHO), CRC stands as the third most common cancer worldwide, accounting for approximately 10% of all cancer cases, and is the second leading cause of cancer-related deaths globally (WHO, 2023).

The incidence and survival rates of CRC have been improved in the last decade (Bray et al., 2018; Murphy et al., 2024; Sharma et al., 2022; Sung et al., 2021). These results are linked to advancements in various disciplines, including the extensive incorporation of colonoscopy in national screening initiatives (Gupta et al., 2020; Hewett & Rex, 2015; Kahi et al., 2016; Rex et al., 2017; Shaukat & Levin, 2022), precise preoperative imaging and staging (Bhoday et al., 2017; So et al., 2017), neoadjuvant chemotherapy (Kong et al., 2021; Lin et al., 2021), and the acceptance of total mesorectal excision (TME) as the established standard for rectal cancer surgery (de Lacy et al., 2018). Therefore, in the contemporary landscape, survival following early CRC is generally anticipated.

In the context of innovation reshaping clinical practice and influencing patient outcomes, there is a growing necessity to comprehend the mechanisms underlying novel medications (Guler et al., 2019). Therefore, this review synthesises the existing literature on targeted therapies, emphasising the molecular biomarkers implicated, and elucidating their associated pathways. This comprehensive exploration delves into pivotal biomarkers for predicting treatment responses, focusing on chemotherapy (MSI, KRAS, and BRAF) and immunotherapy (PD-L1 and TMB). Understanding the roles of these biomarkers unravels their potential implications for shaping the trajectory of CRC treatment.

Chemotherapy is the primary therapeutic approach for individuals with metastatic colon cancer. Additionally, targeted agents like bevacizumab or cetuximab, tailored to the RAS gene mutational status and tumour location, are concurrently applied to augment the anti-tumour efficacy of chemotherapy agents. Despite these efforts, the overall prognosis for patients remains grim (Wu, 2018). In recent times, a transformative development in colon cancer treatment has emerged through immunotherapy. In a clinical investigation, pembrolizumab, an anti-programmed cell death 1 (PD-1) agent, demonstrated notable enhancements in progression-free survival (16.5 vs 8.2 months) and demonstrated fewer side effects than traditional chemotherapy when used as a primary treatment for patients diagnosed with microsatellite instability-high (MSI-H) or mismatch repair-deficient (dMMR) metastatic mCRC (André et al., 2020).

11.4.1 Chemotherapy (KRAS, BRAF)

The treatment modalities for CRC include a diverse array of interventions, from surgical resection to chemotherapy, targeted therapy, immunotherapy, gene therapy, and combination therapies. Current chemotherapy treatments for mCRC include single-agent therapy (mainly fluoropyrimidine 5-fluorouracil (5-FU)-based) and multi-agent regimens such as FOLFOX (leucovorin, 5-FU, and oxaliplatin), FOLFIRI (leucovorin, 5-FU, and irinotecan), and CAPOX/XELOX (capecitabine and oxaliplatin) (Sakata & Larson, 2022).

11.4.1.1 KRAS Molecular Function, Mutation, and Role in CRC

The RAS-encoded proteins are a family of GTPase-related proteins involved in cell signal transduction. RAS genes, including KRAS, NRAS, and HRAS, are mutated in up to 50% of sporadic CRC cases (Requena & Garcia-Buitrago, 2020). A mutation in RAS genes can cause cell proliferation and tumour formation. Mutations of KRAS, a downstream pathway of EGFR, predominate in CRC. Thus, inhibiting EGFR can suppress the activation of KRAS and its downstream pathways. Cetuximab (Epitol), a monoclonal antibody specifically used for blocking EGFR, is one of the most representative targeted therapy drugs and is currently approved for first-line monotherapy in CRC patients (Moiseyenko et al., 2018). Patients with KRAS-mutant CRC have a poorer prognosis than those with KRAS wild-type CRC, especially in the metastatic setting, and are unlikely to benefit from anti-EGFR therapy. EGFR inhibitors, particularly cetuximab and panitumumab, show effectiveness only in individuals with KRAS/NRAS wild-type metastatic CRC and are commonly considered interchangeable (Biller & Schrag, 2021; Douillard et al., 2014).

The process of GTP binding to KRAS is recognised for its potential to enhance effector interaction, initiating various downstream pathways including the fibrosarcoma (RAF)-mitogen-activated protein kinase (MEK)-extracellular-signal-regulated kinase (ERK) and phosphatidylinositol 3-kinase (PI3K)-protein kinase B (AKT)-mechanistic target of rapamycin (mTOR) pathways, which promote cellular growth and viability.

On the other hand, the activity of GDP-bound KRAS is diminished, hindering its role of stimulating signal transduction. When KRAS mutations occur, GTP hydrolysis is disrupted, and/or nucleotide exchange is enhanced, leading to the continuous activation of downstream pathways including BRAF, MEK, and ERK by KRAS. This activation triggers the proliferation and metastasis of tumour cells. A comprehensive understanding of the KRAS mutational status of a primary tumour is essential in the management of metastatic disease, as it is a crucial predictor of resistance to monoclonal antibodies of the epidermal growth factor receptor (Bokemeyer et al., 2009; Douillard et al., 2010).

11.4.1.2 BRAF Mutations and Their Role in CRC

Mutations in BRAF refer to alterations in B-type RAF (BRAF), resulting in sustained activation of the MAPK pathway, independent of KRAS status. An estimated 12% of CRC occurrences exhibit a mutation in the BRAF gene. Among them, around 90% have shown a missense mutation at the V600 E position (Requena & Garcia-Buitrago, 2020).

BRAF mutations at position V600E are characterised by missense mutations in codon 600, leading to the substitution of glutamic acid with valine. KRAS mutations are commonly observed in all CRC subtypes, whereas BRAFV600E mutations are highly represented in tumours with MSI and CIMP-high tumours (De Roock et al., 2010; Saridaki et al., 2010). Patients with BRAF mutations do not derive any therapeutic benefit from targeted therapy unless the administration of a BRAF inhibitor (encorafenib) is initiated. Studies have reported that the concurrent treatment of EGFR inhibitors with BRAF inhibitors extends the overall survival rate of these patients (Al-Salama, 2021; Mauri et al., 2021).

Angiogenesis is considered a pivotal hallmark of cancer and is regulated by a balance between pro- and anti-angiogenic factors. The vascular endothelial growth factor (VEGF) acts as the key angiogenic factor in CRC (Seeber et al., 2018). Studies have shown a correlation between CRC progression and metastasis and an increased level of VEGF (Guba et al., 2004). Bevacizumab, a humanised IgG monoclonal antibody, was approved by the FDA as the first VEGF-targeted therapeutic for metastatic CRC and remains the only VEGF-targeted agent approved by the FDA for use as both first- and second-line treatment for this type of cancer (Hurwitz et al., 2004; Xie et al., 2020).

Several studies have reported that in patients with KRAS, NRAS wild-type left-sided tumours may benefit from EGFR inhibition, using either cetuximab or panitumumab along with chemotherapy (Douillard et al., 2014). EGFR inhibitors may prolong the survival rate by 6 months compared to chemotherapy alone (Sakata & Larson, 2022). The VEGF inhibitor bevacizumab can be considered a substitute for EGFR inhibitors in right-sided tumours or as a second-line treatment (Sakata & Larson, 2022).

11.4.2 Immunotherapy (PD-L1, TMB)

Immunotherapy has emerged as a promising avenue in the treatment of CRC, especially among patients exhibiting high microsatellite instability (MSI-H) or deficient mismatch repair (dMMR). Immune checkpoint inhibitors (ICIs) have garnered attention for their efficacy in managing advanced or metastatic CRC cases with MSI-H or dMMR markers, typically as a subsequent treatment following other interventions. Pembrolizumab, nivolumab, and ipilimumab stand out as some commonly employed ICIs in CRC therapy (Huyghe et al., 2019).

11.4.2.1 Programmed Death-Ligand 1 (PD-L1)

PD-L1 is a negative immune checkpoint pathway that inhibits immune responses, and its upregulation contributes to many cancers (Lin et al., 2020). PD-L1 also interacts with the programmed cell death 1 (PD-1) receptor on immune cells, suppressing the body's immune response. The gene responsible for encoding PD-1, known as Pdcd1, was first identified by Tsuku Honjo's research team (1992) while investigating molecules associated with programmed cell death in T cells (Ishida et al., 1992).

PD-L1 plays a crucial role in the tumour microenvironment (TME). Tumour cells increase PD-L1 expression to trigger immune evasion because PD-L1 binds to PD-1 on the surfaces of T cells, deactivating them. These deactivated T-cells become exhausted, losing their ability to proliferate and act as effector cells in the TME. Cancer cells exploit these effects to evade immune surveillance, leading to adaptive immune resistance (Lin et al., 2020). Besides its interaction with PD-1, recent research suggests that PD-L1 can bind to CD80 (albeit with lower affinity than PD-1). This binding enables CD80 to function as a receptor for PD-L1, transmitting inhibitory signals to T cells (Rutering et al., 2015; Zhang et al., 2023).

On the other hand, immune checkpoint inhibitors (ICIs) disrupt the interaction between PD-1 and PD-L1, reactivating the body's immune response against tumours (Hou et al., 2021). Theoretically, higher PD-L1 expression in tumour tissues correlates with a more positive response to ICI treatment. Thus, PD-L1 expression may become a predictive biomarker for the response to immunotherapy in various solid tumours, such as NSCLC, melanoma, and renal cell cancer (Aguilar et al., 2019; Chen et al., 2018; Wang et al., 2016).

A positive PD-L1 expression has been documented in over half of individuals diagnosed with colon cancer (with a defined threshold of 10%) (Ahn et al., 2021; Patel & Kurzrock, 2015). Despite the correlation between high PD-L1 expression and improved prognosis in colon cancer patients (Li et al., 2019; Wang et al., 2021; Yang et al., 2019), existing clinical data indicate that relying solely on PD-L1 expression is insufficient for accurately predicting the response to immunotherapy in colon cancer.

Atezolizumab and pembrolizumab play key roles in the treatment of CRC with immunotherapy, especially in the context of PD-L1 biomarker immunotherapy (Chen et al., 2023). PD-1/PD-L1 inhibitors, including pembrolizumab and atezolizumab, have shown promising clinical benefits in metastatic CRC. These inhibitors have demonstrated encouraging response rates, especially in patients with dMMR-MSI-H tumours (Zhang et al., 2022). The use of PD-1/PD-L1 inhibitors stands for a new direction in mCRC treatment, with ongoing trials exploring their efficacy and safety profiles (Zhang et al., 2022).

11.5 RESEARCH ADVANCEMENT IN COLORECTAL CANCER

11.5.1 Omics Technologies

Omics techniques include genomics, transcriptomics, proteomics, metabolomics, glycomics, and volatolomics. The application of omics analyses can be extended to various sample types including breath, stool, urine, blood, bowel lavage fluid, and tumour tissue to detect biomarkers for the diagnosis, prognosis, and treatment responses of CRC as illustrated in Figure 11.3 (Alorda-Clara et al., 2022).

Blood	Genomics	**Tissues**	Genomics
	Epigenomics		Transcriptomics
	Transcriptomics		Proteomics
	Proteomics		Glycomics
	Metabolomics		Multi-omics
	Glycomics		
Urine	Genomics	**Stool**	Genomics
	Transcriptomics		Transcriptomics
	Metabolomics		Proteomics
			Metabolomics
Bowel lavage fluid	Genomics	**Breath**	Volatolomics
	Proteomics		
	Mircobiome		

Figure 11.3 Sample types for the omics analyses.

Metagenomics techniques were used to distinguish between microbiomes linked to CRC and control samples. A study analysed the faecal microbiome samples of 74 CRC patients and 54 control samples. As a result, butyryl-CoA dehydrogenase from F. nucleatum was identified as a potential biomarker for the early diagnosis of CRC (Yu et al., 2017).

Another study used methylation-specific PCR (MSP) and real-time MSP (qMSP) genomics techniques to detect DNA methylation biomarkers in the stool samples of patients with CRC. The authors detected four potential methylated biomarkers (COL4A1, COL4A2, TLX2, and ITGA4) in the stool samples of CRC patients, demonstrating high accuracy in detecting colorectal neoplasms in the stool (Liu et al., 2020).

DNA methylation is a type of alteration that occurs in DNA molecules, and the modified DNA can be used as a biomarker for detecting and screening CRC. Some examples of DNA methylation biomarkers include septin 9, homolog 1, vimentin, p16, mutL homolog 1, adenomatous polyposis coli (APC) promoter hypermethylation, and death-associated protein kinase 1 (DAPK1) (Beniwal et al., 2023).

11.5.1.1 Transcriptomics

Gene expression analysis may detect promising biomarkers of enormous clinical significance. Ahmed et al. conducted a microarray expression study to evaluate the potential of microRNAs (miRNAs) as diagnostic markers for CRC screening. They found that 12 miRNAs (miR-7, miR-17, miR-20a, miR-21, miR-92a, miR-96, miR-106a, miR-134, miR-183, miR-196a, miR-199a-3p, and miR214) were overexpressed in the stool of CRC patients. Moreover, the study showed a decreased expression of eight miRNAs (miR-9, miR-29b, miR-127-5p, miR-138, miR-143, miR-146a, miR-222, and miR-938) in the stool of CRC patients (Ahmed et al., 2018).

11.5.1.2 Proteomics

Protein biomarkers can be identified by several techniques such as two-dimensional difference gel electrophoresis (2D-DIGE), two-dimensional gel electrophoresis coupled with liquid chromatography/mass spectrometry (2-DE-MS), and liquid chromatography-mass spectrometry (LC-MS) (Boja & Rodriguez, 2014). Two-dimensional gel electrophoresis and mass spectrometry were used to show the higher abundance of ACTBL2 in human CRC tissue specimens (Ghazanfar et al.,

2017). Protein panels were isolated from the serum of CRC patients, using liquid chromatography-tandem mass spectrometry (LC-MS/MS) and a machine-learning statistical model. Five invasive blood biomarkers (SOD3, LRG1, HPX, ITIH4, and EGFR) showed the potential to detect CRC with 70% sensitivity at over 89% specificity of a human population. Additionally, four proteins (ITIH3, GC, CRP, and CD44) showed the potential to distinguish metastasis CRC cases with regional lymph node invasion from the localised CRC (Ivancic et al., 2020).

11.5.1.3 Metabolomics

Metabolomics is the study of small molecules known as metabolites and their interactions within cells, body fluids, and tissues.

Mallafré-Muro et al. (2021) applied the Preferred Reporting Items for Systematic Reviews and Meta-Analyses (PRISMA) guidelines in their study to evaluate the metabolomes and volatilomes of CRC patients' urine samples. Some 244 compounds were identified in the urine samples of CRC patients and most of these compounds were organic acids and derivatives. The compounds included 40 amino acids and their derivatives, five dipeptides, and nine dicarboxylic and tricarboxylic acids and their derivatives. In addition, the authors detected 22 lipids including five prenol lipids and 15 fatty acyls; 30 benzenoids including three naphthalene, eight phenols, and six hippuric acids and derivatives; and 20 nucleosides including five pyrimidines and 14 purine nucleosides. Of those, 163 compounds were metabolites.

Four upregulated biomarker metabolites (L-dopa, 3-hydroxybutyric acid, N1, N12-diacetylspermine, and L-histidinol) were detected in the urine samples of CRC patients. By contrast, hippuric acid, pyruvic acid, tartaric acid, and hydroquinone were downregulated in the urine samples of these patients.

11.5.1.4 Glycomics

Glycomics is the study of all glycan structures and functions that describe molecules with N- and O- glycosidic bonds including glycolipids, proteoglycans, and glycoproteins (Alorda-Clara et al., 2022). The glycome can be analysed in CRC samples using flow cytometry, microarrays, chromatography, mass spectrometry, and enzyme-linked immunosorbent assay (Holst et al., 2015). N-glycan was analysed in the plasma of 40 CRC patients, and numerous glycome alterations linked with CRC were found. The results indicated an increase in glycans with no galactose residues (agalactosylation), tri- and tetra-galactosylated glycans (galactosylation), tri- and tetra-sialylated structures (sialylation), and an increase in highly branched glycans (GlcNAc antennae). Furthermore, the results revealed a decrease in mono- and di-galactosylated structures, mono-sialylated glycans, galactosylated and sialylated bi-antennary GlcNAc glycans, and neutral core fucosylated glycans with one or two galactose residues (Doherty et al., 2018).

11.5.1.5 Volatolomics

Volatolomics is the study of volatilomes and volatile organic compounds (VOCs) formed in a biological system. Volatilome analysis is used to diagnose CRC.

VOCs can be detected in breath, urine, or skin (Alorda-Clara et al., 2022; Bosch et al., 2019). VOCs can be determined using gas chromatography-mass spectrometry (GC-MS), electronic nose (eNose), and proton transfer reaction-mass spectrometry (PTR-MS) (Alorda-Clara et al., 2022, Van De Goor et al., 2017; Van Keulen et al., 2020). Wang et al. (2014) compared the concentrations of VOCs in the exhalations of 20 CRC patients and 20 healthy controls. They found that the exhalations of CRC patients contained significantly higher concentrations of 2,2-dimethyl decane, cyclohexanone, 4-ethyl-1-octyn-3-ol, dodecane, ethylaniline, trans-2-dodecen-1-ol, cyclooctylmethanol, and 3-hydroxy-2,4,4-trimethylpentyl 2-methyl propanoate, but significantly lower concentrations of 6-t-butyl-2,2,9,9-tetramethyl-3,5-decadien-7-yne than those in the exhalations of healthy controls.

The VOC profiles of the urine samples of 83 CRC patients and 50 healthy controls were analysed using field asymmetric ion mobility spectrometry (FAIMS), a novel screening tool. As a result, the researchers found that the sensitivity and specificity for CRC detection were 88% and 60%, respectively (Arasaradnam et al., 2014).

11.5.2 Ongoing Clinical Trials

Several clinical trials are currently underway, aimed at finding biomarkers related to the diagnosis, prognosis, and treatment of CRC. The goal is to find molecular markers to aid in early detection, predict disease progression, and assist in guiding personalised treatment strategies for CRC patients. Diverse protein-, DNA-, and RNA-based biomarkers are being studied to optimise diagnostic tools, prognostic capabilities, and treatment selection according to the individual tumour characteristics (Alese et al., 2023; Alves Martins et al., 2019). KRAS gene mutations, for example, have been used to predict responses to specific therapies for CRC (Alese et al., 2023; Ogunwobi et al., 2020). Ultimately, clinical trials help bridge the gap between research findings and clinical practice, paving the way for more effective and tailored approaches to treat CRC.

For instance, investigations into cetuximab treatment efficacy in CMS4 patients with RAS wild-type CRC have shown prolonged overall survival outcomes compared to conventional therapies. These trials underscore the potential of targeted therapies guided by specific biomarkers to improve patient outcomes (Ohnmacht et al., 2023). The evolution of precision medicine in metastatic CRC involves a comprehensive molecular profiling strategy that integrates genetic, epigenetic, and microbiomics factors to tailor treatment decisions for enhanced efficacy and personalised care (Puccini et al., 2022). Furthermore, ongoing research endeavours are dedicated to identifying novel biomarkers – beyond established markers like MSI or RAS/BRAF mutations – to refine patient selection for optimised treatment strategies in CRC (Ogunwobi et al., 2020).

11.6 CONCLUSION

CRC treatment relies on personalised, multidisciplinary approaches involving surgery, chemotherapy, targeted therapy, and immunotherapy. Early-stage CRC is typically treated with surgical resection, whereas advanced stages benefit from chemotherapy and targeted therapies, such as EGFR and BRAF inhibitors. Immunotherapy, particularly with immune checkpoint inhibitors like pembrolizumab and nivolumab, has shown promise in MSI-H/dMMR metastatic CRC, offering improved survival and fewer side effects. The integration of omics technologies and liquid biopsies further enhances diagnosis, prognosis, and treatment strategies, paving the way for precision medicine in CRC care.

REFERENCES

Aceto, N., et al. (2014). Circulating tumor cell clusters are oligoclonal precursors of breast cancer metastasis. *Cell*, *158*(5):1110–1122.

Ahmed, F.E., et al. (2018). RT-qPCR for fecal mature micro RNA quantification and validation. *Methods Mol Biol*, *1765*:203–215.

Aguilar, E. J., et al. (2019). Outcomes to first line pembrolizumab in patients with non-small-cell lung cancer and extremely high PD-L1 expression. *Ann Oncol*, *30*(10):1653–1659.

Ahn, A. R., et al. (2021). Correlation of PIK3CA mutation with programmed death ligand-1 (PD-L1) expression and their clinicopathological significance in colorectal cancer. *Ann Translat Med*, *9*(18):1406.

Alese, O. B., et al. (2023). Update on emerging therapies for advanced colorectal cancer. *American Society of Clinical Oncology Educational Book*, *43*:e389574.

Al-Maghrabi, J., et al. (2018). Immunohistochemical staining of cytokeratin 20 and cytokeratin 7 in colorectal carcinomas: four different immunostaining profiles. *Saudi J Gastroenterol*, *24*(2):129–134.

Alorda-Clara, M., et al. (2022). Use of omics technologies for the detection of colorectal cancer biomarkers. *Cancers*, *14*(3):817.

Al-Salama, Z. T. (2021). Encorafenib: a review in metastatic colorectal cancer with a BRAF V600E mutation. *Drugs*, *81*(7):849–856.

Alves Martins, B. A., et al. (2019). Biomarkers in colorectal cancer: the role of translational proteomics research. *Frontiers Oncol*, *9*:1284.

André, T., et al. (2020). Pembrolizumab in microsatellite-instability-high advanced colorectal cancer. *New Engl J Med*, *383*(23):2207–2218.

Arasaradnam, R.P., et al. (2014). Detection of colorectal cancer (CRC) by urinary volatile organic compound analysis. *PLoS One*, *9*(9):e108750.

Arshad, S., et al. (2022). Cell free DNA; diagnostic and prognostic approaches to oncology. *Adv Cancer Biol Metastasis*, *5*:100052.

Baptistella, A.R., et al. (2016). Heterogeneous expression of A33 in colorectal cancer: possible explanation for A33 antibody treatment failure. *Anticancer Drugs*, *27*(8):734–737.

Barbareschi, M., et al. (2003). CDX-2 homeobox gene expression is a reliable marker of colorectal adenocarcinoma metastases to the lungs. *Am J Surg Pathol*, *27*(2):141–149.

Beniwal, S.S., et al. (2023). Current status and emerging trends in colorectal cancer screening and diagnostics. *Biosensors*, *13*(10):926.

Bhoday, J., et al. (2017). How should imaging direct/orient management of rectal cancer? *Clin Colon Rectal Surg*, *30*(05):297–312.

Biller, L. H., Schrag, D. (2021). Diagnosis and treatment of metastatic colorectal cancer: a review. *JAMA*, *325*(7):669.

Boja, E. S., Rodriguez, H. (2014). Proteogenomic convergence for understanding cancer pathways and networks. *Clinical Proteom*, *11*:1–17.

Bokemeyer, C., et al. (2009). Fluorouracil, leucovorin, and oxaliplatin with and without cetuximab in the first-line treatment of metastatic colorectal cancer. *J Clin Oncol*, *27*(5):663–671.

Bosch, S., et al. (2019). Fecal volatile organic compounds for early detection of colorectal cancer: where are we now? *J Cancer Res Clin Oncol*, *145*:223–234.

Bray, F., et al. (2018). Global cancer statistics 2018: GLOBOCAN estimates of incidence and mortality worldwide for 36 cancers in 185 countries. *CA*, *68*(6):394–424.

Capuozzo, M., et al. (2023). Circulating tumor cells as predictive and prognostic biomarkers in solid tumors. *Cells*, *12*(22):2590.

Chan, S.C.H., Liang, J.Q. (2022). Advances in tests for colorectal cancer screening and diagnosis. *Expert Rev Mol Diagn*, *22*(4):449–460.

Chen, G., et al. (2018). Exosomal PD-L1 contributes to immunosuppression and is associated with anti-PD-1 response. *Nature*, *560*(7718):382–386.

Chen, H., Sharon, E. (2013). IGF-1R as an anticancer target: trials and tribulations. *Chinese J Cancer*, 32(5):242–252.

Chen, X., et al. (2023). Anti-PD-1/PD-L1 therapy for colorectal cancer: clinical implications and future considerations. *Translat Oncol*, *40*:101851.

Christou, N., et al. (2019). Circulating tumour cells, circulating tumour DNA and circulating tumour miRNA in blood assays in the different steps of colorectal cancer management, a review of the evidence in 2019. *Biomed Res Int*, *4*: 5953036.

Chu, J., et al. (2020). Population-based screening for BRAFV600E in metastatic colorectal cancer reveals increased prevalence and poor prognosis. *Clin Cancer Res*, *26*:4599–4605.

Cronin, K. A., et al. (2018). Annual report to the nation on the status of cancer, part I: National statistics. *Cancer*, *124*(13):2785–2800.

De la Haba-Rodríguez, J., et al. (2002). Soluble dipeptidyl peptidase IV (CD-26) in serum of patients with colorectal carcinoma. *Neoplasma*, *49*(5):307–311.

De Lacy, F. B., et al. (2018). The Future of rectal cancer surgery: a narrative review of an international symposium. *Surg Innovat*, *25*(5):525–535.

De Roock, W., et al. (2010). Effects of KRAS, BRAF, NRAS, and PIK3CA mutations on the efficacy of cetuximab plus chemotherapy in chemotherapy-refractory metastatic colorectal cancer: A retrospective consortium analysis. *Lancet Oncol*, *11*(8):753–762.

Doherty, M., et al. (2018). Plasma N-glycans in colorectal cancer risk. *Sci Rep*, *8*(1):8655.

Douillard, J. Y., et al. (2014). Results from PRIME: Randomized phase III study of panitumumab with FOLFOX4 for first-line treatment of metastatic colorectal cancer. *Ann Oncol*, *25*(7):1346–1355.

Douillard, J. Y., et al. (2010). Randomized, phase III trial of panitumumab with infusional fluorouracil, leucovorin, and oxaliplatin (FOLFOX4) versus FOLFOX4 alone as first-line treatment in patients with previously untreated metastatic colorectal cancer: the PRIME study. *J Clin Oncol*, *28*(31):4697–4705.

Eissing, M. et al. (2018). PTEN Hamartoma tumor syndrome and immune dysregulation. *Transl Oncol* 2018 Nov 30; *12*(2):361–367. Doi: 10.1016/j.tranon.2018.11.003

Eshghifar, N. et al. (2017). Tumor suppressor genes in familial adenomatous polyposis. *Gastroenterol Hepatol Bed Bench* 2017 Winter; *10*(1):3–13.

Fang, Z., et al. (2021). Identification of a prognostic gene signature of colon cancer using integrated bioinformatics analysis. *World J Surg Oncol*, *19*(1):13.

Ferrari, A., et al. (2021). Towards novel non-invasive colorectal cancer screening methods: a comprehensive review. *Cancers*, *13*(8):1820.

Garcia-Carbonero, N., et al. (2020). KRAS and BRAF mutations as prognostic and predictive biomarkers for standard chemotherapy response in metastatic colorectal cancer: a single institutional study. *Cells*, *9*:219.

Gekas, C., et al. (2016). β-Catenin is required for T-cell leukemia initiation and MYC transcription downstream of Notch1. *Leukemia, 30*(10):2002–2010.
Ghazanfar, S., et al. (2017). Identification of actin beta-like 2 (ACTBL2) as novel, upregulated protein in colorectal cancer. *J Proteomics, 152*:33–40.
Griffin, M. (2023). New target for colorectal cancer: GPA33 [internet]. Available at: https://kactusbio.com/blogs/news/new-target-for-colorectal-cancer-gpa33 (accessed Apr 2024).
Guba, M., et al. (2004). Vascular endothelial growth factor in colorectal cancer. *Int J Colorect Dis, 19*(6):510–517.
Gude, S.S., et al. (2023). Colorectal cancer diagnostic methods: the present and future. *Cureus, 15*(4):e37622.
Guler, I., et al. (2019). Precision medicine for metastatic colorectal cancer: An evolving era. *Expert Rev Gastroenterol Hepatol, 13*(10):919–931.
Gupta, S., et al. (2020). Spotlight: US multi-society task force on colorectal cancer recommendations for follow-up after colonoscopy and polypectomy. *Gastroenterol, 158*(4):1154.
He, K., Gan, W. (2022). Wnt/β-catenin signaling pathway in the development and progression of colorectal cancer. *Cancer Manag Res* 2023 May 23; *15*:435–448. Doi: 10.2147/CMAR.S411168
Hernandez, B. Y., et al. (2005). CK20 and CK7 protein expression in colorectal cancer: demonstration of the utility of a population-based tissue microarray. *Human Pathol, 36*(3):275–281.
Hewett, D. G., Rex, D. K. (2015). The big picture: does colonoscopy work? *Gastrointest Endosc Clin North Am, 25*(2):403–413.
Holst, S., et al. (2015). Glycosylation characteristics of colorectal cancer. *Adv Cancer Res, 126*:203–256.
Hong, Y., et al. (2016). Circulating tumor cell clusters: What we know and what we expect (Review). *Int J Oncol, 49*:2206–2216.
Hou, J., et al. (2020). Clinical significance of epithelial-mesenchymal transition typing of circulating tumour cells in colorectal cancer. *Colorectal Dis, 22*(5):581–587.
Hou, W., et al. (2021). Immune check point inhibitors and immune-related adverse events in small cell lung cancer. *Frontiers Oncol, 11*:604227.
Hurwitz, H., et al. (2004). Bevacizumab plus irinotecan, fluorouracil, and leucovorin for metastatic colorectal cancer. *New Engl J Med, 350*(23):2335–2342.
Huyghe, N., et al. (2019). Immunotherapy with immune checkpoint inhibitors in colorectal cancer: What is the future beyond deficient mismatch-repair tumours? *Gastroenterol Rep, 8*(1):11–24.
Ishida, Y., et al. (1992). Induced expression of PD-1, a novel member of the immunoglobulin gene superfamily, upon programmed cell death. *EMBO J, 11*(11):3887–3895.
Ivancic, M.M., et al. (2020). Noninvasive detection of colorectal carcinomas using serum protein biomarkers. *J Surg Res, 246*:160–169.
Joosse, S.A., et al. (2015). Biology, detection, and clinical implications of circulating tumor cells. *EMBO Mol Med, 7*(1):1–11.
Ju, S., et al. (2022). Detection of circulating tumor cells: opportunities and challenges. *Biomark Res,10*(1):58.
Kahi, C. J., et al. (2016). Colonoscopy surveillance after colorectal cancer resection: recommendations of the US multi-society task force on colorectal cancer. *Gastroenterol, 150*(3):758–768. e11
Kang, S., et al. (2018). The significance of microsatellite instability in colorectal cancer after controlling for clinicopathological factors. *Medicine, 97*(1):e0019.
Kong, J. C., et al. (2021). Total neoadjuvant therapy in locally advanced rectal cancer: a systematic review and metaanalysis of oncological and operative outcomes. *Ann Surg Oncol, 28*(12):7476–7486.
Lalos, A., et al. (2021). Prognostic significance of CD8+ T-cells density in stage III colorectal cancer depends on SDF-1 expression. *Sci Rep*. 2021 Jan 12; *11*:775. Doi: 10.1038/s41598-020-80382-2
Li, Y., et al. (2019). The prognostic and clinicopathological roles of PD-L1 expression in colorectal cancer: a systematic review and meta-analysis. *Frontiers Pharmacol, 10*:139.
Liang, P.S., Dominitz, J.A. (2019). Colorectal cancer screening: is colonoscopy the best option? *Med Clin North Am, 103*(1):111–123.
Lin, H., et al. (2021). Meta-analysis of neoadjuvant chemotherapy versus neoadjuvant chemoradiotherapy for locally advanced rectal cancer. *World J Surg Oncol, 19*(1):141.
Lin, X., et al. (2020). Progress in PD-1/PD-L1 pathway inhibitors: from biomacromolecules to small molecules. *Eur J Med Chem, 186*:111876.
Liu, X., et al. (2020). High-yield methylation markers for stool-based detection of colorectal cancer. *Dig Dis Sci, 65*(6):1710–1719.

Mallafré-Muro, C., et al. (2021). Comprehensive volatilome and metabolome signatures of colorectal cancer in urine: a systematic review and meta-analysis. *Cancers*, *13*(11):2534.
Marcuello, M., et al. (2019). Circulating biomarkers for early detection and clinical management of colorectal cancer. *Mol Aspects Med*, *69*:107–122.
Mauri, G., et al. (2021). The evolutionary landscape of treatment for BRAFV600E mutant metastatic colorectal cancer. *Cancers*, *13*(1):137.
Moiseyenko, V. M., et al. (2018). First-line cetuximab monotherapy in KRAS/NRAS/BRAF mutation-negative colorectal cancer patients. *Clinical Drug Investigat*, *38*(6):553–562.
Murer, P., et al. (2020). A novel human monoclonal antibody specific to the A33 glycoprotein recognizes colorectal cancer and inhibits metastasis. *MAbs*, *12*(1):1714371.
Murphy, C. C., et al. (2024). Declines in colorectal cancer incidence and mortality rates slow among older adults. *Clin Gastroenterol Hepatol*, *22*(2):416–419.e5.
NIH, National Library of Medicine, (2024). GPA33 glycoprotein A33 [Homo sapiens (human)] [internet, updated 2024]; available at: https://www.ncbi.nlm.nih.gov/gene/10223 [accessed Apr 2024].
Nikolouzakis, T.K., et al. (2018). Improving diagnosis, prognosis, and prediction by using biomarkers in CRC patients (Review). *Oncol Rep*, *39*(6):2455–2472.
Oellerich, M., et al. (2021). Liquid biopsies: donor-derived cell-free DNA for the detection of kidney allograft injury. *Nat Rev Nephrol*, *17*(9):591–603.
Ogawa, H., et al. (2002). Expression of CD10 by stromal cells during colorectal tumor development. *Hum Pathol*; *33*(8):806–811. Doi: 10.1053/hupa.2002.125773
Ogunwobi, O. O., et al. (2020). Biomarkers in colorectal cancer: current research and future prospects. *Int J Molec Sci*, *21*(15):5311.
Oh, H.H., Joo, Y.E. (2020). Novel biomarkers for the diagnosis and prognosis of colorectal cancer. *Intest Res*,*18*(2):168–183.
Ohnmacht, A. J., et al. (2023). The oncology biomarker discovery framework reveals cetuximab and bevacizumab response patterns in metastatic colorectal cancer. *Nature Communicat*, *14*(1):5391.
Opstelten, R., et al. (2021). GPA33 is expressed on multiple human blood cell types and distinguishes CD4+ central memory T cells with and without effector function. *Eur J Immunol*, *51*(6):1377–1389.
Papavassiliou, K. A., et al. (2024). Targeting the TGF-β signaling axis in metastatic colorectal cancer: where do we stand? *Int J Mol Sci* 2023 Dec 4; *24*(23):17101. Doi: 10.3390/ijms242317101
Patel, S. P., Kurzrock, R. (2015). PD-L1 expression as a predictive biomarker in cancer immunotherapy. *Molec Cancer Therapeut*, *14*(4):847–856.
Petit, J., et al. (2019). Cell-Free DNA as a diagnostic blood-based biomarker for colorectal cancer: a systematic review. *J Surg Res*, *236*:184–197.
Petrik, J., et al. (2022). Circulating tumor cells in colorectal cancer: detection systems and clinical utility. *Int J Mol Sci*, *23*(21):13582.
Pickhardt, P.J. (2016). Emerging stool-based and blood-based non-invasive DNA tests for colorectal cancer screening: the importance of cancer prevention in addition to cancer detection. *Abdom Radiol*, *41*(8):1441–1444.
Platet, N., et al. (2017). The tumor suppressor CDX2 opposes pro-metastatic biomechanical modifications of colon cancer cells through organization of the actin cytoskeleton. *Cancer Lett*, *386*:57–64.
Pös, Z., et al. (2020). Technical and methodological aspects of cell-free nucleic acids analyzes. *Int J Mol Sci*, *21*(22):8634.
Puccini, A., et al. (2022). Biomarkers in metastatic colorectal cancer: status quo and future perspective. *Cancers*, *14*(19):4828.
Requena, D. O., Garcia-Buitrago, M. (2020). Molecular insights into colorectal carcinoma. *Arch Med Res*, *51*(8):839–844.
Rex, D. K., et al. (2017). Colorectal cancer screening: recommendations for physicians and patients from the U.S. multi-society task force on colorectal cancer. *Gastroenterol* *153*(1):307–323.
Rutering, J., et al. (2015). Improved method for isolation of neonatal rat cardiomyocytes with increased yield of C-Kit+ cardiac progenitor cells. *J Stem Cell Res Ther*, *5*(6):1–8.
Saadi, S., et al. (2023). Circulating tumor cells in colorectal cancer: a review of detection methods and clinical relevance. *Contemp Oncol*, *27*(3):123–131.
Sahin, I. H., et al. (2021). Mismatch repair (MMR) gene alteration and BRAF V600E mutation are potential predictive biomarkers of immune checkpoint inhibitors in MMR-deficient colorectal cancer. *The Oncologist*, *26*(8):668–675. Doi: 10.1002/onco.13741

Sakata, S., Larson, D. W. (2022). Targeted therapy for colorectal cancer. *Surg Oncol Clin North Am*, *31*(2):255–264.
Salvi, S., et al. (2016). Cell-free DNA as a diagnostic marker for cancer: current insights. *Onco Targets Ther*, 9:6549–6559.
Saridaki, Z., et al. (2010). BRAF mutations, microsatellite instability status and cyclin D1 expression predict metastatic colorectal patients' outcome. *Br J Cancer*, *102*(12):1762–1768.
Satelli, A., et al. (2015). Epithelial-mesenchymal transitioned circulating tumor cells capture for detecting tumor progression. *Clin Cancer Res*, *21*(4):899–906.
Sedlackova, T., et al. (2013). Fragmentation of DNA affects the accuracy of the DNA quantitation by the commonly used methods. *Biolog Proced Online*, *15*:1–8.
Seeber, A., et al. (2018). Anti-angiogenics: their value in colorectal cancer therapy. *Oncol Res Treatm*, *41*(4):188–193.
Sharma, R., et al. (2022). Global, regional, and national burden of colorectal cancer and its risk factors, 1990–2019: a systematic analysis for the Global Burden of Disease Study 2019. *Lancet Gastroenterol Hepatol*, *7*(7):627–647.
Shaukat, A., Levin, T. R. (2022). Current and future colorectal cancer screening strategies. *Nature Rev. Gastroenterol Hepatol*, *19*(8):521–531.
Siegel, R. L., et al. (2020). Cancer statistics, 2020. *CA*, 70(1):7–30.
SinoBiology (n.d.). GPA33/A33 general information: GPA33/A33 protein, antibody, and gene [internet]. Available at: https://www.sinobiological.com/resource/a33 (accessed Apr 2024).
So, J. S., et al. (2017). Accuracy of preoperative local staging of primary colorectal cancer by using computed tomography: reappraisal based on data collected at a highly organized cancer center. *Ann Coloproctol*, *33*(5):192–196.
Sung, H., et al. (2021). Global cancer statistics 2020: GLOBOCAN estimates of incidence and mortality worldwide for 36 cancers in 185 countries. *CA*, *71*(3):209–249.
Tabone, T., Heath, J. K. (2008). GPA33 glycoprotein A33 (transmembrane), In: Atlas Genet Cytogenet Oncol Haematol, [internet] available at: https://atlasgeneticsoncology.org/gene/40735/gpa33-(glycoprotein-a33-(transmembrane) (accessed Apr 2024).
Tan, C. R. et al. (2016). Circulating tumor cells versus circulating tumor DNA in colorectal cancer: pros and cons. *Curr Colorectal Cancer Rep*, *12*(3):151–161.
Vacante, M., et al. (2018). Biomarkers in colorectal cancer: Current clinical utility and future perspectives. *World J Clin Cases*, *6*(15):869–881. Doi: 10.12998/wjcc.v6.i15.869
Van De Goor, R. M. G. E., et al. (2017). Feasibility of electronic nose technology for discriminating between head and neck, bladder, and colon carcinomas. *Eur Arch Oto-Rhino-Laryngol*, *274*:1053–1060.
Van Keulen, K. E., et al. (2020). Volatile organic compounds in breath can serve as a non-invasive diagnostic biomarker for the detection of advanced adenomas and colorectal cancer. *Aliment Pharmacol Therapeut*, *51*(3):334–346.
Vymetalkova, V., et al. (2018). Circulating cell-free DNA and colorectal cancer: a systematic review. *Int J Mol Sci*, *19*(11):3356.
Wang, C., et al. (2014). Noninvasive detection of colorectal cancer by analysis of exhaled breath. *Anal Bioanal Chem*, *406*(19):4757–4763.
Wang, S., et al. (2021). Clinicopathological and prognostic significance of PD-L1 expression in colorectal cancer: a meta-analysis. *Int J Colorect Dis*, *36*(1):117–130.
Wang, X., et al. (2016). PD-L1 expression in human cancers and its association with clinical outcomes. *OncoTargets Thet*, *9*:5023–5039.
WHO (2023) Colorectal cancer, [internet] available at: https://www.who.int/news-room/fact-sheets/detail/colorectal-cancer (accessed Jan 2024).
Wong, S. C., et al. (2009). Clinical significance of cytokeratin 20-positive circulating tumor cells detected by a refined immunomagnetic enrichment assay in colorectal cancer patients. *Clin Cancer Res 15* (3): 1005–1012. 10.1158/1078-0432.CCR-08-1515
Wu, C. (2018). Systemic therapy for colon cancer. *Surg Oncol Clin North Ame*, *27*(2):235–242.
Xie, Y.-H., et al. (2020). Comprehensive review of targeted therapy for colorectal cancer. *Signal Transduct Targeted Ther*, *5*(1):22.
Yang, L., et al. (2019). Prognostic and clinicopathological value of PD-L1 in colorectal cancer: Asystematic review and meta-analysis. *OncoTargets Ther*, *12*:3671–3682.
Yao, T., et al. (2023). Optimization of screening strategies for colorectal cancer based on fecal DNA and occult blood testing. *Eur J Public Health*, *33*(2):336–341.

Yu, J., et al. (2017). Metagenomic analysis of faecal microbiome as a tool towards targeted non-invasive biomarkers for colorectal cancer. *Gut, 66*(1):70–78.
Zhang, X., et al. (2022). Clinical benefits of PD-1/PD-L1 inhibitors in patients with metastatic colorectal cancer: a systematic review and meta-analysis. *World J Surg Oncol, 20*(1):93.
Zhang, Y., et al. (2023). Blockade of trans PD-L1 interaction with CD80 augments antitumor immunity. *Proc Nat Acad Sci USA, 120*(16):e2205085120.
Zhong, Y., et al. (2021). Role of phenotypes of circulating tumor cells in the diagnosis and treatment of colorectal cancer. *Cancer Manag Res, 13*:7077–7085.
Zygulska, A.L., Pierzchalski, P. (2022). Novel diagnostic biomarkers in colorectal cancer. *Int J Mol Sci, 23*(2):852.

12 Survivorship and Supportive Care

Praveen Halagali, Vamshi Krishna Tippavajhala, Himanshu Sharma and Harpreet Singh

12.1 INTRODUCTION

12.1.1 Background and Significance

At a five-year survival rate of 75.0%, colorectal cancer (CRC) ranks second in South Korea's cancer incidence rate, surpassing the country's overall cancer survival rate of 70.4%. A greater number of people have survived cancer thanks to early case diagnosis and technological advancements in cancer therapy. Treatment for CRC might differ depending on the stage of the disease. However, this results in permanent bodily alterations and several body related issues. Surgery is the mainstay of care for CRC patients (Russell et al., 2015). Depending on the location and stage, more chemotherapy and radiation therapy may be required. Long-term side effects from CRC surgery may include digestive problems, intestinal adhesions, and sexual dysfunction. For example, 6–24 months following surgery, 71.3% of patients who had rectal cancer suffered an excessive bowel movement (Schlairet et al., 2010). Common adverse effects of chemotherapy include baldness, vomiting, and nausea. When the therapy is stopped, these adverse effects ultimately go away. However, the quality of life (QoL) is worse in patients who develop peripheral neuropathy following surgery and oxaliplatin, the standard colon cancer treatment (Besora et al., 2018; Neibart et al., 2020). Moreover, side effects from radiation therapy include fecal incontinence and sexual dysfunction in individuals. CRC survivors face a variety of issues, as demonstrated by prior research a lowered quality of life, a restricted social life, unpleasant feelings like worry and melancholy, and financial hardships (Drury et al., 2017; Scheele et al., 2015). Survivors of CRC face greater obstacles than their healthy counterparts, including adjustments to their duties at home, cultural issues, and financial difficulties (Kotronoulas et al., 2017). In addition, survivors still worry about a recurrence. Because of this, they require a lot of emotional assistance to overcome their fear of recurrence (Hui, 2014). The need to handle late complications, avoid secondary cancer, and address psychosocial issues after treatment ends has grown along with the number of survivors of CRC . The quality of life of survivors has also drawn more attention. However, throughout treatment, the majority of medical personnel prioritize attending to patients' bodily demands, such as peripheral neuropathy and bowel movement abnormalities (Mayer et al., 2017). The term "supportive care" describes the essential services that individuals impacted by cancer require to satisfy their physical, social, emotional, spiritual, and informational requirements. Cancer survivors have information requirements about their course of care, prognosis, recovery, monitoring, sexuality, and body image (Ahn et al., 2017) When their therapy is finished, cancer survivors feel relieved. They are still unsure of what will happen next, though. Cancer survivors therefore have unfulfilled emotional, spiritual, or social needs at least somewhat. Reduced quality of life is common in cancer survivors who have unfulfilled physical demands as a result of side effects from therapy, such as tingling, exhaustion, and baldness. Accurate identification of patients' needs and concerns is necessary in order to provide a systematic approach to supporting any health challenges that survivors of cancer may encounter in the future. To guarantee that survivors receive proper care, factors impacting their requirements for supportive care must be recognized. We can create a supportive care program that is effective if we have a thorough grasp of the requirements of survivors (Mosher et al., 2016; Vodermaier & Linden, 2019).

The demands of cancer survivors for supportive care are impacted by emotional disorders such as sadness and anxiety. In actuality, clinically significant levels of anxiety and despair are experienced by CRC survivors. The QoL of survivors of CRC is significantly correlated with unmet supportive care demands. After cancer therapy, a survivor's QoL also varies over time. In light of their varying emotional states and quality of life, CRC survivors' requirements for supportive care must be investigated (Averyt & Nishimoto, 2014; Santin et al., 2015). The needs of CRC survivors for supportive care have been examined in a number of research. Multiple programs have been developed and implemented as part of ongoing attempts to manage survivors (Gordon et al., 2015; Han et al., 2019; Park et al., 2016). Research on the variables impacting CRC survivors' demands for supportive care is, nonetheless, lacking. One of the national cancer strategies in South Korea is the management of cancer

DOI: 10.1201/9781032704517-12

survivors (Yang, 2014). Cancer-related societies have conducted several research. Nevertheless, the majority of these cancer survivor studies have exclusively looked at breast cancer survivors (Angolkar et al., 2022; Appleton et al., 2013). Few studies has assessed survivors of CRC (Ashing-Giwa, 2005; Au et al., 2012).

12.1.2 Purpose of the Chapter

This study aimed to determine the degree of supportive care requirements and the impact of emotional state and quality of life on these needs in survivors of CRC. The study's conclusions will aid in the creation of initiatives aimed at managing CRC survivors.

12.2 CRC INCIDENCE AND EPIDEMIOLOGY

In 2012, CRC accounted for 1.4 million new cases and around 700.000 deaths worldwide. It is the third most often diagnosed cancer and the fourth largest cause of cancer-related mortality worldwide. By 2030, it is anticipated that the burden of cancer will rise by 60% to over 1.1 million cancer deaths and 2.2 million new cases. The burden of CRC is distributed unevenly, with over two-thirds of cases and over 60% of deaths taking place in nations with high or very high HDIs. There has been a rapid rise in CRC mortality and incidence in many medium-to-high HDI nations, especially in South America, Asia, and Eastern Europe. In contrast, some of the highest indexed HDI countries, including the United States, New Zealand, Australia, and Western European nations have seen a stabilization or decline in the incidence and death rates of CRC. It is believed that cancer prevention and early detection through screening are the causes of the decreasing trends in incidence in these nations. Improvements in chemotherapy, perioperative care, and radiation, in addition to the reasons that have led to decreases in incidence, will have played a part in the consistently declining trends in CRC mortality in many high-income countries (Bailey et al., 2015).

12.3 CRC SURVIVORSHIP

12.3.1 Defining Survivorship in CRC

The word "survivorship" refers to the ability to move on from a variety of traumas and tragedies, such as sexual assault, cancer, the atomic bomb, and the Holocaust. The notion of survivorship differs across and within situations, which further contributes to this uncertainty. There is a stark discrepancy in the definition of cancer survivorship, with various organizations and geographical areas promoting definitions that commence (1) at the time of diagnosis, (2) following the completion of treatment, or (3) after a predetermined amount of time has passed and the person is still free of the disease. The concept of cancer survival also takes into account whether or not terminally ill patients, as well as the friends and family of survivors, are included. highlighted the distinction between cancer "survivorship" and "survivor." The former speaks of the person who has received a cancer diagnosis, whereas the latter speaks of the person's continuing healthcare need (Bains et al., 2012; Baravelli et al., 2009; Brawarsky et al., 2013; Caravati-Jouvenceaux et al., 2011; Carpentier et al., 2013). Fitzhugh classified cancer survivorship into three separate "seasons" acute, protracted, and permanent in his seminal work *Seasons of Survival.* Acute survival starts with diagnosis and is typified by dread, worry, and uncertainty about the future in addition to diagnosis and treatment. At the time of remission or the conclusion of therapy, Extended Survival starts. Each person's experience with extended survival is unique and depends on their circumstances and level of adaptability. This phase is characterized by physical restrictions, dread of recurrence, isolation from healthcare professionals (HCPs), and watchful waiting. Lastly, a developing sense of healing characterizes lasting survival. However, the person's encounter with cancer has permanently altered their social, psychological, and physical makeup (Chambers et al., 2012a; Chambers et al., 2009; Chambers et al., 2012b; Cooper et al., 2008; Cragun et al., 2012). Numerous concept evaluations have looked at adult survival. These studies, however, concentrate on cancer survivability overall, chronic disease survivorship, and breast cancer survivorship among African American women. There hasn't been much research done to expand on the ideas generated by these concept analyses or look into the applicability of these conceptual models among cancer survivors of various kinds. Since survival rates have increased over the previous 20 years, cancer survivorship has just lately gained attention in Europe. As new, culturally varied perspectives on cancer survivorship surface, our grasp of the term will only deepen and expand. Forecast rises in the rate of CRC survival will support ongoing empirical focus on the experiences and results of CRC survivors (Brenner, 2002; Brenner et al., 2012; Custers et al., 2016; Kievit, 2002; Spoorthi Shetty et al., 2023). It is a profoundly transformative event that possesses a duality of universally applicable good and bad features that are specific to each individual" (Hewitt et al., 2006; Hewitt et al., 2007).

12.3.2 The Need for Survivorship Care

Survivorship from CRC is a unique stage of cancer care that addresses patients' long-term concerns. It ensures a smooth transition from treatment-focused cancer control to rehabilitation, recovery, and post-treatment monitoring (Grunfeld, 2006). SCPs are designed to provide comprehensive, interdisciplinary, and multifaceted treatment. A systematic program for detection and monitoring can help doctors and caregivers identify relapsed illness, and other noncancer disease progression, and treat and manage chronic and post-primary therapy complications. To help these survivors move as closely as feasible back to their premorbid condition, the care plan must also address the psychological implications of the disease and identify some of their unmet supportive care requirements (Green et al., 2002; Kattlove & Winn, 2003). The enormous volume of CRC survivors and their distinct requirements need the development and improvement of survivorship care at the institutional level. Due of the numerous follow-up appointments that each patient must attend, there are a lot of patients overall. Moreover, a CRC survivor's demands are very different from those of a typical cancer patient undergoing active therapy. A multidisciplinary team is needed to coordinate and maybe even customize a treatment care plan for every patient due to the range of long-term demands. Lastly, as survivorship care is still in its infancy, investing time and money in this emerging subject will undoubtedly provide a wealth of chances to influence the direction of service growth outside of specific institutions (Birgisson et al., 2005; Mäkelä et al., 1995).

12.3.2.1 Medical Needs

Cancer monitoring is arguably the most important necessity for CRC survivors. One of the biggest worries for cancer survivors is frequently the ongoing possibility of recurrence. They also run the risk of developing second primary CRCs, with 43% of these developing more than two years after the first diagnosis (Pietra et al., 1998; Secco et al., 2002). Survivors of CRC are also more likely to develop noncolorectal second primary malignancies, including breast, prostate, genitourinary, skin, and lung cancers (Chau et al., 2004; Glimelius, 1992; Lafata et al., 2005). Promoting early metastasis identification is a major goal of post-treatment monitoring for CRC patients. It will promote the possibility of curative resection and intervention through early identification. It is important to identify cancer recurrences before the onset of symptoms to improve survival rates. Research has indicated that asymptomatic CRC recurrences could respond better to R0 surgical resection (Earle & Neville, 2004; Phipps et al., 2008). For instance, Chau et al. found in a retrospective review of patients with stage III and II colon cancer that the proportion of asymptomatic patients who underwent potentially curative resection via computed tomography imaging was higher than that of patients whose relapse was identified by symptoms (26.5 vs. 3.1%). Furthermore, vigilant surveillance can identify asymptomatic diseases that are not treatable with curative medicine, enabling early palliative treatment. According to research conducted by the Nordic gastrointestinal tumor adjuvant therapy group, receiving palliative chemotherapy early on improves survival and delays the course of the disease (Earle et al., 2003). According to the American Society of Clinical Oncology's (ASCO) guidelines for the follow-up care of CRC, surveillance plans should be customized for each patient's functional level and risk of recurrence, which is based on the clinical stage and other biological characteristics of their particular disease. For a period of five years, a carcinoembryonic antigen testing, physical examination, and medical history should be done every three to six months. In terms of imaging, it is generally advised to get an annual CT scan (abdominal and chest) every three years. Depending on the results of the prior endoscopy, a surveillance colonoscopy should be done one year following the original procedure and then every three to five years after that. Additionally, they advise against using surveillance tests on patients who are not candidates for systemic treatment or surgery due to serious concomitant illnesses (Owen, 2012; Snyder et al., 2008). Identification of recurrence and patient survival depends on surveillance. The best way to provide cancer monitoring for these survivors has not yet been determined, despite standards having been set for follow-up times and modalities. When they go from active treatment to survivorship, the goal should be to coordinate investigations and promote compliance. Doctors must treat other medical problems in addition to monitoring for cancer recurrence and taking advantage of preventive medicine options. Eighty percent of CRC survivors had at least one comorbidity, indicating a significant burden of comorbidity. These prevalent comorbidities include psychological disorders, musculoskeletal disorders, and cardiovascular illness (Buffart et al., 2012; Lynch et al., 2008). Even among survivors who have gone more than five years without a cancer diagnosis, the co-occurrence of comorbidity and a cancer diagnosis will result in reduced productivity, increased functional restrictions, and a worse quality of life in terms of health (Ibrahim & Al-Homaidh, 2011; Mosher et al., 2009; Peddle et al., 2008). To investigate patterns of care, Earle and Neville compared five-year CRC survivors to matched noncancer controls. They discovered that cancer survivors were less likely to obtain recommended care for long-term problems. Cancer survivors under the care of PCPs and oncology specialists had a higher chance of receiving adequate treatment; PCPs were more likely to propose noncancer-related care, while oncology specialists were more likely to prescribe cancer-related care. The results of research conducted on breast cancer survivors were comparable. Comprehensive instructions about the frequency and severity of follow-up have been provided to survivors of CRC. The SCP still has a lot of holes, though, including who should follow up with patients and if they should be screened for other conditions unrelated to cancer (Ibrahim & Al-Homaidh, 2011; Meyerhardt et al., 2006).

12.3.2.2 Physical Activity in CRC Survivors

According to a study, sedentary activity is defined as "waking activities performed in a seated or reclining posture that require very low energy expenditure." Cancer survivors report engaging in sedentary behavior for up to 70% of the day. Regular and prolonged moderate-to-vigorous physical exercise is linked to improved health-related QoL and less tiredness symptoms in survivors of CRC, according to an increasing body of research. Although many cancer survivors are interested in changing their lifestyle, poor health habits are rather common. The study, which included 753 elderly long-term survivors of CRC, prostate, and breast cancer, also found some risk variables for unhealthy lifestyle choices. Reduced physical functioning and increased role constraints as a result of physical issues were linked to older age, less education, and more comorbidities (Isomura et al., 2006; Nilsen et al., 2008; Wolin et al., 2009). Exercise may also lower the incidence of cancer death and recurrence among survivors of breast and CRC, according to data. Meta-analyses of over 60 studies provide compelling evidence of a preventative impact on colon cancer, with a reduction of 20–25% in risk for those with the greatest degree of physical activity compared to those with the lowest level. Physical activity does not depend on the source since there is a dose-response impact and both leisure and work activities reduce risk comparably. Such relationships were discovered in additional research on European and Asian ethnicities (Larsen et al., 2006; Larsson et al., 2006; Zhang et al., 2006). The general population is not well-informed about the protective effects of physical exercise against colon cancer, despite these compelling findings. Currently, the NCIC Clinical Trials Group in Canada is leading research named challenge to investigate if exercise might increase cancer patients' chances of survival. Aside from the case of cancer, physical activity generally reduces the risk of metabolic syndrome, falls, diabetes, cardiovascular disease, and premature death. It also enhances cardiopulmonary fitness, older adults' cognitive function, sleep quality, and bone density. Combined with calorie restriction, physical activity also helps with weight loss and maintenance. Comes to managing the comorbidities of CRC survivors, all of these are crucial. Physical exercise was shown to be safe both before and after cancer treatment, according to a meta-analysis of physical activity intervention studies in a variety of cancer survivor groups. Exercise has been demonstrated to have potential benefits in reducing the risk of CRC recurrence as well as the comorbidities that cancer patients must deal with. The least amount of time and minimal intensity that patients with CRC should be encouraged to participate in should be further investigated. Also, it has been found that the general population is not well-informed about the benefits of physical activity in preventing CRC. Subsequent studies can be conducted to assess the most effective strategies for raising public awareness and encouraging patient activity (see Table 12.1) (Coups et al., 2008).

12.3.2.3 The Demands of Survivors of CRC in Terms of Behavior

Patients and survivors with potentially fatal illnesses like cancer will also seek information on their condition, the therapy they had, any side effects, the likelihood of a recurrence, and other self-care strategies. Research looked at 619 cancer survivors and the many elements that support information-seeking behavior. Although the majority of survivors prefer to get cancer information from their normal healthcare professional, they discovered that almost two out of every three survivors sought information about the disease. The study demonstrated that by supplying social

Table 12.1: Focusing on physical activity in CRC survivors

S.no	Aspects	Recommendations/ Considerations	Benefits	References
1	Types of physical activity	Aerobic exercises (e.g., walking, cycling)	Improves cardiovascular health	(Speck et al., 2011).
2	Frequency	Include strength training exercises 2–3 times per week	Reduces the risk of recurrence	(Mayer et al., 2007).
3	Duration	Break activity into manageable sessions throughout the day if needed	Boosts energy levels	(Sahoo et al., 2023).
4	Intensity	Gradual progression based on individual fitness levels	Improves mood and mental well-being	(Cassileth et al., 1980).
5	Tailoring exercise	Adapt exercises to individual abilities and preferences	Reduces fatigue and anxiety	(Derdiarian, 1986).
6	Post-surgery considerations	Gradual return to physical activity after surgery, following healthcare provider advice	Facilitates recovery	(Fallowfield et al., 1995).
7	Monitoring and safety	Regularly assess for any adverse effects or discomfort during and after exercise	Improves sleep quality	(Leydon et al., 2000).

support and regulating uncertainty and anxiety, knowledge-seeking also helps with coping. (Rees & Bath, 2000).

Numerous other research on behavioral seeking have confirmed the currently accepted paradigm, which states that most cancer survivors want to know as much as they can about their condition and course of treatment. Not every victim, though, would choose that level of disclosure. The intended time and quantity of information differed depending on whether the person was attempting to keep hope alive or had a blunting approach (avoid scary information. Research has been done on the advantages of knowledge for cancer patients, especially those with breast cancer. Even if there hasn't been much study done on CRC patients and the advantages of having enough knowledge, it's still important to acknowledge the benefits of information searching. These advantages include reduced anxiety and mood disorders, enhanced coping strategies throughout the diagnostic, treatment, and post-treatment phases, and increased patient engagement in decision-making, which translates to higher satisfaction with treatment choices. and enhanced family communication. In a study of survival information requirements and preferences, researchers examined CRC survivors. The study found that awareness of the most important components of diagnosis and treatment was typically accurate among people who did not obtain a survival strategy. When combined, the study's survivors mainly remembered basic details about their diagnosis and course of treatment, but they remembered less about their stage and continuing dangers. Additionally, respondents reported receiving medical information regarding their illness and its follow-up more often than nonmedical information on things like financial and insurance difficulties, support groups, and therapy. The dearth of information about nonmedical concerns, such as practical difficulties and psychosocial assistance, emphasizes the necessity of interdisciplinary care for cancer survivors. (Harrison-Woermke & Graydon, 1993; Luker et al., 1995).

12.3.2.4 Emotional Needs and Quality of Life

Although individuals who have survived CRC often report a high quality of life, receiving a cancer diagnosis can hurt one's social and functional health. Years after the cancer diagnosis, side effects from the therapy, such as physical limits, weariness, depression, and a poor body image, may still exist. These are especially true for patients undergoing radiation therapy, stoma-related surgeries, elderly or female survivors, as well as those who are overweight, obese, or have poor incomes. According to Patients with a diverting ostomy were more likely than others to feel dissatisfied with their physical appearance. (Damian & Tattersall, 1991). One of the most significant effects of cancer survival is QoL, which is the subjective assessment of an individual's happiness with their general health and well-being. This includes social, psychological, functional, and physical functioning. The majority of survivor quality of life reports have demonstrated that these survivors of various cancers had lower mental and physical QoL compared to age-matched persons who were cancer-free. One year following diagnosis and therapy, the most common concerns expressed were tiredness (67%), sleep difficulties (48%), and fear of recurrence (68%). Older long-term survivors utilized more mental health services and had worse psychological outcomes than those without cancer, despite the fact that they typically reported lower levels of worry and sadness. This is a problem for those who have survived CRC since the age-adjusted risk of CRC is more than three times greater in those over 65 than in those between the ages of 50 and 64 (Johnson et al., 1988; Rainey, 1985). In terms of mental state, the CES-Depression Scale indicated that the depressed symptoms of CRC survivors were not particularly high at diagnosis and gradually improved within the first year following therapy. improved social networks, greater community participation, and no restrictions on their regular social activities were associated with improved mental health outcomes for survivors. Even with modest levels of depressive symptoms, 26–44% of long-term survivors of CRC persisted in worrying about future diagnostic tests, symptoms serving as indications of cancer, recurrence of the disease, or obtaining a second malignancy. Anxiety and sadness were linked to concerns about cancer-related health, with 24% of survivors expressing depression levels high enough to require a clinical depression diagnosis. Therefore, it is necessary to evaluate all CRC survivors for psychological distress, sadness, exhaustion, and pain since this might reveal problems that need to be addressed to enhance their quality-of-life following therapy. Furthermore, there is always a chance that those who have survived CRC will have to deal with the long-term effects of having a stoma created following surgery. The quality of life may be impacted by a permanent stoma. Patients who had surgery, radiation therapy, and permanent stoma for locally advanced rectal cancer reported reduced pain, exhaustion, and bowel symptoms after treatment. However, even while global quality of life was unaffected, having a stoma has also been linked to negative body image and greater financial anxiety. There is a strong, albeit small, correlation between intentions to engage in health-related activities and perceived risk and concern in CRC survivors. While it is important to address the emotional needs of CRC survivors, if fear is minimized and appropriately managed by a doctor, it can serve as motivation for individuals to adopt healthier lifestyles. (Johnson & Adelstein, 1991; Meredith et al., 1996).

12.4 TREATMENT-RELATED COMPLICATIONS

Survivors of CRC will also need to manage the ongoing side effects of their therapies. Along with the surgery, patients will need to endure, their therapy comes with several drawbacks and side effects. Some of the patients will experience peripheral neuropathy caused by oxaliplatin as a side effect of chemotherapy. Although their effectiveness is uncertain, opioids, nonsteroidal anti-inflammatory drugs, and referrals to neurology or pain management for persistent symptoms are advised. Bowel dysfunction is yet another typical adverse effect. Up to ten years following therapy, 13–50% of patients report having chronic diarrhea. Diarrhea can have a detrimental impact on quality of life and has been linked to activity limits. Following CRC therapy, antidiarrheal drugs, bulk-forming agents, the use of pads, and diet modification are commonly used to manage gastrointestinal issues. Probiotic supplementation and the avoidance of particular foods have not shown any effect. Doctors should try to ask patients about their bowel habits frequently. Patients might believe that their therapy would inevitably result in atypical bowel symptoms. Unless they are routinely questioned, they might not bring up these issues in their follow-up sessions (see Table 12.2) (Samuelsson et al., 2022; Son et al., 2018).

12.5 PROBLEMS FACED BY CRC PATIENTS

12.5.1 Striking for Self-Assurance

The cancer specialist nurses (CSNs) observed that both the individuals with the CRC diagnosis and their family members suffer feelings of panic, worry, and uncertainty following the diagnosis. As a result, the CSNs work to ensure that family members feel secure and taken care of. The CSNs described how they actively listen and make an attempt to validate and normalize feelings to offer assistance. Moreover, they reassure relatives that the individual with a CRC diagnosis would receive the finest treatment available and request that they put their faith in them to monitor everything honestly and completely, saying, "We are very clear about that what we know we will

Table 12.2: Treatment-related complications in survivorship and supportive care for CRC patients

Complication	Strategies/Interventions	Considerations	Reference
Fatigue	Gradual introduction of physical activity	Regular sleep and relaxation techniques	(Forsberg et al., 1996)
Gastrointestinal issues	Dietary modifications (e.g., fiber intake adjustment)	Hydration and electrolyte balance	(Hamashima, 2002)
Neuropathy	Gentle exercises for nerve function	Medication adjustments as needed	(Sapp et al., 2003)
Emotional distress	Psychosocial support groups and counseling	Integration of mindfulness practices	(Trentham-Dietz et al., 2003)
Pain and Discomfort	Medications for pain management	Alternative therapies (e.g., acupuncture)	(Vacante et al., 2012)

let them know." In addition, the CSNs provide family members with information on other services, such as nutritionists, so they won't have to worry about what to prepare for loved ones who have been diagnosed with CRC. By educating the family about the diagnosis, symptom management, anticipated course of treatment, and significant milestones, the CSNs also provide them with a framework to cling to. However, despite their best attempts to inform family members of what to expect, there are still instances in which a person's physical changes after receiving a CRC diagnosis can come as a shock and be unanticipated, and the intensity of their symptoms can cause anxiety for their family. Being consistent along the trajectory is another method to aim for confidence, according to CSNs. Family members were reassured by the CSNs that they could get in touch with them if they had queries or concerns. The CSNs make an effort to respond to all calls and inquiries, stating that they are aware of the location and method of accessing the extensive healthcare system. They will call and feel comfortable if they have a name, a number, and a known face since that is how it operates (CSN 4). This can occasionally result in family members calling CSNs when their loved one is undergoing surgery or receiving cancer treatment, even though other clinics should have been responsible for their care (Lundy et al., 2009).

12.5.2 Searching for Ways to Support

The CNSs in this study's description of the Swedish CRC structure include a limited number of guaranteed contacts between the CSNs and family members, usually limited to the time of diagnosis. Contact with individuals who receive a CRC diagnosis later in the disease's trajectory is often done over the phone, and family members are only contacted at their request. The CSNs must take advantage of the opportunity to try to give emotional support during the initial meeting to fulfill their obligation. When a doctor notifies a family member of a patient's diagnosis of CRC, CSNs indicate they attempt to offer support through nonverbal communication, even if there isn't a set time for their appointments with the patient and family. Developing a bond with the family members is another tactic they might use to attempt to fulfill their obligation despite their limited resources. According to the CSNs, family members' solid relationship feels supported even in the absence of communication. Moreover, this kind of relationship usually raises the possibility that family members will get in touch with them later on if they need assistance. This was viewed by some CSNs as a sufficient means of providing for their families. For some, however, it was a compromise: "This is what I would have wanted to do, to be able to get in touch with and follow up, for family members as well." Being unable to follow up causes the CSNs to question if the assistance was truly satisfying the requirements, especially since they believed that supportive care needs would evolve over time. For example, family members may first refuse a therapist, but if communication with them becomes irregular or nonexistent, they may later want one (Gray et al., 2011).

12.5.3 Seeking Individualization

The CSNs believed that it was critical to evaluate each individual family member's supportive care requirements due to the variety of the members. Nevertheless, needs evaluations for family members lack the same frameworks as those for individuals with CRC. Rather, the CSNs attempt, with differing degrees of participation, to determine the requirements by employing open-ended inquiries, interpreting nonverbal cues, and focusing on tone to ascertain needs that are both immediate and potentially future. The CSNs stated that to offer sufficient assistance, they need to determine how CRC affects the particular family's daily activities. To accomplish this, they must evaluate the circumstances at home and understand the distribution of household duties among the family members. Additionally, they make an effort to understand the dynamics inside the family. It is crucial to attempt to understand the dynamics of the family because even if they have a caring and supportive bond, not all family members are eager to talk about symptoms or engage in direct caregiving. On whether or not to address any recognized issues with familial relationships, there is disagreement. While some CSNs advocated for a rigid coordinating CSN position, others accepted this as a component of a therapeutic CSN role. Needs related to family relationships were viewed as family matters by the latter. The CSNs plan the treatment of individuals with CRC based on the assessment (Ko et al., 2003).

12.5.4 Balancing between Needs

It is not often simple to meet the requirements of family members since they may conflict with those of CSNs or people who have been diagnosed with CRC. Swedish law prohibits CSNs from sharing health information with relatives unless the individual receiving the cancer diagnosis gives consent. Therefore, CSNs are powerless to intervene if family members pressure them to discuss the prognosis without the person who has been diagnosed with CRC present and the diagnosed person objects. Participants indicated they tried to work around the issue by telling the family about the prognosis, in general, to still try to address the demands of the family members. Participants also discussed the challenge of family members seeking guidance on how to persuade individuals with CRC not to modify their treatment choices: "Otherwise, it is a little difficult if the patient does not want to and the children are persistent." It's not a simple task. However, the patient retains the last say in the matter. Also, family members might not be aware of what is ideal for a person who has been diagnosed with CRC. For example, family members demand eating during rehabilitation or may push for relaxation while the CSNs advise activities. This frustrates people and makes it necessary for CSNs to speak out in support of those who have been diagnosed with CRC. Supporting family members also has the added benefit of helping the CSNs since self-assured and less anxious family members facilitate easier operations. Families with confidence are essential to CSNs because they guarantee that no information is lost and create a nurturing atmosphere at home, which includes keeping an eye on symptoms. This is especially crucial as the majority of people with CRC are elderly, meaning that the CSNs' duty would be lessened because they wouldn't have to worry about them. The CSNs stated that they attempt to convince family members of the recognized advantages when the person with CRC decides not to include any of them (Deimling et al., 2006).

12.6 PSYCHOSOCIAL INTERVENTIONS

Psychosocial interventions are helpful supplements to medical treatment in the context of cancer care. They include therapeutic communication that addresses patients' informational and psychosocial needs in order to help minimize stress, enhance quality of life, lessen depression, and support cancer patients through their diagnosis and recovery. Psychoeducation, family therapy, stress management training, support groups, and cognitive-behavioral therapy are examples of common psychosocial therapies. Systematic studies have demonstrated the beneficial effects of psychosocial therapies, such as self-care, psychoeducational, and social and emotional support, in enhancing the quality of life and reducing stress anxiety, and depression in cancer patients. In addition to being cost-effective, these therapies shorten hospital stays. Supportive therapies have been demonstrated to successfully reduce psychological distress, boost patients' fighting spirit, and minimize fatalism in patients with stomach, colorectal, and breast cancer. These interventions include listening, validation, stress management, and problem-solving. In conclusion, not enough research has been done to determine how colorectal patients' results are affected by pre- and post-operative psychological therapies. The forms and viability of psychological therapies for patients with CRC are not well understood (Paika et al., 2010).

12.7 DETERMINING THE QUALITY OF LIFE

Patients with CRC are linked to several variables. We categorized these elements into five main groups to make the conversation easier to understand sociodemographic traits, elements about health, variables associated with cancer and surgical treatments, factors about lifestyle, and other components.

12.7.1 Socio-demographic Characteristics

The findings regarding CRC QoL and age are debatable. Some researchers found that QoL increased with age, while other studies found that QoL decreased with age. There is debate over the psychological and physical aspects of quality of life. Because wealth has a greater influence on QoL than education, education level does not determine QoL. There is evidence to suggest that lower income is associated with lower quality of life in terms of physical, social, and emotional well-being. Additionally, having a house was a separate predictor of QoL (Abdelsattar et al., 2016).

12.7.2 Health-Related Factors

Compared to patients without cancer, patients with CRC reported much higher rates of comorbid disorders and lower physical and mental QoL. Patients with two or more comorbidities or those who had just been diagnosed experienced even worse outcomes. Overall QoL was significantly impacted by a few particular conditions, including heart disease, anxiety/depression, and urine disorders. For example, concerns about a cancer recurrence or second instance, even five years or more after the initial diagnosis, may account for a portion of the greater than average frequency of depression among people of comparable age in the general population. When it comes to the relationship between body mass index and quality of life, cancer survivors who were overweight or of healthy weight reported greater levels of vitality, overall health, and physical functioning than those who were obese (Kneuertz et al., 2015).

12.7.3 Cancer-Related and Surgical Procedure Factors

The symptoms, treatment options, and length of therapy are all influenced by the stage and location of CRC at diagnosis, which is significant in deciding quality of life. Patients in stage I showed a steadily improving trend in their quality of life, whereas those in stage IV showed a declining tendency. Rather, those with stage II and stage III cancer saw a decline in QoL at first, which was followed by an improvement in QoL ratings. One reason that could apply is how one views (or rethinks) quality of life following a diagnosis of CRC. However, other research found no connection between QoL and tumor stage. Surgery can have an impact on a number of QoL factors because of its physical and psychological aftereffects. Following surgery, patients had a sharp reduction in their QoL, which progressively improved around three months later. Also, elderly folks take longer to regain their quality of life (Vatandoust et al., 2016).

12.7.4 Other Factors

A few lifestyle variables, including exercise, food, alcohol use, and smoking, has been linked to quality of life. Because moderate to intensive physical exercise reduces weariness and discomfort, it is associated with an improved physical quality of life. Probiotic medication and a healthy diet high in fruits and vegetables and low in fat decreased gastrointestinal dysfunction, which can significantly lower quality of life. Alcohol use has been linked to a worse quality of life when smoking was involved. A change in one behavior has a lesser cumulative impact on quality of life than a change in one lifestyle factor (Wang et al., 2015).

12.8 SURVIVORSHIP CHALLENGES FOR EARLY-ONSET COLORECTAL CANCER (EOCRC)

The survival rate of EOCRC patients is found to be on par with, if not better than, that of Late-Onset Colorectal Cancer (LOCRC) patients, despite the latter's aggressive character and advanced stage upon diagnosis. The data from SEERs indicates that those under 50 had a higher stage-specific survival rate than those over 50. However, when survival was broken down by age group, it was shown that people under 35 often had a bad outlook. This may point to a distinct molecular and biological profile that makes the illness more likely to be lethal and aggressive. Age is actually thought to be a poorer prognostic factor, meaning that the prognosis is worse the younger the patient. More young adult cancer patients are surviving the disease as a result of recent advancements in cancer therapy, which have increased these patients' survival rates by over 80% (Lieu et al., 2014).

Young CRC survivors deal with the same problems as their older counterparts, including long-term side effects from therapy, malignancy, fear of recurrence, and other psychological problems (Figure 12.1). A customized survival strategy is necessary because of the distinct survivor experiences that young patients go through at various transitional phases, as revealed by a recent qualitative study. The majority of their psychological suffering stems from having to change their life plan to accommodate the unwanted guest. Numerous physical and psychological symptoms, including pain, persistent tiredness, anorexia, dyspnea, mental disorientation,

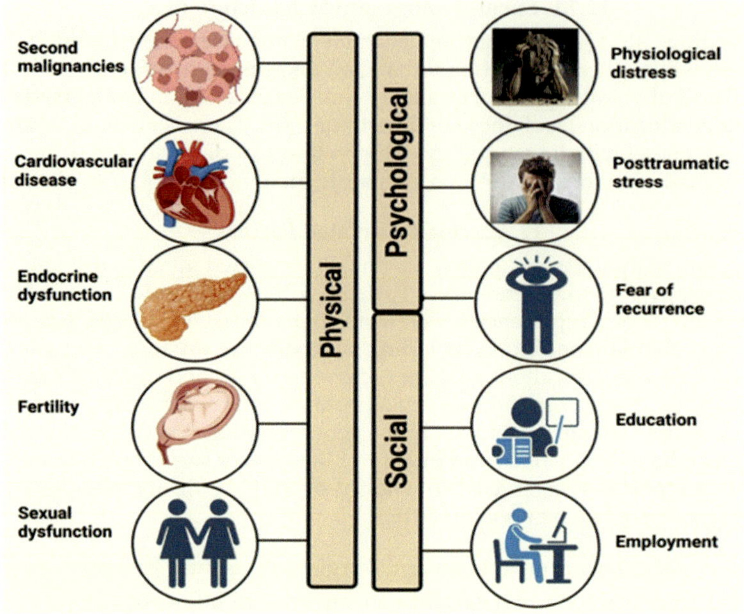

Figure 12.1 Common challenges perceived by young adult survivors of CRC.

intestinal blockage, genitourinary issues, and others, have been linked to either cancer or treatment-related side effects. The care of these patients must take into account several underappreciated concerns, including mental health, employment and financial security, sexual dysfunction, family planning, and psychological disorders. It seems that young adults must bear heavy symptom loads to maintain their independence. As a result, these patients should get proactive intervention and support centered on psychological, spiritual, physical, and mental issues (Liu et al., 2019).

12.9 PREVENTION AND RISK REDUCTION

By avoiding modifiable risk factors and detecting the disease early, CRC may be substantially prevented. EOCRC was believed to have a close connection to genetic disorders. According to epidemiological research, hereditary family cancer syndromes account for less than 5% of EOCRC cases. Less than 1% are brought on by inflammatory bowel disorders, and about 5% are hereditary. The majority are irregular EOCRCs. Sporadic CRC development is thought to be complex. The majority of the blame has been placed on modifiable risk factors such as high-calorie consumption, sedentary behavior, and physical inactivity (Bhupathiraju & Hu, 2016; Nguyen et al., 2018). Together, these variables cause obesity, which is the result of an energy imbalance. This imbalance has been shown to be a substantial risk factor for CRC with an early start, particularly in women. Therefore, it is believed that the widespread adoption of sedentary lifestyles and poor eating habits is to blame for the recent steady growth in EOCRC incidence seen around the world. The rising rates of smoking, diabetes mellitus (DM), and obesity are further contributing factors. It was discovered that DM was linked to a 20–38% higher incidence of CRC. The link was discovered to be very reliant on the length of DM; patients with DM for ≥10 years showed a significant association, but those with DM for <10 years showed no significant association (Figure 12.2) (Larsson et al., 2005).

Examining this assertion closely reveals that the incidence of DM, smoking, and sedentary lifestyles has grown across all age groups, not only young people. As a result, it is unable to adequately account for the rising prevalence of CRC in those under 40. In addition, compared to LOCRC patients, Chen et al. showed that the rate of obesity and overweight is lower in EOCRC patients (Kallergi et al., 2015; Yuhara et al., 2011). Additionally, physicians have frequently noted that EOCRC patients are typically well-integrated, active individuals without any comorbidities. It is impossible to determine with precision what causes what and how risk variables are related because they are all observational research Conversely, a meta-analysis has demonstrated that

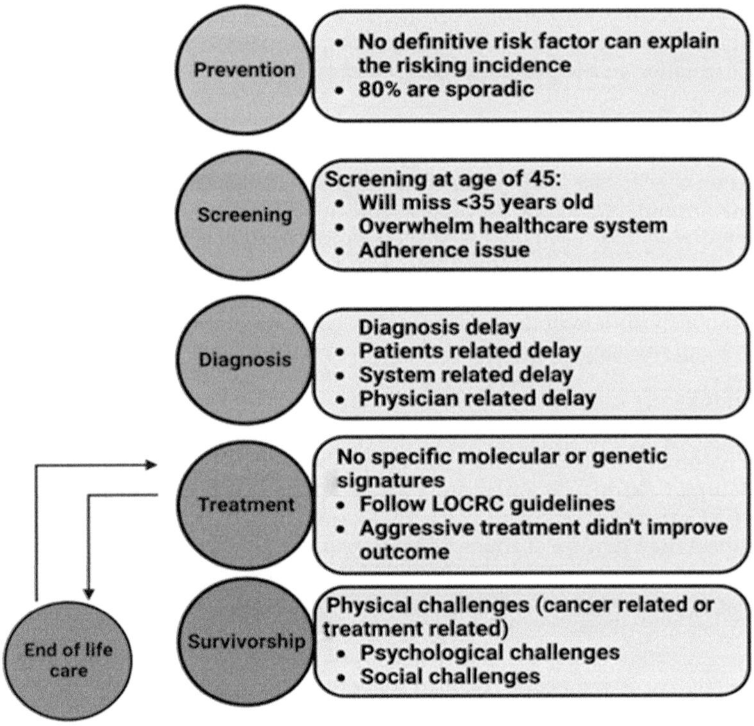

Figure 12.2 Challenges faced by young adult with CRC across the cancer care continuum.

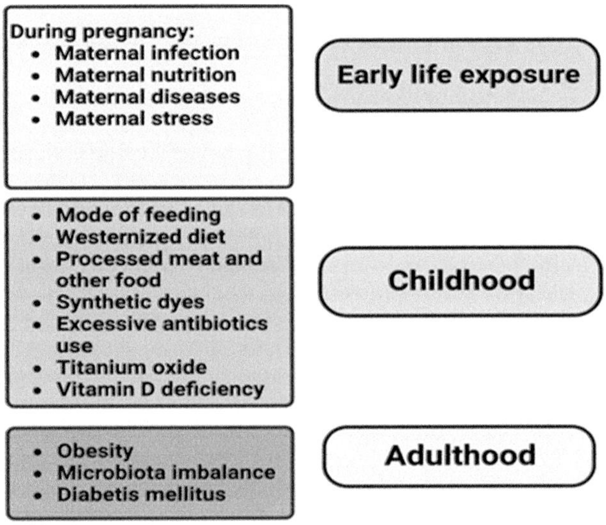

Figure 12.3 The reported accumulative risk factors for EOCRC.

engaging in physical exercise was linked to a noteworthy 30% decrease in the risk of colon cancer (Chen et al., 2017). It is important to note that research is still being done to determine how these characteristics are related to EOCRC. The risk of CRC in adulthood is assumed to be an accumulative risk (Figure 12.3), starting with exposure while still in the womb, delivery method, early life feeding strategy, and ending with all environmental exposure during childhood. These elements are believed to interact to disrupt the colon microbiota, increasing the risk of inflammation and malignant transformation in the process (Puzzono et al., 2021).

12.10 FUTURE PERSPECTIVE

Uptil now, a large portion of the efforts to redesign care for those "living with and beyond" cancer has concentrated on cancer "survivors," or those who have finished their treatment and are trying to find their "new normal," which could involve putting their cancer experience behind them or learning to deal with the long-term effects of their cancer or its treatment. To guarantee that the needs of people who have cancer for a long time but do not survive it are satisfied, additional work will need to be done in the future. The requirements of those with cancer that are treatable but not curable are still not well understood, and better data gathering has only lately begun to hint at the possible magnitude of this population.

12.11 CONCLUSION

Greater resources must be made available to CSNs for them to fulfill their responsibility of assisting family members of CRC patients. CSNs may be able to discover multidimensional requirements by undertaking frequent systematic evaluations of family members' needs while adhering to the CRC patient's treatment plan. This would highlight the need for supportive care that family members may require at times when they have little or no interaction with medical experts. By defining the needs in this way, it may be possible to emphasize the need for support and make the case for setting priorities for work and resources. Then, customized support is organized, arranged, and monitored through the creation of supportive care plans. However, it appears that a reassessment of the job as a whole is essential if CSNs are to lead supportive care. Crucially, administrative or medically focused responsibilities cannot take precedence over essential nursing responsibilities. Critical analysis of the centering of care also seems to be important. Specifically, family members should be involved and supported for both their benefit and the benefit of the person with a CRC diagnosis.

REFERENCES

Abdelsattar, Z. M., et al. (2016). Colorectal cancer outcomes and treatment patterns in patients too young for average-risk screening. *Cancer*, *122*(6):929–934.

Ahn, S., et al. (2017). Quality of life among Korean gastrointestinal cancer survivors. *Eur J Oncol Nursing*, *30*:15–21.

Angolkar, M., et al. (2022). Nanotechnological advancements in the brain tumor therapy: a novel approach. *Therapeut Delivery*, *13*(11):531–557.

Appleton, L., et al. (2013). Patients' experiences of living beyond colorectal cancer: a qualitative study. *Eur J Oncol Nursing*, *17*(5):610–617.

Ashing-Giwa, K. T. (2005). The contextual model of HRQoL: a paradigm for expanding the HRQoL framework. *Quality of Life Res*, *14*(2):297–307.

Au, T. Y., et al. (2012). Health-seeking behaviors and sexuality in rectal cancer survivors in Taiwan: associations with spirituality and resourcefulness. *Oncology Nursing Forum*, *39*:5.

Averyt, J. C., Nishimoto, P. W. (2014). Psychosocial issues in colorectal cancer survivorship: the top ten questions patients may not be asking. *J Gastrointest Oncol*, *5*(5):395–400.

Bailey, C. E., et al. (2015). Functional deficits and symptoms of long-term survivors of colorectal cancer treated by multimodality therapy differ by age at diagnosis. *J Gastrointest Surg*, *19*(1):180–188.

Bains, M., et al. (2012). Helping cancer survivors return to work: what providers tell us about the challenges in assisting cancer patients with work questions. *J Occupat Rehabil*, *22*(1):71–77.

Baravelli, C., et al. (2009). The views of bowel cancer survivors and health care professionals regarding survivorship care plans and post treatment follow up. *J Cancer Survivorship*, *3*(2):99–108.

Besora, S., et al. (2018). Rechallenge with oxaliplatin and peripheral neuropathy in colorectal cancer patients. *J Cancer Res Clin Oncol*, *144*(9):1793–1801.

Bhupathiraju, S. N., Hu, F. B. (2016). Epidemiology of obesity and diabetes and their cardiovascular complications. *Circulation Res*, *118*(11):1723–1735.

Birgisson, H., et al. (2005). Occurrence of second cancers in patients treated with radiotherapy for rectal cancer. *J Clin Oncol*, *23*(25):6126–6131.

Brawarsky, P., et al. (2013). Surveillance after resection for colorectal cancer. *Cancer, 119*(6):1235–1242.
Brenner, H. (2002). Long-term survival rates of cancer patients achieved by the end of the 20th century: a period analysis. *Lancet, 360*(9340):1131–1135.
Brenner, H., et al. (2012). Progress in colorectal cancer survival in Europe from the late 1980s to the early 21st century: the EUROCARE study. *Int J Cancer, 131*(7):1649–1658.
Buffart, L. M., et al. (2012). Self-reported physical activity: its correlates and relationship with health-related quality of life in a large cohort of colorectal cancer survivors. *PLoS ONE, 7*(5):1–9.
Caravati-Jouvenceaux, A., et al. (2011). Health-related quality of life among long-term survivors of colorectal cancer: a population-based study. *Oncologist, 16*(11):1626–1636.
Carpentier, M. Y., et al. (2013). Receipt of recommended surveillance among colorectal cancer survivors: A systematic review. *J Cancer Survivorship, 7*(3):464–483.
Cassileth, B. R., et al. (1980). Information and participation preferences among cancer patients. *Ann Intern Med, 92*(6):832–836.
Chambers, S. K., et al. (2009). Relationship over time between psychological distress and physical activity in colorectal cancer survivors. *J Clin Oncol, 27*(10):1600–1606.
Chambers, S. K., et al. (2012a). Survivor identity after colorectal cancer: antecedents, prevalence and outcomes. *Psycho-Oncol, 21*(9):962–969.
Chambers, S. K., et al. (2012b). A five-year prospective study of quality of life after colorectal cancer. *Quality of Life Res, 21*(9):1551–1564.
Chau, I., et al. (2004). The value of routine serum carcino-embryonic antigen measurement and computed tomography in the surveillance of patients after adjuvant chemotherapy for colorectal cancer. *J Clin Oncol, 22*(8):1420–1429.
Chen, F. W., et al. (2017). Advanced-stage colorectal cancer in persons younger than 50 years not associated with longer duration of symptoms or time to diagnosis. *Clin Gastroenterol Hepatol, 15*(5):728–737.e3.
Cooper, G. S., et al. (2008). Receipt of guideline-recommended follow-up in older colorectal cancer survivors: a population-based analysis. *Cancer, 113*(8):2029–2037.
Coups, E. J., et al. (2008). Awareness of the role of physical activity in colon cancer prevention. *Patient Educat Counseling, 72*(2):246–251.
Cragun, D., et al. (2012). Colorectal cancer survivors' interest in genetic testing for hereditary cancer: Implications for universal tumor screening. *Genet Testing Molec Biomark, 16*(6):493–499.
Custers, J. A. E., et al. (2016). Fear of cancer recurrence in colorectal cancer survivors. *Supportive Care Cancer, 24*(2):555–562.
Damian, D., Tattersall, M. H. N. (1991). Letters to patients: improving communication in cancer care. *Lancet, 338*(8772):923–925.
Deimling, G. T., et al. (2006). Cancer-related health worries and psychological distress among older adult, long-term cancer survivors. *Psycho-Oncol, 15*(4):306–320.
Derdiarian, A. K. (1986). Informational needs of recently diagnosed cancer patients. *Nursing Res, 35*(5):276–281.
Drury, A., et al. (2017). Cancer survivorship: advancing the concept in the context of colorectal cancer. *Eur J Oncol Nursing, 29*:135–147.
Earle, C. C., et al. (2003). Quality of non-breast cancer health maintenance among elderly breast cancer survivors. *J Clin Oncol, 21*(8):1447–1451.
Earle, C. C., Neville, B. A. (2004). Under use of necessary care among cancer survivors. *Cancer, 101*(8):1712–1719.
Fallowfield, L., et al. (1995). No news is not good news: Information preferences of patients with cancer. *Psycho-Oncol, 4*(3), 197–202.
Forsberg, C., et al. (1996). Well-being and its relation to coping ability in patients with colo-rectal and gastric cancer before and after surgery. *Scand J Caring Sci, 10*(1), 35–44.
Glimelius, B. (1992). Expectancy or primary chemotherapy in patients with advanced asymptomatic colorectal cancer: a randomized trial. *J Clin Oncol, 10*(6):904–911.
Gordon, L. G., et al. (2015). A telephone-delivered multiple health behaviour change intervention for colorectal cancer survivors: making the case for cost-effective healthcare. *Eur J Cancer Care, 24*(6):854–861.
Gray, N. M., et al. (2011). Modifiable and fixed factors predicting quality of life in people with colorectal cancer. *Br J Cancer, 104*(11):1697–1703.
Green, R. J., et al. (2002). Surveillance for second primary colorectal cancer after adjuvant chemotherapy: an analysis of intergroup 0089. *Ann Intern Med, 136*(4):261–269.

Grunfeld, E. (2006). Looking beyond survival: how are we looking at survivorship? *J Clin Oncol*, *24*(32):5166–5169.
Hamashima, C. (2002). Long-term quality of life of postoperative rectal cancer patients. *J Gastroenterol Hepatol*, *17*(5):571–576.
Han, S. J., et al. (2019). Quality of life and supportive care needs of back-to-work breast cancer survivors. *Korean J Adult Nursing*, *31*(5):552–561.
Harrison-Woermke, D. E., Graydon, J. E. (1993). Perceived informational needs of breast cancer patients receiving radiation therapy after excisional biopsy and axillary node dissection. *Cancer Nursing*, *16*(6):449–455.
Hewitt, M. E., et al. (2007). Perspectives on post-treatment cancer care: qualitative research with survivors, nurses, and physicians. *J Clin Oncol*, *25*(16):2270–2273.
Hewitt, M., et al. (2006). *From Cancer Patient to Cancer Survivor*, Institute of Medicine and National Research Council, Washington DC.
Hui, D. (2014). Definition of supportive care: Does the semantic matter? *Curr Opinion Oncol*, *26*(4):372–379.
Ibrahim, E. M., Al-Homaidh, A. (2011). Physical activity and survival after breast cancer diagnosis: Meta-analysis of published studies. *Med Oncol*, *28*(3):753–765.
Isomura, K., et al. (2006). Physical activity and colorectal cancer: the fukuoka colorectal cancer study. *Cancer Sci*, *97*(10):1099–1104.
Johnson, I. A., Adelstein, D. J. (1991). The use of recorded interviews to enhance physician-patient communication. *J Cancer Educat*, *6*(2):99–102.
Johnson, J. E., et al. (1988). Reducing the negative impact of radiation therapy on functional status. *Cancer*, *61*(1):46–51.
Kallergi, A., et al. (2015). Diabetes mellitus and colorectal cancer risk. *Curr Colorectal Cancer Rep*, *11*(4):209–216.
Kattlove, H., Winn, R. J. (2003). Ongoing care of patients after primary treatment for their cancer. *CA*, *53*(3):172–196.
Kievit, J. (2002). Follow-up of patients with colorectal cancer: numbers needed to test and treat. *Eur J Cancer*, *38*(7):986–999.
Kneuertz, P. J., et al. (2015). Overtreatment of young adults with colon cancer: more intense treatments with unmatched survival gains. *JAMA Surg*, *150*(5):402–409.
Ko, C. Y., et al. (2003). Evaluating health utility in patients with melanoma, breast cancer, colon cancer, and lung cancer: A nationwide, population-based assessment. *J Surg Res* *114*(1):1–5.
Kotronoulas, G., et al. (2017). A systematic review of the supportive care needs of people living with and beyond cancer of the colon and/or rectum. *Eur J Oncol Nursing*, *29*:60–70.
Lafata, J. E., et al. (2005). Routine surveillance care after cancer treatment with curative intent. *Medical Care*, *43*(6):592–599.
Larsen, I. K., et al. (2006). Lifestyle as a predictor for colonic neoplasia in asymptomatic individuals. *BMC Gastroenterology*, *6*:5.
Larsson, S. C., et al. (2005). Diabetes mellitus and risk of colorectal cancer: a meta-analysis. *J Nat Cancer Instit*, *97*(22):1679–1687.
Larsson, S. C., et al. (2006). Physical activity, obesity, and risk of colon and rectal cancer in a cohort of Swedish men. *Eur J Cancer*, *42*(15):2590–2597.
Leydon, G. M., et al. (2000). Cancer patients' information needs and information seeking behaviour: in depth interview study. *Br Med J*, *320*(7239):909–913.
Lieu, C. H., et al. (2014). Association of age with survival in patients with metastatic colorectal cancer: analysis from the ARCAD clinical trials program. *J Clin Oncol*, *32*(27):2975–2982.
Liu, P. H., et al. (2019). Association of obesity with risk of early-onset colorectal cancer among women. *JAMA Oncol*, *5*(1):37–44.
Luker, K. A., et al. (1995). The information needs of women newly diagnosed with breast cancer. *J Adv Nursing*, *22*(1):134–141.
Lundy, J. J., et al. (2009). Exploring household income as a predictor of psychological well-being among long-term colorectal cancer survivors. *Quality of Life Res*, *18*(2):157–161.
Lynch, B. M., et al. (2008). Prospective relationships of physical activity with quality of life among colorectal cancer survivors. *J Clin Oncol*, *26*(27):4480–4487.
Mäkelä, J. T., et al. (1995). Five-year follow-up after radical surgery for colorectal cancer: results of a prospective randomized trial. *Arch Surg*, *130*(10):1062–1067.
Mayer, D. K., et al. (2017). Defining cancer survivors, their needs, and perspectives on survivorship health care in the USA. *Lancet Oncol*, *18*(1):e11–e18.

Mayer, D. K., et al. (2007). Cancer survivors information seeking behaviors: a comparison of survivors who do and do not seek information about cancer. *Patient Educat Counseling, 65*(3):342–350.
Meredith, C., et al. (1996). Information needs of cancer patients in west Scotland: cross sectional survey of patients' views. *Br Med J, 313*(7059):724–726.
Meyerhardt, J. A., et al. (2006). Impact of physical activity on cancer recurrence and survival in patients with stage III colon cancer: findings from CALGB 89803. *J Clin Oncol, 24*(22):3535–3541.
Mosher, C. E., et al. (2009). Associations between lifestyle factors and quality of life among older long-term breast, prostate, and colorectal cancer survivors. *Cancer, 115*(17):4001–4009.
Mosher, C. E., et al. (2016). Mental health outcomes during colorectal cancer survivorship: a review of the literature. *Psycho-Oncol, 25*(11):1261–1270.
Neibart, S. S., et al. (2020). Quality of life after radiotherapy for rectal and anal cancer. *Curr Colorectal Cancer Rep 16*(1):1–10.
Nguyen, L. H., et al. (2018). Sedentary behaviors, TV viewing time, and risk of young-onset colorectal cancer. *JNCI Cancer Spectrum, 2*(4):pky073.
Nilsen, T. I. L., et al. (2008). Recreational physical activity and cancer risk in subsites of the colon (the Nord-Trøndelag health study). *Cancer Epidemiol Biomark Prevent, 17*(1):183–188.
Owen, N. (2012). Sedentary behavior: understanding and influencing adults' prolonged sitting time. *Preventive Med, 55*(6):535–539.
Paika, V., et al. (2010). Personality variables are associated with colorectal cancer patients' quality of life independent of psychological distress and disease severity. *Psycho-Oncol, 19*(3):273–282.
Park, J. H., et al. (2016). Factors influencing posttraumatic growth in survivors of breast cancer. *J Korean Acad Nursing, 46*(3):454–462.
Peddle, C. J., et al. (2008). Associations between exercise, quality of life, and fatigue in colorectal cancer survivors. *Dis Colon Rectum, 51*(8):1242–1248.
Phipps, E., et al. (2008). Quality of life and symptom attribution in long-term colon cancer survivors. *J Evaluat Clin Pract, 14*(2):254–258.
Pietra, N., et al. (1998). Role of follow-up in management of local recurrences of colorectal cancer: a prospective, randomized study. *Dis Colon Rectum, 41*(9):1127–1133.
Puzzono, M., et al. (2021). The role of diet and lifestyle in early-onset colorectal cancer: a systematic review. *Cancers, 13*(23): 5933.
Rainey, L. C. (1985). Effects of preparatory patient education for radiation oncology patients. *Cancer, 56*(5):1056–1061.
Rees, C. E., Bath, P. A. (2000). The information needs and source preferences of women with breast cancer and their family members: a review of the literature published between 1988 and 1998. *J Adv Nursing, 31*(4):833–841.
Russell, L., et al. (2015). Psychological distress, quality of life, symptoms and unmet needs of colorectal cancer survivors near the end of treatment. *J Cancer Survivorship, 9*(3):462–470.
Sahoo, A. K., et al (2023). Current advances in dry eye disease diagnosis and treatment. *J Med Pharmaceut Allied Sci, 12*(2):5724–5729.
Samuelsson, M., et al. (2022). Cancer specialist nurses' experiences of supporting family members of persons diagnosed with colorectal cancer: a qualitative study. *Eur J Oncol Nursing, 61*:102205.
Santin, O., et al. (2015). Self-reported psychosocial needs and health-related quality of life of colorectal cancer survivors. *Eur J Oncol Nursing, 19*(4):336–342.
Sapp, A. L., et al. (2003). Social networks and quality of life among female long-term colorectal cancer survivors. *Cancer, 98*(8):1749–1758.
Scheele, J., et al. (2015). Quality of life after sphincter-preserving rectal cancer resection. *Clin Colorectal Cancer, 14*(4):e33–e40.
Schlairet, M., et al. (2010). Piloting a needs assessment to guide development of a survivorship program for a community cancer center. *Oncol Nursing Forum, 37*(4):501–508.
Secco, G. B., et al. (2002). Efficacy and cost of risk-adapted follow-up in patients after colorectal cancer surgery: a prospective, randomized and controlled trial. *Eur J Surg Oncol, 28*(4):418–423.
Snyder, C. F., et al. (2008). Preventive care for colorectal cancer survivors: a 5-year longitudinal study. *J Clin Oncol, 26*(7):1073–1079.
Son, H., et al. (2018). Effect of psychosocial interventions on the quality of life of patients with colorectal cancer: a systematic review and meta-analysis. *Health Quality Life Outcomes, 16*(1):119.
Speck, R. M., et al. (2011). Erratum to: an update of controlled physical activity trials in cancer survivors: a systematic review and meta-analysis. *J Cancer Survivorship, 5*(1): 112.

Spoorthi Shetty, S., et al. (2023). Oral insulin delivery: Barriers, strategies, and formulation approaches: A comprehensive review. *Int J Biolog Macromolec, 242*:1–27.

Trentham-Dietz, A., et al. (2003). Health-related quality of life in female long-term colorectal cancer survivors. *Oncologist, 8*(4):342–349.

Vacante, M., et al. (2012). Centenarians and supercentenarians: a black swan. Emerging social, medical and surgical problems. *BMC Surgery, 12*(Suppl.1): S36.

Vatandoust, S., et al. (2016). Metastatic colorectal cancer in young adults: a study from the South Australian population-based registry. *Clin Colorectal Cancer, 15*(1):32–36.

Vodermaier, A., Linden, W. (2019). Social support buffers against anxiety and depressive symptoms in patients with cancer only if support is wanted: a large sample replication. *Supportive Care Cancer, 27*(7):2345–2347.

Wang, M. J., et al. (2015). The prognostic factors and multiple biomarkers in young patients with colorectal cancer. *Scientific Reports, 5*:10645.

Wolin, K. Y., et al. (2009). Physical activity and colon cancer prevention: a meta-analysis. *Br J Cancer, 100*(4):611–616.

Yang, Y. H. (2014). A path analysis on factors influencing second primary cancer screening practices in stomach, colon, and breast cancer survivors. *J Korean Acad Nursing, 44*(2):139–148.

Yuhara, H., et al. (2011). Is diabetes mellitus an independent risk factor for colon cancer and rectal cancer. *Am J Gastroenterol, 106*(11):1911–1921.

Zhang, Y., et al. (2006). Occupational and leisure-time physical activity and risk of colon cancer by subsite. *J Occupat Environm Med, 48*(3):236–243.

13 Colorectal Cancer in Distinct Demographics: Addressing the Challenges in Young Adults and the Elderly

Roy Rillera Marzo

13.1 INTRODUCTION

Colorectal cancer (CRC) stands as a formidable public health challenge worldwide, significantly contributing to the global burden of cancer-related morbidity and mortality. As the third most commonly diagnosed cancer globally, CRC's impact on individuals and healthcare systems is profound and multifaceted, necessitating comprehensive strategies for its management and prevention (Sung et al., 2021). This chapter seeks to explore CRC in detail, emphasizing the necessity of demographic-specific research and intervention strategies to effectively combat this disease.

13.1.1 Overview of Colorectal Cancer

CRC originates in the epithelial cells lining the colon or rectum, part of the human digestive system responsible for the final stages of digestive processing. The transformation of benign polyps into cancer over time highlights the importance of early detection through regular screening. The disease's etiology is multifactorial, with lifestyle factors such as high consumption of red and processed meats, sedentary lifestyle, obesity, smoking, and heavy alcohol use being significant contributors. Moreover, genetic predispositions and familial syndromes like Lynch syndrome and familial adenomatous polyposis (FAP) significantly elevate risk levels, underscoring the need for targeted prevention and screening strategies within these high-risk populations (WCRF, 2018).

13.1.2 Rationale for Focusing on Distinct Demographics

The epidemiology of CRC reveals shifting patterns, particularly a disturbing rise in incidence among individuals younger than 50, contradicting long-standing perceptions of CRC as predominantly an ailment of the elderly (Siegel et al., 2023). This demographic trend challenges the current screening paradigms and calls for a re-evaluation of risk assessments and prevention measures tailored for this younger cohort. Conversely, the elderly demographic, while continuing to account for the majority of CRC cases, presents unique management challenges due to the prevalence of comorbidities, altered pharmacodynamics, and an increased risk of treatment-related complications. These observations necessitate a nuanced examination of CRC across life stages, advocating for tailored approaches that account for the biological and physiological diversities among different age groups (Sawicki et al., 2021).

13.1.3 Objectives of the Chapter

This chapter aims to delineate the complex landscape of CRC, highlighting its epidemiological shifts, elucidating the multifactorial risk factors, and exploring diagnostic and therapeutic challenges unique to various demographic groups. Specifically, it seeks to:

a. **Illuminate the shifting epidemiology of CRC**, with a focus on the rising incidence among young adults and the persistent prevalence in the elderly, exploring the implications of these trends for public health and clinical practice.

b. **Examine the distinct risk factors, diagnostic challenges, and therapeutic modalities** pertinent to different age groups, emphasizing the necessity for demographic-specific strategies in managing CRC.

c. **Discuss the broader implications of these epidemiological and clinical observations for screening guidelines, public health policies, and clinical protocols**, aiming to enhance early detection, optimize treatment outcomes, and ultimately alleviate the CRC burden across all age demographics.

By addressing these objectives, the chapter contributes to the ongoing discourse on CRC, advocating for refined prevention, diagnosis, and treatment paradigms that reflect the intricate needs of diverse populations. Through an in-depth analysis informed by current research and clinical insights, this narrative aspires to offer valuable perspectives for clinicians, researchers, and policymakers striving to mitigate the impact of this pervasive disease.

DOI: 10.1201/9781032704517-13

13.2 EPIDEMIOLOGICAL TRENDS IN CRC ACROSS DIFFERENT AGE GROUPS

The global burden of CRC represents a major public health challenge, with its incidence and mortality rates reflecting a complex interplay of genetic, environmental, and lifestyle factors. This section delves deeper into the worldwide incidence and prevalence of CRC, highlights the concerning rise in cases among young adults, and discusses the ongoing high incidence and unique management challenges in the elderly population.

13.2.1 Global Incidence and Prevalence

Colon cancer is the second leading cause of cancer-related deaths worldwide. In 2020, more than 1.9 million new cases of colorectal cancer and more than 930,000 deaths due to colorectal cancer were estimated to have occurred worldwide (WHO, 2023).

The distribution of CRC cases globally shows a marked disparity, with developed countries reporting higher incidence rates. This variation is largely attributed to lifestyle factors prevalent in these regions, such as diets rich in processed meats and low in fiber, combined with sedentary lifestyles and increased obesity rates. Moreover, the aging global population further contributes to the rise in CRC cases, as age remains a significant risk factor for the disease (Bray et al., 2018).

13.2.2 Trends in Young Adults

A notable epidemiological shift has been observed in CRC incidence, with a significant rise in cases among individuals younger than 50 years in high-income countries. This trend is particularly alarming as it contrasts with the stable or declining rates in the older population, likely a result of effective screening programs. The increase in young-onset CRC cases is thought to be driven by a blend of genetic predispositions, environmental exposures, and lifestyle choices distinct from those traditionally associated with CRC risk in older adults. Young-onset CRC often presents at a more advanced stage, highlighting a gap in awareness and the absence of screening recommendations for this age group. The trend is further complicated by the distinct molecular characteristics of CRC in young adults, suggesting different pathophysiological underpinnings compared to those in older patients (Siegel et al., 2023).

13.2.3 Trends in the Elderly Population

Despite the rise in young-onset CRC, the majority of cases occur in the elderly, with individuals aged 65 and older representing approximately 70% of all new diagnoses. In many developed countries, incidence rates among the elderly have been stabilizing or decreasing, thanks in part to widespread screening that facilitates early detection and removal of precancerous lesions. However, managing CRC in the elderly is fraught with challenges, including a higher prevalence of comorbidities and the physiological impacts of aging, which necessitate a nuanced approach to treatment to balance efficacy against potential risks (Sawicki et al., 2021).

The increasing demand for CRC care, spurred by the growing elderly population and the rising incidence among young adults, calls for healthcare systems to adapt and evolve. This includes the implementation of prevention strategies, early detection initiatives, and management approaches that are tailored to meet the distinct needs of these demographic groups.

13.3 CRC IN YOUNG ADULTS

The landscape of CRC is undergoing a significant shift, with an increasing incidence among young adults under 50 years of age. This demographic trend presents unique clinical challenges and raises important questions regarding the underlying causes and optimal management strategies for this population.

13.3.1 Characteristic Features and Aggressiveness

CRC in young adults often presents with aggressive features that distinguish it from cases typically diagnosed in older individuals. Notably, tumors found in younger patients tend to be diagnosed at a later stage and exhibit histological patterns associated with poorer outcomes, such as mucinous and signet-ring cell carcinomas. Furthermore, genomic analysis has revealed a higher prevalence of microsatellite instability and mutations in pivotal genes like BRAF and KRAS in young-onset CRC, contributing to the aggressive nature of these cancers. These molecular characteristics not only influence the behavior of the disease but also have significant implications for treatment selection and prognosis (Marx et al., 2023; Gao et al., 2022).

13.3.2 Challenges in Diagnosis and Late Presentation

The diagnosis of CRC in young adults is fraught with challenges, primarily due to a low index of suspicion and the absence of routine screening in this age group. Symptoms such as abdominal pain, rectal bleeding, and changes in bowel habits are often mistakenly attributed to more benign conditions, leading to delayed diagnosis and advanced disease at presentation. The lack of established screening guidelines for individuals under 50 exacerbates this issue, resulting in many young patients being diagnosed at stages where treatment options are more limited and less effective (Akimoto et al., 2021).

13.3.3 Cancer-Specific Survival Rates and Treatment Outcomes

Despite the aggressive presentation of CRC in young adults, research indicates that when matched for stage and treatment, survival rates can be comparable to those observed in older populations. This underscores the critical importance of early detection and the application of aggressive, tailored treatment strategies. However, the intensive nature of treatments often required for young adults can lead to significant long-term side effects, including fertility concerns, sexual dysfunction, and an increased risk of secondary malignancies, highlighting the need for comprehensive survivorship care planning (Siegel et al., 2023).

13.3.4 Financial and Psychological Impact on Young Adults

The financial and psychological impacts of a CRC diagnosis can be particularly profound for young adults. At a stage in life often characterized by career development, family planning, or further education, the burdens of cancer treatment can lead to significant financial strain and emotional distress. Costs associated with care can precipitate financial insolvency, while the emotional toll of battling a life-threatening illness can lead to anxiety, depression, and a diminished quality of life. Access to supportive care services, including mental health support and financial counseling, is essential for addressing these challenges (Keyvani et al., 2023).

13.3.5 The Need for Enhanced Screening and Early Detection

Given the rising incidence of CRC among young adults and the associated diagnostic challenges, there is a growing consensus on the need for enhanced screening and early detection efforts targeted at this population. Strategies to improve early detection rates include raising awareness among both healthcare providers and young adults about the symptoms of CRC and the potential benefits of early screening. Research into lowering the screening age and developing targeted screening strategies for high-risk individuals, such as those with a family history of CRC or known genetic predispositions, could lead to earlier diagnosis, improved treatment outcomes, and a reduction in the disease burden among young adults (Engel-Nitz et al., 2023).

13.4 CRC IN THE ELDERLY

The management of CRC in elderly patients presents unique challenges, necessitating a nuanced approach that accounts for the interplay of comorbidities, cognitive functioning, frailty, and the specifics of treatment outcomes.

13.4.1 Interaction with Comorbidities and Performance Status

The treatment landscape for elderly CRC patients is significantly influenced by the presence of comorbidities, which are prevalent in this population. Diseases such as cardiovascular disorders, diabetes, and chronic kidney disease not only complicate the choice of treatment modalities but also affect surgical outcomes and overall survival rates. The assessment of a patient's performance status, through tools like the ECOG scale or Karnofsky Performance Status, provides crucial insights into their ability to endure the rigors of cancer treatment. This assessment is vital for tailoring treatment approaches that balance the curative or palliative intent of therapy with the patient's overall health and quality of life expectations. The presence of multiple comorbidities often necessitates adjustments in treatment plans, such as modifying surgical techniques or chemotherapy dosing, to minimize adverse effects and optimize outcomes (Papaccio et al., 2023).

13.4.2 Challenges in Cognitive Functioning and Social Support

Cognitive decline and social support systems play critical roles in the treatment and recovery of elderly CRC patients. Cognitive impairments, ranging from mild cognitive deficits to advanced dementia, can severely limit a patient's capacity to understand their diagnosis, consent to treatment, and adhere to treatment regimens. The role of social support is equally critical, as strong familial or community support networks are associated with better treatment adherence, management of side effects, and overall patient well-being. Conversely, social isolation can exacerbate the challenges faced during treatment, leading to delays in receiving care, reduced treatment efficacy, and a decline in mental health. Addressing these issues requires a multidisciplinary approach, incorporating medical, nursing, social, and familial resources to support the patient's journey through treatment and recovery (Suzuki et al., 2021).

13.4.3 Personalized Treatment Approaches and Consideration of Frailty

Frailty stands out as a key consideration in managing CRC in the elderly, characterized by decreased physiological reserves and increased vulnerability to stressors. This condition necessitates a careful evaluation of treatment options to avoid over-treatment or under-treatment. The Comprehensive Geriatric Assessment (CGA) is an invaluable tool in this context, enabling healthcare providers to develop personalized treatment plans that consider the patient's frailty, comorbidities, and individual health goals. Such personalized plans may include less invasive surgical options, tailored chemotherapy regimens, or supportive care focused on maintaining independence and quality of life. This patient-centric approach emphasizes the importance of treating the patient holistically, rather than focusing solely on the cancer (Salaffi et al., 2023).

Personalized treatment approaches for the elderly must carefully navigate the spectrum of frailty, balancing the therapeutic benefits against potential risks. The concept of frailty, encompassing more than just chronological age, highlights a patient's vulnerability to adverse outcomes. The CGA serves as a cornerstone for evaluating frailty, enabling a holistic view of the patient that includes physical health, cognitive status, social circumstances, and nutritional status. This assessment guides oncologists in making informed decisions, potentially steering towards treatments that prioritize quality of life. Innovations in minimally invasive surgical techniques and the advent of targeted therapies offer promising avenues for providing effective treatment while minimizing the burden on frail patients. Moreover, the integration of palliative care from the onset of treatment planning emphasizes symptom management and supports maintaining patient autonomy and dignity (Salaffi et al., 2023).

13.4.4 Postoperative Outcomes and Predictors

The prognosis and outcomes following CRC treatment in the elderly are determined by a complex interplay of factors, including the disease's stage at diagnosis, the chosen surgical interventions, and the patient's specific health characteristics such as age, comorbidities, and frailty status. Predictors of poor postoperative outcomes include advanced age, high ASA scores, the necessity for emergency surgery, and preoperative functional limitations. The adoption of Enhanced Recovery After Surgery (ERAS) protocols has been shown to significantly improve postoperative results, reducing hospital stays and minimizing complications. These protocols underscore the value of a comprehensive, multidisciplinary approach to perioperative care, ensuring that elderly patients receive optimized, individualized treatment that addresses the full spectrum of their health needs (Voutsadakis, 2023).

Enhanced Recovery After Surgery (ERAS) protocols represent a paradigm shift in postoperative care for elderly CRC patients, emphasizing early mobilization, optimized pain management, and nutritional support to facilitate recovery. These protocols have been instrumental in reducing the length of hospital stays and improving patient satisfaction without compromising surgical outcomes. Furthermore, predictive models, incorporating factors such as ASA scores and preoperative functional status, are increasingly used to stratify risk and tailor postoperative care plans. Such models enable healthcare teams to anticipate complications and implement pre-emptive strategies, thereby improving outcomes. As the field advances, there is a growing emphasis on leveraging predictive analytics and personalized medicine to refine these models further, ensuring that each patient receives the most appropriate care based on their specific risk profile and preferences (Voutsadakis, 2023).

13.4.5 Advancements in Geriatric Oncology

The field of geriatric oncology is rapidly evolving, with ongoing research focusing on better understanding the biology of cancer in the elderly and developing age-specific treatment guidelines. Key areas of interest include the identification of biomarkers that can predict treatment response and toxicity in older adults and the development of novel therapeutic agents that are both effective and well-tolerated by this population. Additionally, clinical trials are increasingly incorporating geriatric assessment measures to ensure that study findings are relevant to older patients. These efforts aim to fill the current knowledge gaps and ultimately improve the standard of care for elderly CRC patients.

13.4.6 Ethical Considerations in Treatment Decisions

Ethical considerations play a crucial role in treating elderly CRC patients, particularly when balancing the potential benefits and harms of aggressive treatments. Decisions about whether to pursue, withhold, or withdraw treatment must be made in close consultation with patients and their families, respecting patient autonomy and informed consent. Advance care planning and discussions about goals of care are essential components of the treatment planning process, ensuring that interventions align with the patient's values and preferences.

13.5 DIAGNOSTIC METHODOLOGIES IN DISTINCT DEMOGRAPHIC GROUPS

Diagnostic methodologies for CRC must be nuanced and flexible to effectively address the diverse needs of distinct demographic groups. As CRC incidence patterns shift, with rising rates among younger adults and ongoing challenges in treating the elderly, the approach to diagnosis similarly requires adaptation. This necessitates a comprehensive evaluation of standard diagnostic techniques, alongside considerations for age-specific adjustments that can enhance early detection and improve outcomes.

13.5.1 Standard Diagnostic Techniques

Standard diagnostic techniques for CRC play a pivotal role in early detection and management. Colonoscopy, with its ability to visualize the entire colon and remove polyps, remains the cornerstone of CRC diagnosis. Its effectiveness is complemented by flexible sigmoidoscopy and CT colonography, which serve as less invasive alternatives for individuals unable or unwilling to undergo colonoscopy. Fecal occult blood tests (FOBT) and fecal immunochemical tests (FIT) offer non-invasive screening options, targeting early-stage detection where treatment outcomes are most favorable. These methodologies are underpinned by a growing understanding of CRC pathogenesis, emphasizing the importance of early detection in improving patient outcomes (Rintala et al., 2023).

13.5.2 Adjustments in Diagnostic Approaches for Young Adults

The increasing incidence of CRC among individuals under 50 highlights the need for adjusted diagnostic approaches. This demographic shift necessitates heightened awareness among healthcare providers and the public regarding CRC symptoms in young adults. Symptoms such as unexplained weight loss, rectal bleeding, or persistent changes in bowel habits warrant prompt evaluation, typically through colonoscopy, to facilitate early detection. Moreover, genetic counseling and testing become crucial for those with a family history of CRC, offering a proactive approach to identifying at-risk individuals. This emphasis on genetic predisposition and early symptom recognition aims to bridge the gap in screening guidelines that traditionally exclude younger adults (Singh et al., 2023).

13.5.3 Tailored Diagnostics for the Elderly

In elderly patients, diagnostic strategies must account for the prevalence of comorbidities and the increased risk of procedure-related complications. Non-invasive tests, such as FIT, are often preferred for those with limited life expectancy or significant health concerns. When invasive procedures like colonoscopy are necessary, comprehensive pre-procedural assessments are vital to mitigating risks. CT colonography emerges as a valuable alternative in this context, providing a less invasive diagnostic option that minimizes stress and discomfort for elderly patients. This tailored approach reflects a broader shift towards personalized medicine, seeking to optimize diagnostic accuracy while minimizing patient burden (Diamantis et al., 2021).

13.5.4 Advancements in Diagnostic Technologies

Recent advancements in diagnostic technologies, such as liquid biopsy and molecular testing, offer promising avenues for early CRC detection. Liquid biopsy, through the analysis of circulating tumor DNA (ctDNA) in blood samples, provides a non-invasive method for detecting CRC and monitoring treatment response. Molecular testing of tumor tissue can identify specific genetic mutations, guiding targeted therapy decisions. These innovations underscore the potential for personalized diagnostic pathways, enhancing early detection and enabling tailored treatment strategies.

13.5.5 Psychological and Financial Implications

The psychological toll of a CRC diagnosis is significant, with the potential to affect every aspect of a patient's life. Young adults, often in the prime of their lives, may experience profound disruptions to their personal and professional development. The shock of a cancer diagnosis can derail plans for education, career advancement, and family life, leading to feelings of isolation, depression, and anxiety. The loss of normalcy and the fear of mortality are heavy burdens, compounded by the physical side effects of treatment. Psychological support, including counseling and support groups, plays a crucial role in helping young adults navigate these challenges, offering a space for expression and shared experiences that can foster resilience and coping strategies.

For the elderly, the psychological impact of CRC diagnosis is intertwined with the challenges of aging. Concerns about independence, the burden on family caregivers, and the existential dread of facing one's mortality in the later stages of life can exacerbate feelings of depression and anxiety. Elderly patients may also grieve the loss of their health and fear the potential for pain and disability. Psychological interventions tailored to the elderly, which consider the cognitive and emotional aspects of aging, are essential in providing the support needed to navigate the diagnosis and treatment of CRC.

The financial implications of CRC are equally daunting and can lead to significant stress and anxiety. Young adults, who may not have established financial stability or comprehensive health insurance, can find themselves facing overwhelming medical bills. The costs associated with diagnostic tests, surgical procedures, chemotherapy, and radiation therapy, along with indirect costs such as lost wages and travel expenses for treatment, can lead to financial toxicity. This economic burden can hinder adherence to treatment plans and impact decision-making, potentially compromising treatment outcomes. Financial counseling services, assistance programs, and insurance navigation are critical resources that can help young adults manage the costs of care.

For elderly patients, the financial burden of CRC treatment may be influenced by fixed incomes, retirement savings, and the complexities of Medicare or other insurance coverage. Out-of-pocket expenses, even for those with insurance, can be significant, particularly for medications, long-term care, and supportive services not fully covered by insurance plans. The financial strain can exacerbate existing health disparities, limiting access to high-quality care and support services. Elderly patients benefit from comprehensive financial counseling that addresses insurance benefits, medication assistance programs, and eligibility for government and community resources designed to support older adults with cancer.

13.5.6 Addressing Psychological and Financial Needs

Addressing the psychological and financial needs of CRC patients requires an integrated approach that spans the continuum of care. Healthcare providers, social workers, financial counselors, and support organizations must collaborate to identify at-risk patients and provide the necessary resources and support. Early intervention, tailored to the specific needs and circumstances of each patient, can improve quality of life, treatment adherence, and ultimately, health outcomes. Initiatives aimed at raising awareness of these resources, alongside policy efforts to reduce the financial burden of cancer care, are crucial in ensuring that all patients have access to the support they need during their cancer journey.

13.6 THERAPEUTIC STRATEGIES AND MANAGEMENT

The management of CRC encompasses a spectrum of therapeutic strategies, each tailored to meet the unique needs of individual patients across various stages of the disease. Effective treatment requires a multidisciplinary approach, incorporating surgical interventions, chemotherapy, radiation therapy, and personalized care plans that consider the patient's age, health status, and specific tumor characteristics.

13.6.1 Surgical Interventions

Surgical resection is the foundational treatment for CRC, offering the best opportunity for a cure, particularly in the early stages of the disease. The evolution of surgical techniques has significantly improved patient outcomes, transitioning from traditional open surgeries to minimally invasive approaches such as laparoscopic and robotic-assisted surgeries. These advanced techniques are associated with shorter hospital stays, reduced postoperative pain, and quicker return to normal activities, without compromising the oncological outcomes. Total mesorectal excision (TME) for rectal cancer, often preceded by neoadjuvant chemoradiation in more advanced stages, exemplifies the tailored approach to surgical treatment, aiming to reduce the risk of local recurrence and preserve sphincter function, when possible. The decision-making process for the surgical approach takes into consideration not only the tumor characteristics but also the patient's overall health status, including comorbidities and functional status, to ensure the chosen method aligns with the patient's recovery capabilities and long-term quality of life (Siegel et al., 2023; Sabbagh et al., 2023).

13.6.2 Chemotherapy and Radiation Therapy

Adjuvant chemotherapy and radiation therapy are integral to the comprehensive management of CRC, particularly for patients with stage III and certain high-risk stage II diseases. The selection of chemotherapeutic agents has become increasingly personalized, guided by the tumor's molecular profile to optimize effectiveness and minimize toxicity. For example, the presence of specific genetic mutations in the tumor, such as those in the RAS or BRAF genes, can influence the choice of chemotherapeutic agents and targeted therapies. Radiation therapy, particularly in rectal cancer, is utilized in the neoadjuvant setting to reduce tumor size and improve surgical outcomes. This combined modality approach underscores the importance of a multidisciplinary team in planning and implementing the treatment strategy to achieve the best possible outcomes while considering the patient's preferences and overall health (Maurel et al., 2023).

13.6.3 Tailored Therapeutic Strategies for Young Adults

The increasing incidence of CRC among young adults necessitates therapeutic strategies that address the unique challenges and aggressive nature of their disease. Fertility preservation becomes a critical consideration for young patients undergoing treatment, necessitating early referral to reproductive specialists before initiating therapy. Emerging treatments, such as targeted therapies and immunotherapies, have shown promise in improving outcomes for young adults, especially those whose tumors exhibit specific molecular characteristics. Moreover, the provision of comprehensive psychosocial support is essential for addressing the wide range of emotional and mental health challenges that young patients face, including coping with the diagnosis, managing treatment side effects, and navigating the impact of the disease on their personal and professional lives (Boardman et al., 2020).

13.6.4 Special Considerations in Treating Elderly Patients

The treatment of elderly CRC patients requires careful consideration of several factors, including the prevalence of comorbidities, the increased risk of treatment-related toxicity, and the overall goals of care. Comprehensive geriatric assessment (CGA) plays a vital role in evaluating the patient's health status beyond the tumor characteristics, identifying frailty and other vulnerabilities that may impact the patient's ability to tolerate standard treatment regimens. Modifying treatment plans for the elderly may involve adjustments such as lower doses of chemotherapy, consideration of less invasive surgical options, or the prioritization of palliative care measures aimed at maintaining independence and quality of life. This personalized approach emphasizes the need to balance the potential benefits of treatment against the risks and the patient's preferences and values (Ma et al., 2023; Liu et al., 2023)

13.7 DISCUSSION

The management of CRC presents unique challenges and necessitates distinct approaches across different age groups, underlining the complexity of its treatment landscape. This section explores the varying challenges encountered, the implications for healthcare policies and protocols, and envisions future directions in research and clinical practice to enhance outcomes for CRC patients universally.

13.7.1 Comparison of Challenges and Approaches in Different Age Groups

The management of CRC highlights a spectrum of challenges unique to each age group. In young adults, the disease's aggressive nature and its tendency to be diagnosed at a later stage necessitate a proactive approach in both awareness and screening practices. Unlike their older counterparts, young adults may not benefit from current screening guidelines that typically start at age 50, leading to diagnoses during more advanced disease stages. This situation underscores the importance of genetic screening and lifestyle counseling as part of a younger individual's healthcare routine (Siegel et al., 2023).

Conversely, elderly patients present a different set of challenges, including a higher burden of comorbidities and a greater sensitivity to treatment side effects. The management strategies in this group require a delicate balance between aggressive treatment of the cancer and the preservation of quality of life, highlighting the role of comprehensive geriatric assessments in guiding treatment decisions (Papaccio et al., 2023).

Across all age groups, the integration of multidisciplinary care teams is essential for addressing the broad spectrum of needs presented by CRC patients. These teams can offer comprehensive support, from diagnosis through treatment and survivorship, tailoring interventions to each patient's specific circumstances. For young adults, this might include fertility preservation services and psychosocial support, while for the elderly, it may focus on managing comorbidities and incorporating palliative care principles.

As CRC incidence continues to evolve, with increasing cases among younger individuals and the ongoing challenge of treating an aging population, healthcare systems must adapt. This includes revising screening guidelines to reflect changing epidemiology, enhancing genetic and lifestyle counseling, and ensuring that treatment plans are informed by a thorough understanding of each patient's unique health status and life context.

13.7.2 Implications for Policy and Healthcare Protocols

For young adults, the current screening recommendations, which typically commence at age 50, may not be sufficient to catch early-stage diseases in this demographic. The American Cancer Society has already responded by lowering the recommended screening age to 45 for those at average risk, but even this adjustment may not address the needs of those with genetic predispositions or other risk factors that necessitate earlier intervention. Incorporating genetic and lifestyle risk assessments into routine healthcare visits for young adults could significantly enhance early detection rates. Such assessments would allow for personalized screening schedules, tailored to the individual's specific risk profile.

Moreover, public health education plays a crucial role in altering the trajectory of CRC in younger populations. Education campaigns designed to raise awareness of CRC's risk factors, such as diet, physical inactivity, and smoking, as well as its potential symptoms, are essential. These campaigns should target not only the general public but also healthcare providers, who can sometimes attribute symptoms indicative of CRC in young adults to less serious conditions due to the historical rarity of the disease in this age group. Enhancing awareness can lead to earlier diagnoses, significantly improving treatment outcomes and survival rates (Meester et al., 2015).

For the elderly, the integration of comprehensive geriatric assessments (CGA) into cancer care planning represents a critical step towards personalized medicine. These assessments evaluate more than just the cancer itself, considering the patient's overall functional status, comorbid conditions, cognitive function, and social support systems. This holistic approach ensures that treatment decisions are made in the context of the patient's broader health and life circumstances, aiming to maintain or improve quality of life alongside cancer treatment.

Additionally, there's a pressing need for the development and integration of less invasive treatment options into the care protocols for elderly CRC patients. Advances in surgical techniques targeted therapies, and radiation treatment modalities that offer effective cancer control with reduced side effects and recovery times are particularly beneficial for this group. Support for research in these areas is crucial, as is the dissemination of findings to ensure that new, less invasive options are rapidly incorporated into clinical practice.

The importance of public health education extends across all demographic groups but is particularly pertinent for young adults and the elderly. For young adults, education can empower individuals to seek medical advice when they experience potential CRC symptoms and to adopt lifestyle changes that reduce their risk of developing the disease. For the elderly, education can help to dispel misconceptions about the risks and benefits of CRC screening and treatment, encouraging more informed discussions with healthcare providers.

Public health campaigns should utilize multiple channels, including social media, community events, and healthcare settings, to disseminate information on CRC prevention, the importance of screening, and the advancements in treatment options. Such initiatives can build a culture of awareness and proactive health management, ultimately leading to earlier detection, more effective treatments, and better overall outcomes for individuals affected by CRC.

13.8 SUMMARY OF KEY FINDINGS

The exploration of CRC across different demographic groups reveals significant disparities in incidence, diagnosis, treatment responses, and outcomes. Key findings from the discussions include.

13.8.1 Increasing Incidence in Young Adults

Recent studies have documented a concerning rise in CRC incidence among individuals younger than 50. This shift challenges the traditional view of CRC as a condition predominantly affecting older adults and signals a critical need for public health strategies focused on younger populations. Initiatives may include lowering the recommended age for routine screening and implementing targeted awareness campaigns that educate on CRC's risk factors and symptoms. These efforts aim to improve early detection rates, ultimately enhancing treatment success and survival outcomes for this demographic (Siegel et al., 2023).

13.8.2 Diagnostic Challenges and Late Presentation

Young adults with CRC often face significant delays in diagnosis. Symptoms such as abdominal pain or rectal bleeding can be mistakenly attributed to more common, less serious conditions in younger individuals, leading to advanced disease stages at diagnosis. This contrasts with the elderly, who benefit from routine screening but must navigate the complexities of comorbid conditions that can complicate both the diagnosis and treatment process. Addressing these diagnostic challenges requires increasing awareness among healthcare providers and patients alike, emphasizing the importance of timely and thorough evaluations when symptoms arise (Akimoto et al., 2021).

13.8.3 Tailored Treatment Approaches

The management of CRC underscores the importance of personalized treatment strategies that account for the diverse needs of patients across the age spectrum. For young adults, treatments need to consider the potential impact on fertility and the long-term side effects that can affect quality of life. In contrast, treatment plans for elderly patients often require modifications to minimize adverse effects, considering the prevalence of frailty and other comorbidities. These tailored approaches necessitate a multidisciplinary effort, integrating insights from oncology, genetics, and supportive care services to optimize outcomes for each patient (Maurel et al., 2023).

13.8.4 Implications for Policy and Healthcare Protocols

The evolving epidemiology of CRC demands flexible healthcare policies that can respond to the needs of both younger and older populations. For young adults, revising screening guidelines to initiate at a younger age could play a critical role in addressing the rising incidence rates. For the elderly, the integration of comprehensive geriatric assessments into the care planning process can ensure that treatments are appropriately aligned with each patient's unique health status and life goals. These policy adaptations, alongside continued research into less invasive and more effective treatment modalities, are essential for improving CRC care and outcomes across all demographics (Meester et al., 2015).

13.8.5 Final Thoughts on Addressing CRC in Distinct Demographics

Early detection remains a critical factor in the successful management of CRC. For young adults, this includes an urgent need to reassess and potentially lower the age at which routine screening begins, reflecting the shifting epidemiology of the disease. Education plays a vital role in this demographic, emphasizing the importance of recognizing symptoms and seeking medical advice promptly. Healthcare providers should be trained to consider CRC as a possible diagnosis in younger patients presenting with relevant symptoms, overcoming the current bias that CRC is predominantly an older person's disease.

In contrast, for the elderly, the focus shifts towards the customization of treatment plans that account for the broader health context of the patient, including comorbidities and overall functional status. This necessitates not only medical interventions tailored to the physiological realities of aging but also a supportive care framework that addresses the psychological, social, and logistical challenges faced by older adults undergoing cancer treatment.

Future research should aim to dissect the complex interplay of genetic, lifestyle, and environmental factors that contribute to the risk and progression of CRC, particularly in younger adults. This effort should be paralleled by the development of more sophisticated, non-invasive screening tools that can facilitate early detection without adding undue burden to patients. Personalized medicine, driven by advances in genomic and biomarker research, offers the promise of treatments tailored to the individual characteristics of a patient's cancer, maximizing efficacy while minimizing side effects.

A concerted global effort to enhance public awareness about the signs and symptoms of CRC, alongside education on modifiable risk factors, can empower individuals to take proactive steps toward cancer prevention and early diagnosis. Public health campaigns should be inclusive, targeting diverse populations to ensure messages reach those most at risk.

Policies must evolve to ensure equitable access to CRC screening and treatment, addressing disparities that may exist based on geography, socioeconomic status, or healthcare system limitations. Insurance coverage for genetic counseling and testing, particularly for individuals with a family history of CRC or related cancers, should be expanded to facilitate risk assessment and management from an early age.

13.9 CONCLUSION

CRC challenges the medical community to adapt its approaches to prevention, diagnosis, and treatment in response to changing disease dynamics. The path to improved outcomes is multifaceted, requiring innovation in medical research, flexibility in clinical practice, and a commitment to patient education and support. By embracing these principles, healthcare systems worldwide can advance in the fight against CRC, offering hope and healing to patients across all demographics.

REFERENCES

Akimoto, N., et al. (2021). Rising incidence of early-onset colorectal cancer: a call to action. *Nature Rev Clin Oncol, 18*(4):230–243.

Boardman, L. A., et al. (2020). AGA clinical practice update on young adult–onset colorectal cancer diagnosis and management: expert review. *Clin Gastroenterol Hepatol, 18*(11):2415–2424.

Bray, F., et al. (2018). Global cancer statistics 2018: GLOBOCAN estimates of incidence and mortality worldwide for 36 cancers in 185 countries. *CA: A Cancer Journal for Clinicians, 68*(6), 394–424.

Diamantis, A., et al. (2021). Quality of life deterioration and colorectal cancer staging in elderly patients Which comes first?. *Journal of B.U.ON., 26*(4):1266–1270.

Engel-Nitz, N. M., et al. (2023). Colorectal screening among average risk individuals in the United States, 2015–2018. *Preventive Med Rep, 31*:102082.

Gao, Z., et al. (2022). Abstract 5873: Distinct prevalence and spectrum of germline cancer susceptibility gene mutations between early-onset and late-onset colorectal cancer. *Cancer Res, 82*(12 Suppl):5873.

Keyvani, V., et al. (2023). Epidemiological trends and risk factors of gynecological cancers: an update. *Med Oncol, 40*/3:93.

Liu, X. Y., et al. (2023). Effects of preoperative bicarbonate and lactate levels on short-term outcomes and prognosis in elderly patients with colorectal cancer. *BMC Surg, 23*(1):127.

Ma, Q., et al. (2023). Oncological and surgical outcomes of radical surgery in elderly colorectal cancer patients with intestinal obstruction. *Frontiers Surg, 10*:1251461.

Marx, O., et al. (2023). Molecular genetics of early-onset colorectal cancer. *World J Biolog Chem, 14*(2):13–27.

Maurel, J. M., et al. (2023). Practical notes on the current ESMO consensus guidelines for the management of patients with metastatic colorectal cancer in resource-constrained environments of low- to middle-income countries. *Oncol, 101*(Suppl 1):5–8.

Meester, R. G. S., et al. (2015). Public health impact of achieving 80% colorectal cancer screening rates in the United States by 2018. *Cancer, 121*(13):2281–2285.

Papaccio, F., et al. (2023). Proteotranscriptomic analysis of advanced colorectal cancer patient derived organoids for drug sensitivity prediction. *J Experim Clin Cancer Res, 42*(1):8.

Rintala, S., et al. (2023). A synthesis of evidence for cancer-specific screening interventions: A Preventive Medicine Golden Jubilee Review. *Preventive Med, 167*:107395.

Sabbagh, S., et al. (2023). Abstract 5550: Sociodemographic disparities in access to biomarker testing in patients with advanced colorectal cancer. *Cancer Res, 83*(7 Suppl):5550.

Salaffi, F., et al. (2023). Inflammaging and frailty in immune-mediated rheumatic diseases: how to address and score the issue. *Clin Rev Allerg Immunol, 64*.2:206–221.

Sawicki, T., et al. (2021). A review of colorectal cancer in terms of epidemiology, risk factors, development, symptoms and diagnosis. *Cancers, 13*(9):2025.

Siegel, R. L., et al. (2023). Colorectal cancer statistics, 2023. *CA, 73*(3):233–254.

Singh, A. K., et al. (2023). Detection of germline variants with pathogenic potential in 48 patients with familial colorectal cancer by using whole exome sequencing. *BMC Med Genomics, 16*(1):126.

Sung, H., et al. (2021). Global Cancer Statistics 2020: GLOBOCAN estimates of incidence and mortality worldwide for 36 cancers in 185 countries. *CA, 71*(3):209–249.

Suzuki, Y., et al. (2021). O14-2 Evaluation of comprehensive geriatric assessment in elderly patients with pancreaticobiliary cancer. *Ann Oncol*, *32*:101817.
Voutsadakis, I. A. (2023). Presentation, molecular characteristics, treatment, and outcomes of colorectal cancer in patients older than 80 years old. *Medicina*, 59/9:1574.
WCRF. (2018). *World Cancer Research Fund/American Institute for Cancer Research. Continuous Update Project Expert Report. Diet, nutrition, physical activity and prostate cancer*. Continuous Update Project Expert Report [internet], available at: www.wcrf.org/wp-content/uploads/2021/02/prostate-cancer-report.pdf
World Health Organization. (2023). *Tracking universal health coverage: 2023 global monitoring report*. World Health Organization.

14 Future Perspective and Emerging Therapies

S. M. Mahfooz Hasan, Mohammad Waseem, Satya N. Sankwar and Zainab Siddiqui

14.1 INTRODUCTION

Colorectal cancer (CRC) arises when the lining of the colon or rectum undergoes malignant transformation, often starting as benign polyps in the mucosal layer (Huck & Bohl, 2016). Sporadic CRC is the most common type accounting for 70–80% of CRCs. It is caused either due to genetic mutations, environmental exposures, or lifestyle factors (Marcellinaro et al., 2023; Mármol et al., 2017). Sporadic CRC is a heterogeneous disease with distinct subtypes. The most common subtype involves chromosomal instability, while others include microsatellite instability (MSI) and CpG island methylator phenotype (Marcellinaro et al., 2023). Mutations in oncogenes like KRAS, HRAS, NRAS, PIK3CA, c-MYC, SMAD2/4, and tumor suppressor genes (PTEN and p53) also play a significant role in the pathogenesis of CRC (Marcellinaro et al., 2023; Mármol et al., 2017). Hereditary non-polyposis colon cancer (Lynch syndrome) and familial adenomatous polyposis coli account for 15–30% of the cases (Bhattacharya et al., 2024; Marcellinaro et al., 2023). Mismatch repair gene germline mutations are found in Lynch syndrome, while familial adenomatous polyposis coli is found to be associated with adenomatous polyposis coli (APC) gene mutation (Bhattacharya et al., 2024). Understanding these subtypes and mutations is crucial for tailoring treatment strategies and advancing personalized medicine in CRC.

14.2 SIGNIFICANCE OF MEASURING VOLATILE ORGANIC COMPOUNDS AND CONTROLLING THEIR PRODUCTION BY TARGETED THERAPY

Volatile organic compounds (VOC) are usually defined as carbon-based organic compounds that can be directly detected in breath, feces, blood, saliva, and urine (Janfaza et al., 2019; Oxner et al., 2023). The ongoing research into VOCs as biomarkers for cancer shows potential for non-invasive diagnostic methods that could improve early detection and, subsequently, survival rates. The Cancer Odor Database is a comprehensive, web-based database that compiles information on VOCs related to cancer (Janfaza et al., 2019). VOCs are categorized as noninvasive biomarkers that can be used for the diagnosis of CRC (de Boer et al., 2014). A recent study has indicated that using urinary VOCs can detect CRC with significant sensitivity (84%) and specificity (70%) (van Liere et al., 2023). Mezmale et al. used headspace solid-phase microextraction gas chromatography-mass spectrometry to evaluate volatile signatures formed by VOCs emitted from CRC tissue. Both common (138 VOCs) and differential VOC emissions between cancerous and non-cancerous tissues, reflecting shared and distinct metabolic activities were found. Only four compounds (1-propanol, pyridine, isoprene, methyl thiolacetate) showed increased emissions from CRC tissues (Mezmale et al., 2023).

14.2.1 Targeting Metabolic Cycles: Generators of Volatile Organic Compounds

Several studies have put forward the concept of VOCs production due to dysregulated metabolism in cancer cells (Peng et al., 2010; de Boer et al., 2014; Oxner et al., 2023). The "Warburg effect," is one of the metabolic abnormalities seen in cancer cells that allows preference for glycolysis over oxidative phosphorylation even in the presence of sufficient oxygen. Numerous pathways and mutations, including those in the TP53 gene, RAS, APC, and increased PI3K/AKT/mTORC1 pathway, have been shown to affect glycolysis in cancer cells (Zhong et al., 2022). In a study headspace VOCs were analyzed to observe the metabolic changes followed with the inhibition of glycolysis. Therefore, 3-BrPA and 2-DG (2-deoxy-D-glucose) reduced the production of VOCs during glycolysis in lung cancer cell lines (NCI-H460, PC-9, A549) and non-diseased lung cells. Notably, in comparison to normal cells, all the studied lung cancer cells showed increased amounts of acetoin in their headspace (Chu et al., 2024). In another study, it was found that the 2,3-butanediol dehydrogenase converts acetoin to 2,3-butanediol. Later it was found to be elevated in CRC (Lamaudière et al., 2023). Propofol appears to inhibit aerobic glycolysis in CRC cells by interfering with the NMDAR-CAMKII-ERK signaling pathway. Apart from this it also inhibits HIF-1alpha (Chen et al., 2018). In another study, it was reported that propofol downregulates serine and arginine-rich factor 1 via circRNA poly(A) binding protein nuclear 1/ microRNA (miRNA)-638 axis in CRC cells (Zhao & Liu, 2021). This suggests that propofol might exert an anti-cancer effect by altering how these cells metabolize glucose, potentially reducing their growth and

 DOI: 10.1201/9781032704517-14

survival. Nicotinamide phosphoribosyltransferase (NAMT) is an enzyme involved in the NAD+ salvage pathway and catalyzes the conversion of nicotinamide into nicotinamide mononucleotide, a precursor of NAD+ (Tang et al., 2023). As seen in Figure 14.1, NAD+ is required for many metabolic activities, including glycolysis, Kreb's cycle, and oxidative phosphorylation. Elevated NAMT levels have been observed in various cancers, including CRC (Tolstikov et al., 2014; Tang et al., 2023). This elevation supports tumor growth by boosting NAD+ levels, which in turn enhances cancer cell survival, proliferation, and resistance to cellular stress. The first-generation NAMT inhibitors, FK866 and CHS828, demonstrated robust antitumor activity in preclinical models but failed to show objective responses in clinical trials. Issues such as severe dose-limiting toxicities (like thrombocytopenia and gastrointestinal issues), poor pharmacokinetic properties, and retinal toxicity hindered their clinical development (Wei et al., 2022). Clinical trials are ongoing to evaluate NAMT inhibitors alone and in combination with conventional chemotherapeutic agents in CRC patients. These trials aim to determine the safety, efficacy, and optimal dosing of NAMT-targeted therapies. A compound 6c, a naphthalimide derivative inhibits mitochondrial complex III activity and the expression of isocitrate dehydrogenase 2. The combinational therapy of compound 6c + mitoxantrone shows more profound effects on CRC tumorigenesis than individual therapies (Ge et al., 2021).

The monitoring of VOCs in CRC patients could help in the identification of metabolites that could assess mucosal inflammation, polyps, and early-stage colorectal tumors. Thus, leading to improved treatment strategies. Apart from these further studies are required to identify therapeutic targets that are involved in the dysregulated metabolic pathways such as glycolysis, anaerobic respiration, Kerb's cycle, amino acid metabolism, and fatty acid synthesis in CRC.

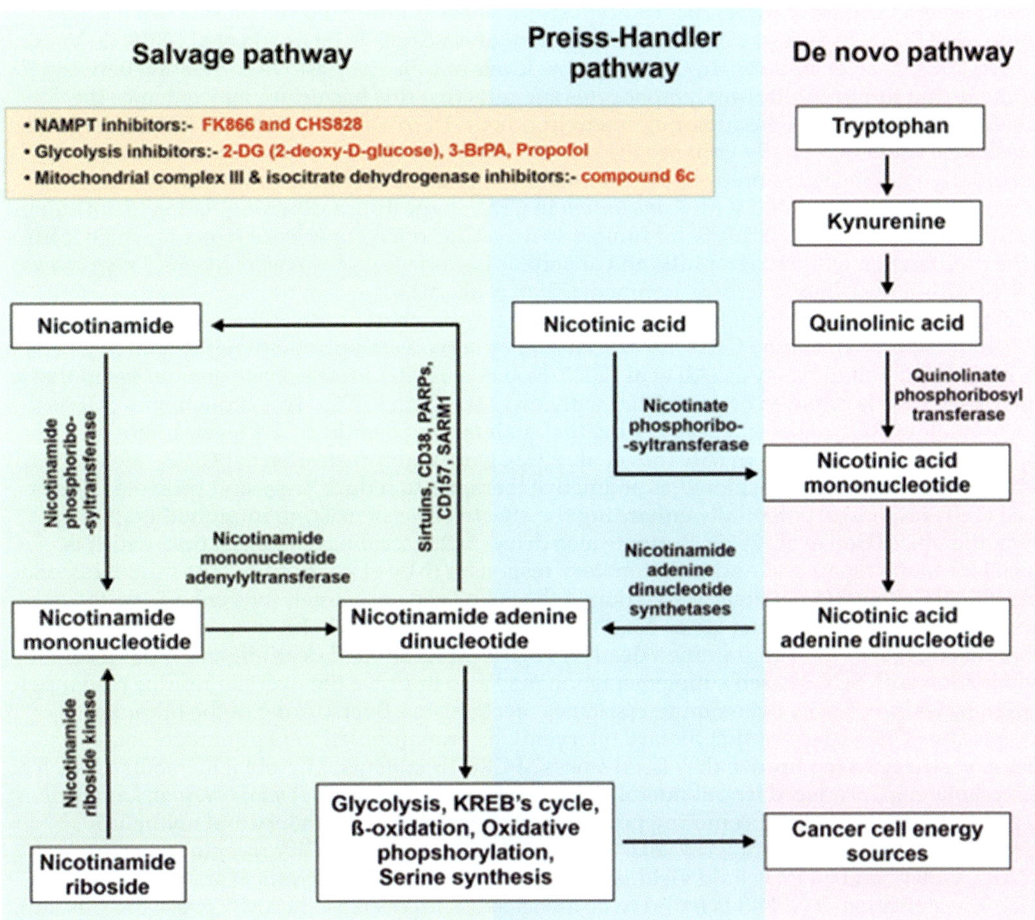

Figure 14.1 Role of nicotinamide dinucleotide in cancer cell survival.

14.3 ROLE OF GUT MICROBIOTA IN COLORECTAL CANCER DEVELOPMENT AND TREATMENT

Gut microbiota is involved in the pathogenesis of CRC by regulating tumor initiation, proliferation, apoptosis, and immune modulation (Hamada et al., 2018; Wong & Yu, 2023; Zhao et al., 2023). These microbes have a role in the remodeling of the tumor microenvironment and can impact cancer development both directly and indirectly by altering the immune responses (de Visser & Joyce, 2023). Several microbes and toxins are released by them which are found to be involved in the development of CRC. For example, *Escherichia coli*, *Fusobacterium nucleatum*, and *Helicobacter pylori* cause inflammation, and DNA damage (Zhao et al., 2023). The findings of Engevik et al. on *Fusobacterium nucleatum* subspecies *polymorphum* highlight the bacterium's ability to interact with the host immune cells through the release of outer membrane vesicles (OMVs) (Engevik et al., 2021). These OMVs are key players in bacterial communication with host cells and can carry various bioactive molecules that influence host immune responses (Li et al., 2023). When *F. nucleatum* OMVs interact with Toll-like receptor 4 on colonic epithelial cells, they trigger a signaling cascade involving the activation of nuclear factor kappa Beta (NF-κB). Pro-inflammatory cytokines are then transcriptionally triggered by NF-κB. Hence, contributing to a pro-inflammatory environment in the gut. OMVs containing CagA also activate the Toll-like receptor 4/ NF-κB mediated signaling pathway (Zhao et al., 2023). Such a response can play a significant role in the pathogenesis of conditions like inflammatory bowel disease and other forms of intestinal inflammation leading to the transformation of precancerous form to cancer.

The existence and growth of cancer cells are largely dependent on the tumor microenvironment, which is a complex ecosystem made up of cancer cells, immune cells (such as T and B lymphocytes, natural killer cells, and tumor-associated macrophages), stromal cells, and a network of microvessels (de Visser & Joyce, 2023). Recent studies have highlighted the presence of bacteria within the TME, which might regulate the behavior of cancer cells (Hamada et al., 2018; de Visser & Joyce, 2023; Zhao et al., 2023). In CRC with low levels of MSI, the positive correlation between *F. nucleatum* and tumor-infiltrating lymphocytes suggests that this bacterium may enhance the immune response within the tumor microenvironment (Hamada et al., 2018). Therefore, targeting *F. nucleatum* could potentially enhance the effectiveness of existing therapeutic approaches, particularly in MSI CRC. Moreover, a decreased number of CD3+ T cells in the tumor microenvironment has been associated with *F. nucleatum* in CRC, suggesting a downregulation of antitumor immunity (Hamada et al., 2018). Solid tumors such as CRC often experience hypoxia which leads to the proliferation of microaerophilic and anaerobic bacteria (e.g., *Bacteroides fragilis*, *Enterococcus faecalis*) within the tumor microenvironment (Zhao et al., 2023).

Colon bacteria ferment undigested dietary fibers and starch to produce short-chain fatty acids (SCFAs) (Figure 14.2). These SCFAs are crucial as they serve as the primary energy source for colon cells through aerobic glycolysis (Ali et al., 2022; Hou et al., 2022). Studies have demonstrated that a high risk of CRC is found to be associated with lower levels of SCFAs. This reduction in SCFAs also promotes CRC progression, suggesting that maintaining healthy SCFA levels might be important for colorectal health (Alvandi et al., 2022). Direct administration of SCFAs, such as sodium butyrate, could be explored as adjunctive therapies to reduce Tregs and promote effector T cell (Teff) responses, potentially enhancing the effectiveness of existing immunotherapies or chemotherapies (Hou et al., 2022). Butyrate also drives M2 macrophage polarization, which is crucial for tissue repair and anti-inflammatory responses (Ney et al., 2023). SCFAs inhibit histone deacetylases, promoting histone acetylation of the Foxp3 gene promoter, thus enhancing the expression of the transcription factor FoxP3 in Tregs (Ney et al., 2023). Immune checkpoint inhibitors (ICI), i.e. PD1 (programmed death 1) /PDL1 (programmed death-ligand 1) or CTLA-4 in combination with SCFA-based supplementation may also improve the overall immune response to malignancies, potentially overcoming resistance mechanisms. Such tuning of the immune response by SCFAs suggests that dietary interventions or manipulation of gut microbiota may offer new strategies to improve the effectiveness of CRC treatments. Microbial metabolites, such as deoxycholic acid, produced by gut microbes (*Clostridium*, *Enterococcus*, *Bifidobacterium*, *Lactobacillus*), may also influence CRC by promoting processes like vasculogenic mimicry and epithelial-mesenchymal transition (Fang et al., 2021; Zhao et al., 2023). Several SCFA receptors such as HCAR2, GPR41, and GPR43 could yield novel therapeutic options (Carretta et al., 2021; Hou et al., 2022). The activation of GPR43 is found to be associated with decreased cAMP levels and Ca^{2+} flux. HCAR2 and GPR43 interact with beta-arrestin via ERK1/2 and inhibit the NF-κB pathway (Carretta et al., 2021). *Clostridium butyricum* strains MIYAIRI 588 (CBM588) clinical trials have shown its impact in reducing the progression-free survival in metastatic renal cancer patients

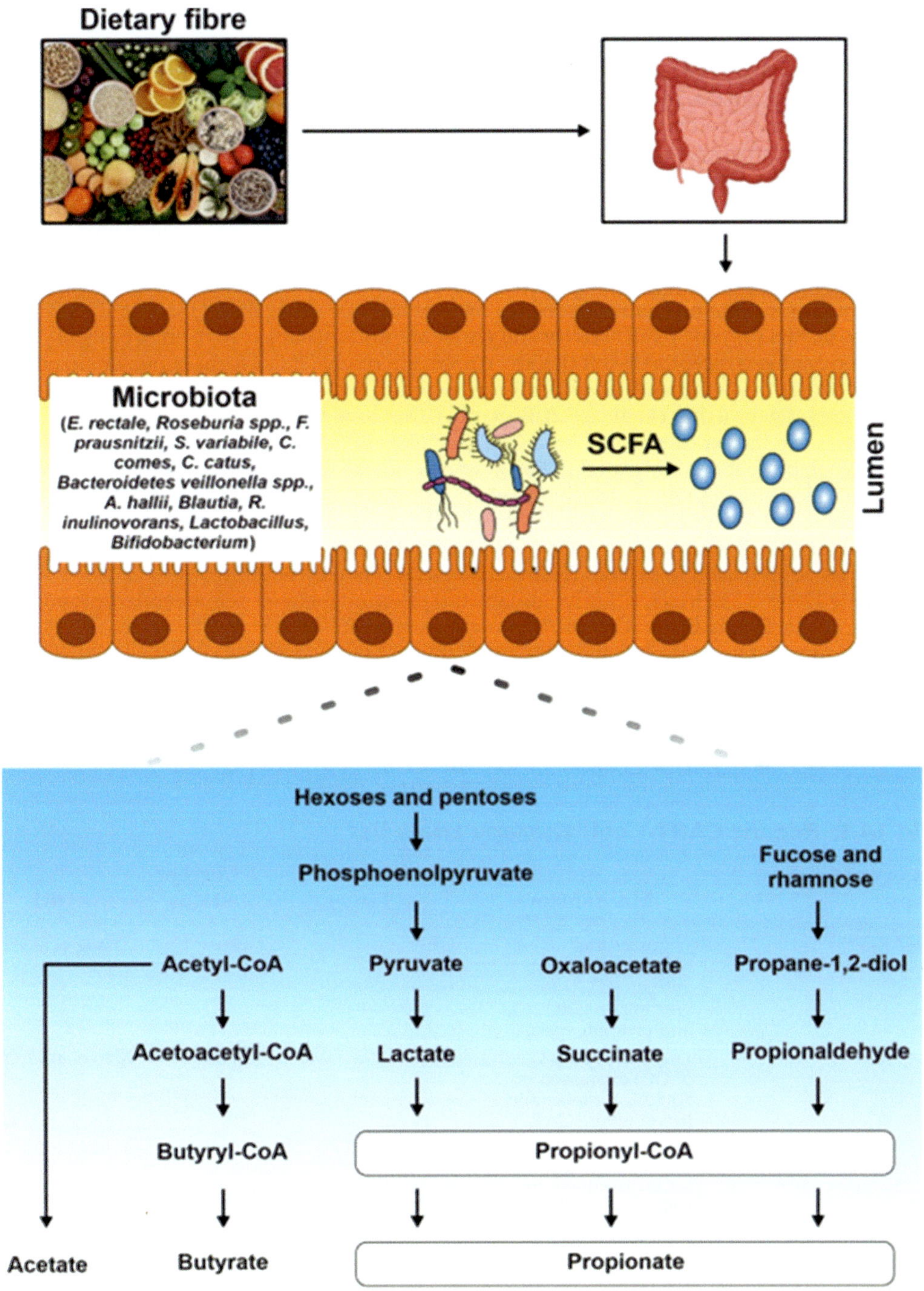

Figure 14.2 Production of SCFAs by gut microbiota.

(Zhao et al., 2023). MSI CRC cells treated with SCFAs (butyrate and propionate) lead to the activation of CD8+ T cells and increased major histocompatibility complex class I surface expression via the cGAS/STING signaling pathway (Mowat et al., 2023). This highlights the complex role that the microbiome plays in modulating cancer biology and immune interactions, emphasizing the need for further research to fully understand these dynamics and their implications for treatment strategies.

14.4 IMMUNOTHERAPY

Immunotherapy proves to be a promising treatment for certain CRC subtypes, particularly MSI-high tumors (Ghazi et al., 2022). The antagonists of PD1 or PDL1 has emerged as a successful approach in various cancers including MSI-high metastatic CRC (Han et al., 2020; Ghazi et al., 2022). Nivolumab, pembrolizumab, JQ1, atezolizumab, avelumab, and cemiplimab are the PD1/PDL1 inhibitors commonly administered to the cancer patients (Han et al., 2020). The FDA approved pembrolizumab (anti-PD1), in 2017 was a second-line treatment for MSI-high metastatic CRC. On the basis of the Keynote-177 research, it was included in the first-line treatment group (Ghazi et al., 2022). The tumor microenvironment resident cells often secrete soluble immunosuppressive proteins/enzymes like PDL1, indoleamine 2,3-dioxygenase 1, and interleukin (IL)-10. Apart from this it also harbors regulatory myeloid-derived suppressor cells, T cells, tumor-associated macrophages, and tumor-associated neutrophils (Tie et al., 2022; Ghazi et al., 2022). Hence, these factors prevent T-cell activity and create an immunosuppressive environment. To this CAR (chimeric antigen receptor) T-cell therapy comes as a rescue. It is a form of personalized treatment where a cancer patient's own T-cells are genetically modified to express CARs that target specific tumor antigens (Ghazi et al., 2022). Recently CD6-based CAR T-cell therapy was found to effectively regulate the expression of CD44, CD133, CD166, EPCAM, and vimentin via activation of CD166 (CD6 receptor). It demonstrated potent cytotoxic effects and significant cytokine release in CRC stem cells leading to their apoptosis (He et al., 2023). Few CAR T-cell therapies have been enlisted in Table 14.1. Despite its success, CAR T-cell therapy faces many challenges that include the risk of secondary T-cell malignancies, toxicities like cytokine release syndrome, and neurotoxicity which emphasizes the need to explore their mechanisms (Zhou et al., 2024).

CheckMate 120 and 142 trials, ICI (nivolumab and ipilimumab) have demonstrated promising results for CRC, including those with BRAF mutations and MSI-high status (Singh et al., 2019; Lenz et al., 2022). Immune checkpoint blockade is a valuable option for a subset of metastatic CRC patients, although it's not universally the first-line treatment for all cases of metastatic CRC (Yao et al., 2023). IL-15 and IL-17a play crucial roles in shaping immune responses in the tumor microenvironment, with IL-15 enhancing antitumor immunity and IL-17a potentially contributing to immune resistance (Chen et al., 2024). Research suggests that concurrently targeting both IL-17a

Table 14.1: Recent CART-T cell therapy trials list

Category of CR-T therapy	Mechanism	Target	Current status	URL
CEA CAR-T	Recognizes the carcinoembryonic antigen (CEA) for binding on CRC cells and promote lysis.	CEA	Phase 1	Zhang et al., 2024
GGC CAR-T	Guanylate cyclase C (GGC) expressed in 70-80% of metastatic lesion where CAR-T cells specifically bind and leads to the target killing of cancerous cells.	GGC	Phase 1	Xiao et al., 2023
Combination of CAR-T cell therapy and checkpoint inhibitors	Checkpoint inhibitor blocks the PD1 pathway which increases the T cell activity.	CD-19 and PD-1	Phase I	Masucci et al., 2023
Enzyme secreting CAR-T cells/Epithelial cell adhesion molecule (EpCAM)	Secrete enzymes that degrades extracellular matrix which enhances T cell penetration in solid tumors	Extracellular matrix	Preclinical research	Zhang et al., 2019
HER2-specific CAR-T cells	HER2-specific CAR-T cells exhibited strong cytotoxicity and cytokine-secreting ability against CRC cells in vitro	Human epidermal growth factor receptor 2 (HER2)	Preclinical research	Xu et al., 2021

and PD1 can enhance antitumor immunity by increasing the presence of cytotoxic T lymphocytes, and reducing the number of immunosuppressive myeloid-derived suppressor cells (Liu et al., 2021; Chen et al., 2024). It shows promising efficacy in mouse tumor models, and the P65/NRF1/miRNA15b axis appears to be the critical pathway involved in this response (Liu et al., 2021). IL-15 is known for stimulating the activation, proliferation, and survival of CD8+ T cells and natural killer cells, making it a promising agent in cancer immunotherapy (Chen et al., 2024). The N-803 (IL-15 agonist) induces anti-tumor responses across various tumors, including hepatocellular carcinoma and CRC (Chen et al., 2024). In a metastatic murine CT26 colon carcinoma model, a combination of anti-PDL1, anti-CTLA-4, and IL-15 led to the enhanced cytotoxic activity of T cells, increased secretion of IFN-γ, and a reduction in immunosuppressive signals (like PD1 and IL-10), resulting in prolonged survival of mice with metastasis (Yu et al., 2010). The hIL15-ABD consisting of human IL-15 and albumin binding domain (ABD) was reported to significantly reduce tumor growth in the murine colon cancer and melanoma models. Furthermore, the use of hIL15-ABD and rat antibody (10F.9G2) against murine PDL1, also showed a stronger anti-tumor effect than either monotherapy alone (Hsu et al., 2021). The use of CD80-Fc in PDL1+ CT26 tumors improves tumor-infiltrating T-cell activity and inhibits tumor progression, indicating the importance of immune checkpoint pathways in CRC (Horn et al., 2018; Chen et al., 2024).

14.5 CHEMOTHERAPEUTIC AGENTS

For CRC surgical removal is often the primary treatment when the tumor is resectable. However, when the cancer is non-resectable or has metastasized, other treatments like chemotherapy, radiotherapy, and immunotherapy become crucial. The lack of specificity makes these therapies cytotoxic, which leads to secondary complications such as nausea, vomiting, hair loss, anemia, and increased probability of getting infections (Yao et al., 2023). The commonly administrated chemotherapeutic agents in CRC are 5-fluorouracil (FLU), cetuximab, panitumumab, irinotecan (IRN), oxaliplatin (OXP), capecitabine (CAP), etc. (Arredondo et al., 2013; García-Foncillas et al., 2019; Blondy et al., 2020; Yao et al., 2023; Nishizawa et al., 2022).

Long-term exposure to FLU results in the development of FLU-resistant CRC cells. It has also been observed that the levels of the covalent complex between 5-fluorodeoxyuridylate (FdUMP) and thymidylate synthase (TS) are elevated in such resistant cells. Therefore, targeting autophagy could enhance the effectiveness of FLU by preventing the removal of FdUMP-TS, potentially reversing resistance as evaluated by Nishizawa and co-workers (Nishizawa et al., 2022). It has also been reported that in FLU-resistant cells, there is a lower expression of mitochondrial F1F0-ATP synthase subunits (Galber et al., 2020; Blondy et al., 2020). Oligomycin A, an ATP synthase inhibitor, can counteract FLU's cytotoxic effects, highlighting the role of ATP synthase in resistance mechanisms (Blondy et al., 2020). FKBP4 (FK506-binding protein 4), a co-chaperone protein identified in the mitochondrial inner membrane space interacts with COA6 (Cobalt-chaperone for cytochrome c oxidase) and regulates the assembly of the COA6/SCO1/SCO2 complex. The COA6/SCO1/SCO2 complex influences the assembly and function of cytochrome c oxidase. Moreover, the deletion of COA6 was found to improve the sensitivity of FLU in colorectal adenocarcinoma cells (Zhu et al., 2022). FLU treatment has also been linked to the increased production of reactive oxygen species. Hence results in the activation of the NRF-2 protein, which further induces the transcription of antioxidant enzymes (SOD1, SOD2, glutathione peroxidase, catalases) in cancer stem cells (Blondy et al., 2020). However FLU is associated with concerning side effects such as fever, mucositis, and thrombocytopenia (Evans et al., 1981). The alternative drug such as CAP, a prodrug of FLU, is often administrated to patients with metastatic CRC. It can be used alone or in combination with FLU which has been shown to improve the treatment response, progression-free survival, and survival in CRC patients (Alzahrani et al., 2023). The addition of OXP and IRN to the FLU regimen has substantially improved survival outcomes in metastatic CRC patients (Kumar et al., 2023). Pt(II) compounds, including OXP, can induce nucleolar stress even when DNA damage response pathways (such as those dependent on ATM/ATR) are inhibited. The nucleolar stress can be induced by cisplatin with the cell cycle arrest at the G1/S checkpoint, implying that the timing within the cell cycle may influence the response to platinum compounds (Pigg et al., 2024). According to the Gruppo Oncologico dell'Italia Meridionale trial, overall survival rates for FOLFIRI (folinic acid, fluorouracil, and IRN) and FOLFOX (5-fluorouracil, leucovorin, OXP) were found to be 14 and 15 months, respectively. The GERCOR Trial demonstrated longer overall survival when patients received FOLFOX first followed by FOLFIRI (21.5 months) compared to FOLFIRI followed by FOLFOX (20.6 months), indicating the potential benefit of such sequential regimens (Kumar et al., 2023). Regorafenib, a multi-kinase inhibitor has been approved for the treatment of advanced

CRC and other malignancies. It targets multiple kinases involved in angiogenesis and tumor growth, offering clinical benefits despite the challenges in preclinical studies (García-Aranda & Redondo, 2019). Cetuximab (Erbitux) and Panitumumab (Vectibix) target the EGFR and are used in the treatment of CRC cases with wild-type KRAS genes (García-Foncillas et al., 2019). While vemurafenib and dabrafenib target mutations in the BRAF gene, which can be found in a subset of CRC patients. MEK inhibitor trametinib targets the MAPK pathway and is used in patients with BRAF V600-mutant CRC in addition to BRAF inhibitors (Shan et al., 2024).

Dagenborg and colleagues studied T-cell density in the intratumoral region of colorectal liver metastases after neoadjuvant chemotherapy (OXP or IRN). A transient period of <9.5 weeks (as computed between neoadjuvant chemotherapy completion and CRC resection) was found to have significantly higher T-cell densities in patients receiving neoadjuvant chemotherapy than non-recipient (Dagenborg et al., 2020). Arredondo et al. conducted a trial involving stage III colon cancer patients who received preoperative CAPOX (CAP + OXP) followed by adjuvant CAPOX; a significant reduction in tumor volume (69.5%) was observed without disease progression during therapy (Arredondo et al., 2013). In another study, the authors further evaluated the significance of FLU along with CAPOX and found a tumor volume reduction of 62.5% (Arredondo et al., 2017). In stage IIIB colon cancer, a triplet regimen of FOLFOXIRI (FLU, leucovorin, OXP, and IRN) followed by resection and adjuvant therapy showed a 91.3% tumor reduction rate. However, there were significant toxicities (grade 3 and 4), and the 2-year recurrence rate was 26.1%, with an overall survival rate of 95.7% (Zhou et al., 2016). Bevacizumab, an anti-VEGF agent, has been combined with other chemotherapy drugs in trials and has been shown to increase survival in patients with CRC and liver metastases (Bisschop et al., 2017). Overall, neoadjuvant chemotherapy is a promising strategy that can improve resectability, reduce tumor burden, and potentially enhance survival rates. However, the side effects, such as grade 3 and 4 toxicities observed in some studies, highlight the need for careful patient selection and management.

14.6 MOLECULAR TARGETS AND THEIR THERAPIES

Genetic abnormalities contribute to the complex heterogeneity of CRC (Sadien et al., 2024; Cao et al., 2024). The Consensus Molecular Subtype classification system was developed to better understand the molecular diversity of CRC (Rejali et al., 2023; Singh et al., 2019). This system categorizes CRC tumors into four distinct subtypes based on their gene expression profiles, helping to identify different underlying biological mechanisms and potentially guiding personalized treatment approaches. It categorizes tumors into four distinct subtypes based on their gene expression profiles. The first type is CMS1 (immune subtype), distinguished by a high density of immune cell infiltration and MSI. The second subtype, CMS2, represents the classical subtype with activated Wnt and Myc signaling pathways. This subtype is often associated with the typical features of CRC, including chromosomal instability. Traditional chemotherapy regimens, such as FOLFOX or FOLFIRI, are often effective for CMS2 tumors (Singh et al., 2019). Metabolic dysregulation and a high CpG island methylator phenotype characterize CMS3 or the metabolic subtype. Finally, the epithelial-mesenchymal transition and stromal invasion define the CMS4 (mesenchymal). This subtype frequently exhibits poor prognosis and characteristics of the tumor microenvironment remodeling (Rejali et al., 2023; Singh et al., 2019). In metastatic CRC, anti-VEGF therapies, FLT4 inhibitors (sorafenib, telatinib, sunitinib, and pazopanib), and Src inhibitors represent a range of targeted treatment options that address different aspects of tumor biology and progression (Singh et al., 2019). Each subtype has unique molecular and clinical characteristics, which can influence treatment strategies and patient outcomes.

Poly [ADP-ribose] polymerase-1 (PARP-1) is a crucial enzyme involved in the repair of single-strand DNA breaks through the base excision repair pathway. It plays a significant role in maintaining genomic stability. Inhibition of PARP-1 has gained attention for its potential in treating various cancers, particularly those with specific genetic vulnerabilities (Graham et al., 2016; Wang et al., 2017). In a study, it was demonstrated that CRC cell lines deficient in MLH1 or MSH3 (involved in mismatch repair system), respond better to a combination of PARP-1 inhibitors and IRN (topoisomerase I inhibitor) (Wang et al., 2017). This combination exploits the impaired DNA repair pathways in these cells, enhancing the therapeutic efficacy. Both poorly and moderately differentiated cells exhibit a higher percentage of positive PARP-1 expression in CRC tissue samples. Moreover, the patients with positive PARP-1 expression are also associated with a reduced recurrence-free survival of 34.7 months. The XRCC2 inhibitor is known to downregulate the expression of PARP-1 as both these proteins form a complex to initiate CRC cell proliferation (Xu et al., 2024).

The Wnt canonical pathway, also known as the Wnt/β-catenin signaling pathway, is a crucial regulatory mechanism involved in various cellular processes, including cell proliferation, differentiation, and embryonic development (Figure 14.3). Its dysregulation is implicated in several cancers including CRC (Cheng et al., 2019; Zhong et al., 2021; He and Gan, 2023). β-catenin is a central signaling molecule in this pathway. When Wnt ligand is not present, β-catenin is targeted for degradation by the "destruction complex," which includes APC, protein phosphatase 2A, axis inhibitor (Axin), glycogen synthase kinase 3 beta, and casein kinase-1 (Zhong et al., 2021; He & Gan, 2023). When Wnt ligands bind to frizzled receptors and co-receptors, this interaction inhibits the

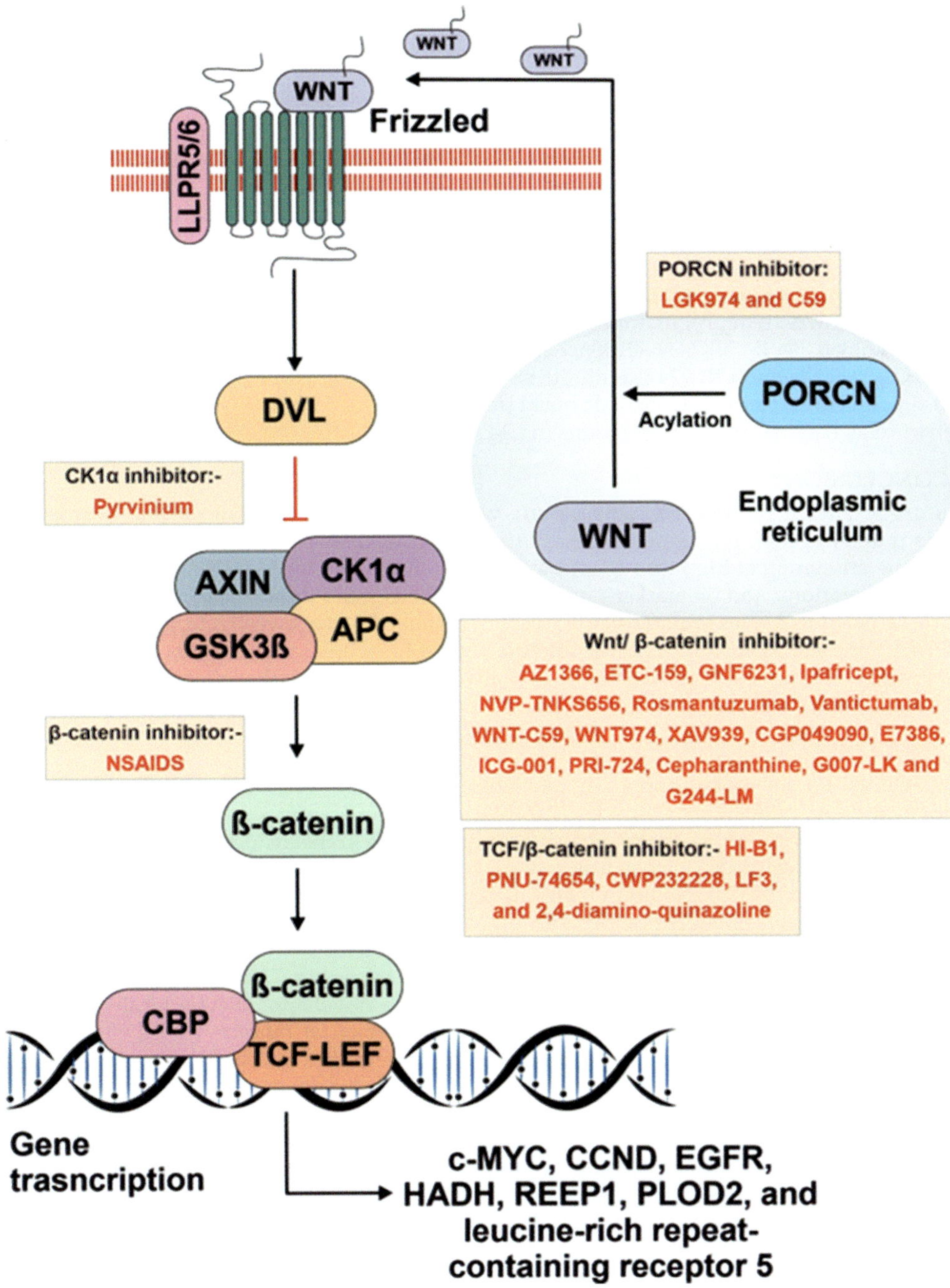

Figure 14.3 Wnt/β-catenin signalling pathway and its inhibitors.

activity of the destruction complex. This complex's component axin interacts with LRP5 or LRP6, to stabilize β-catenin. After nuclear translocation, β-catenin along with the transcription factors lymphoid enhancer binding factor/T-cell factor (TCF) causes the activation of the target genes (c-MYC, CCND, EGFR, HADH, REEP1, PLOD2, and leucine-rich repeat-containing receptor 5) (Zhong et al., 2021; He & Gan, 2023). Cepharanthine effectively inhibits Wnt/ β-catenin signaling in various APC-mutant CRC cell lines (SW480, SW620, LoVo) (Su et al., 2024). Loss-of-function mutations in RNF43, particularly the G659Vfs frame-shift mutation, can lead to the accumulation of frizzled receptors on the cell surface. This accumulation likely sensitizes cells to Wnt ligands, resulting in hyperactivation of the Wnt/β-catenin pathway—one of the hallmarks of many cancers, including CRC (Zhong et al., 2021). The inhibition of WNT secretion through targeting PORCN (WNT-specific O-acyltransferase) is a promising therapeutic strategy, particularly in the context of RNF43 loss-of-function associated tumorigenesis. Inhibitors like LGK974 and C59 have been investigated for their ability to block PORCN activity in various cancers (Cheng et al., 2019; Zhong et al., 2021).

Non-steroidal anti-inflammatory drugs also inhibit β-catenin-mediated target gene expression in CRC cells (Cheng et al., 2019; Rashid et al., 2023). G007-LK and G244-LM (small-molecule inhibitors) downregulate the Wnt/β-catenin signaling pathway by preventing the degradation of Axin, which is induced by poly(ADP-ribosylation) (Katoh, 2018; Cheng et al., 2019). Pyrvinium, a casein kinase 1 alpha inhibitor, has shown effectiveness in treating CRC cell lines (HCT116 and SW480), which harbors mutations in the APC gene or β-catenin (Cheng et al., 2019). HI-B1, PNU-74654, CWP232228, LF3, and 2,4-diamino-quinazoline are known to inhibit the TCF/β-catenin pathway (Katoh, 2018; Cheng et al., 2019; Amerizadeh et al., 2022). Several β-catenin inhibitors of its upstream and downstream regulators have been identified such as rosmantuzumab, ICG-001, GNF6231, AZ1366, NVP-TNKS656, E7386, ETC-159, ipafricept, vantictumab, WNT-C59, WNT974, XAV939, CGP049090, and PRI-724 (Katoh, 2018). Therefore, further investigation of these compounds in clinical settings, may provide novel treatment options for patients with APC mutations and other related signaling dysregulations in CRC.

14.7 CONCLUSION

The future of CRC therapies is evolving rapidly, with a focus on overcoming current treatment limitations and improving patient outcomes through innovative approaches. One of which includes the utilization of high throughput techniques that aid in the identification of genetic mutations, alterations, and biomarkers specific to CRC stages and its subtypes. This helps in tailoring targeted therapies and predicting responses to treatment. Investigating new ICI beyond PD1/PDL1, such as LAG-3, and CTLA-4 inhibitors could be beneficial to patients. Also targeting RTKs, Wnt/β-catenin signaling, PARP-1, NAMT, glycolysis, and oxidative phosphorylation will show promising results.

BIBLIOGRAPHY

Ali, Q., et al. (2022). Microbial short-chain fatty acids: a bridge between dietary fibers and poultry gut health - A review. *Animal Biosci*, *35*(10):1461–1478.

Alvandi, E., et al. (2022). Short-chain fatty acid concentrations in the incidence and risk-stratification of colorectal cancer: a systematic review and meta-analysis. *BMC Med*, *20*(1):323.

Alzahrani, S. M., et al. (2023). 5-Fluorouracil and capecitabine therapies for the treatment of colorectal cancer (Review). *Oncol Rep*, *50*(4):175.

Amerizadeh, F., et al. (2022). Inhibition of the Wnt/b-catenin pathway using PNU-74654 reduces tumor growth in in vitro and in vivo models of colorectal cancer. *Tissue Cell*, *77*: 101853.

Arredondo, J., et al. (2017). Mid-term oncologic outcome of a novel approach for locally advanced colon cancer with neoadjuvant chemotherapy and surgery. *Clin Translat Pncol*, *19*(3):379–385.

Arredondo, J., et al. (2013). Preliminary outcome of a treatment strategy based on perioperative chemotherapy and surgery in patients with locally advanced colon cancer. *Colorectal Dis*, *15*(5): 552–557.

Bhattacharya, P., et al. (2024). Lynch syndrome (hereditary nonpolyposis colorectal cancer). In: *StatPearls [Internet]*. Treasure Island (FL): StatPearls Publishing. Available from: https://www.ncbi.nlm.nih.gov/books/NBK431096/ (accessed July 2024)

Bisschop, C., et al. (2017). Short-course radiotherapy followed by neoadjuvant bevacizumab, capecitabine, and oxaliplatin and subsequent radical treatment in primary stage IV rectal cancer: long-term results of a phase II study. *Ann Surg Oncol*, *24*(9):2632–2638.

Blondy, S., et al. (2020). 5-Fluorouracil resistance mechanisms in colorectal cancer: From classical pathways to promising processes. *Cancer Sci*, *111*(9):3142–3154.

Cao, Q., et al. (2024). Epigenetic alteration in colorectal cancer: potential diagnostic and prognostic implications. *Int J Molec Sci*, 25(6):3358.
Carretta, M. D., et al. (2021). Participation of short-chain fatty acids and their receptors in gut inflammation and colon cancer. *Frontiers Physiol*, *12*:662739.
Chen, X., et al. (2024). Anti-PD-1/PD-L1 therapy for colorectal cancer: clinical implications and future considerations. *Translat Oncol*, *40*:101851.
Chen, X., et al. (2018). Propofol disrupts aerobic glycolysis in colorectal cancer cells via inactivation of the NMDAR-CAMKII-ERK Pathway. *Cellular Physiol Biochem*, *46*(2):492–504.
Cheng, X., et al. (2019). Therapeutic potential of targeting the Wnt/β-catenin signaling pathway in colorectal cancer. *Biomedicine & pharmacotherapy = Biomedecine & pharmacotherapie*, *110*:473–481.
Chu, Y., et al. (2024). Controlling glycolysis to generate characteristic volatile organic compounds of lung cancer cells. *Sci Rep*, *14*:16561.
Dagenborg, V. J., et al. (2020). Neoadjuvant chemotherapy is associated with a transient increase of intratumoral T-cell density in microsatellite stable colorectal liver metastases. *Cancer Boil Ther*, *21*(5):432–440.
de Boer, N. K., et al (2014). The scent of colorectal cancer: detection by volatile organic compound analysis. *Clin Gastroenterol Hepatol*, *12*(7):1085–1089.
de Visser, K. E., Joyce, J. A. (2023). The evolving tumor microenvironment: from cancer initiation to metastatic outgrowth. *Cancer Cell*, *41*(3):374–403.
Engevik, M. A., et al. (2021). *Fusobacterium nucleatum* secretes outer membrane vesicles and promotes intestinal inflammation. *mBio*, *12*(2):e02706–20.
Evans, R. M., et al. (1981). Effect of excess folates and deoxyinosine on the activity and site of action of 5-fluorouracil. *Cancer Res*, *41*(9 Pt 1):3288–3295.
Fang, Y., et al. (2021). The roles of microbial products in the development of colorectal cancer: a review. *Bioengineered*, *12*(1):720–735.
Galber, C., et al. (2020). The role of mitochondrial ATP synthase in cancer. *Biolog Chem*, *401*(11):1199–1214.
García-Aranda, M., Redondo, M. (2019). Targeting receptor kinases in colorectal cancer. *Cancers*, *11*(4):433.
García-Foncillas, J., et al. (2019). Distinguishing features of cetuximab and panitumumab in colorectal cancer and other solid tumors. *Frontiers Oncol*, *9*:849.
Ge, C., et al. (2021). Suppression of oxidative phosphorylation and IDH2 sensitizes colorectal cancer to a naphthalimide derivative and mitoxantrone. *Cancer Letters*, *519*, 30–45.
Ghazi, B., et al. (2022). CAR T-cells for colorectal cancer immunotherapy: ready to go?. *Frontiers Immunol*, *13*:978195.
Graham, D.M., et al. (2016). Molecular subtypes and personalized therapy in metastatic colorectal cancer. *Curr Colorectal Cancer Rep*, *12*:141–150.
Hamada, T., et al. (2018). *Fusobacterium nucleatum* in colorectal cancer relates to immune response differentially by tumor microsatellite instability status. *Cancer Immunol Res*, *6*(11):1327–1336.
Han, Y., et al. (2020). PD-1/PD-L1 pathway: current researches in cancer. *Am J Cancer Res*, *10*(3):727–742.
He, S., et al. (2023). CD166-specific CAR-T cells potently target colorectal cancer cells. *Translat Oncol*, *27*:101575.
He, K., Gan, W. J. (2023). Wnt/β-Catenin Signaling Pathway in the Development and Progression of Colorectal Cancer. *Cancer Management and Research*, *15*, 435–448.
Horn, L. A., et al. (2018). Soluble CD80 protein delays tumor growth and promotes tumor-infiltrating lymphocytes. *Cancer Immunol Res*, *6*(1):59–68.
Hou, H., et al. (2022). Gut microbiota-derived short-chain fatty acids and colorectal cancer: ready for clinical translation? *Cancer Letters*, *526*:225–235.
Hsu, F. T., et al. (2021). Preclinical evaluation of recombinant human IL15 protein fused with albumin binding domain on anti-PD-L1 immunotherapy efficiency and anti-tumor immunity in colon cancer and melanoma. *Cancers*, *13*(8):789.
Huck, M. B., Bohl, J. L. (2016). Colonic polyps: diagnosis and surveillance. *Clin Colon Rectal Surg*, *29*(4):296–305.
Janfaza, S., et al. (2019). Digging deeper into volatile organic compounds associated with cancer. *Biol Meth Protoc*, *4*(1):bpz014.
Katoh M. (2018). Multi-layered prevention and treatment of chronic inflammation, organ fibrosis and cancer associated with canonical WNT/β-catenin signaling activation (Review). *Int J Molec Med*, *42*(2):713–725.

Kumar, A., et al. (2023). Current and emerging therapeutic approaches for colorectal cancer: a comprehensive review. *World J Gastrointest Surg*, *15*(4):495–519.
Lamaudière, M. T. F., et al. (2023). The colorectal cancer microbiota alter their transcriptome to adapt to the acidity, reactive oxygen species, and metabolite availability of gut microenvironments. *mSphere*, *8*(2):e0062722.
Lenz, H. J., et al. (2022). First-line nivolumab plus low-dose ipilimumab for microsatellite instability-high/mismatch repair-deficient metastatic colorectal cancer: the phase II CheckMate 142 study. *J Clin Oncol*, *40*(2):161–170.
Li, D., et al. (2023). Bacterial outer membrane vesicles in cancer: biogenesis, pathogenesis, and clinical application. *Biomed Pharmacother*, *165*:115120.
Liu, C., et al. (2021). Blocking IL-17A enhances tumor response to anti-PD-1 immunotherapy in microsatellite stable colorectal cancer. *J Immunother Cancer*, *9*(1):e001895.
Marcellinaro, R., et al. (2023). Colorectal cancer: current updates and future perspectives. *J Clin Med*, *13*(1):40.
Mármol, I., et al. (2017). Colorectal carcinoma: a general overview and future perspectives in colorectal cancer. *Int J Molec Sci*, *18*(1):197.
Masucci, C., et al. (2023). Case report: severe cutaneous adverse event associated with checkpoint inhibition in the setting of CAR T-cell therapy: beyond CRS. *Frontiers Oncol*, *13*:1171031.
Mezmale, L., et al. (2023). Identification of volatile markers of colorectal cancer from tumor tissues using volatilomic approach. *Molecules (Basel, Switzerland)*, *28*(16):5990.
Mowat, C., et al. (2023). Short chain fatty acids prime colorectal cancer cells to activate antitumor immunity. *Frontiers Immunol*, *14*:1190810.
Ney, L. M., et al. (2023). Short chain fatty acids: key regulators of the local and systemic immune response in inflammatory diseases and infections. *Open Biol*, *13*(3):230014.
Nishizawa, N., et al. (2022). Regulation of 5-fluorodeoxyuridine monophosphate-thymidylate synthase ternary complex levels by autophagy confers resistance to 5-fluorouracil. *FASEB bioAdvances*, *5*(1):43–51.
Oxner M, et al. (2023). The versatility and diagnostic potential of VOC profiling for noninfectious diseases. *BME Front*, *10*(4):0002.
Peng, G., et al. (2010). Detection of lung, breast, colorectal, and prostate cancers from exhaled breath using a single array of nanosensors. *Br J Cancer*, *103*(4):542–551.
Pigg, H. C., et al. (2024). The unique Pt(II)-induced nucleolar stress response and its deviation from DNA damage response pathways. *J Biol Chem*, *300*(11):107858.
Rashid, G., et al. (2023). Non-steroidal anti-inflammatory drugs and biomarkers: a new paradigm in colorectal cancer. *Frontiers Med*, *10*:1130710.
Rejali, L., et al. (2023). Principles of molecular utility for CMS classification in colorectal cancer management. *Cancers*, *15*(10):2746.
Sadien, I. D., et al. (2024). The genomics of sporadic and hereditary colorectal cancer. *Ann Royal Coll Surg Engl*, *106*(4):313–320.
Shan, K. S., et al. (2024). Molecular targeting of the BRAF proto-oncogene/mitogen-activated protein kinase (MAPK) pathway across Cancers. *Int J Molec Sci*, *25*(1):624.
Singh, M. P., et al. (2019). Molecular subtypes of colorectal cancer: an emerging therapeutic opportunity for personalized medicine. *Genes Dis*, *8*(2):133–145.
Su, G., et al. (2024). Cepharanthine suppresses APC-mutant colorectal cancers by down-regulating the expression of β-catenin. *Natural Products Bioprospect*, *14*(1):18.
Tang, H., et al. (2023). Recent advances of targeting nicotinamide phosphoribosyltransferase (NAMPT) for cancer drug discovery. *Eur J Med Chem*, *258*:115607.
Tie, Y., et al. (2022). Immunosuppressive cells in cancer: mechanisms and potential therapeutic targets. *J Hematol Oncol*, *15*(1):61.
Tolstikov, V., et al. (2014). Metabolomics analysis of metabolic effects of nicotinamide phosphoribosyltransferase (NAMPT) inhibition on human cancer cells. *PloS one*, *9*(12):e114019.
van Liere, E. L. S. A., et al. (2023). Urinary volatile organic compounds for colorectal cancer screening: A systematic review and meta-analysis. *Eur J Cancer*, *186*:69–82.
Wang, L., et al. (2017). PARP1 in carcinomas and PARP1 inhibitors as antineoplastic drugs. *Int J Molec Sci*, *18*(10):2111.
Wei, Y., et al. (2022). Review of various NAMPT inhibitors for the treatment of cancer. *Frontiers Pharmacol*, *13*:970553.
Wong, C.C., Yu, J. (2023).Gut microbiota in colorectal cancer development and therapy. *Nat Rev Clin Oncol*, *20*:429–452.

Xiao, L., et al. (2023). A phase 1 dose escalation study of GCC19CART: a novel coupled CAR therapy for patients with metastatic colorectal cancer. *Ann Oncol, 34*(suppl 2):S410–S457.
Xu, J., et al. (2021). HER2-specific chimeric antigen receptor-T cells for targeted therapy of metastatic colorectal cancer. *Cell Death Dis, 12*(12):1109.
Xu, K., et al. (2024). PARP1 bound to XRCC2 promotes tumor progression in colorectal cancer. *Discover Oncol, 15*(1):238.
Yao, S., et al. (2023). From organ preservation to selective surgery: how immunotherapy changes colorectal surgery? *Surgery Open Sci, 15*:44–53.
Yu, P., et al. (2010). Simultaneous blockade of multiple immune system inhibitory checkpoints enhances antitumor activity mediated by interleukin-15 in a murine metastatic colon carcinoma model. *Clin Cancer Res, 16*(24):6019–6028.
Zhang, B. L., et al. (2019). Preclinical evaluation of chimeric antigen receptor-modified T cells specific to epithelial cell adhesion molecule for treating colorectal cancer. *Human Gene Ther, 30*(4):402–412.
Zhang, H., et al. (2024). Phase I trial of hypoxia-responsive CEA CAR-T cell therapy in patients with heavily pretreated solid tumor via intraperitoneal or intravenous transfusion. *J Clin Oncol, 42*;suppl 16:abstr 3514.
Zhao, A., Liu, Y. (2021). Propofol suppresses colorectal cancer development by the circ-PABPN1/miR-638/SRSF1 axis. *Analyt Biochem, 631*:114354.
Zhao, L. Y., et al. (2023). Role of the gut microbiota in anticancer therapy: from molecular mechanisms to clinical applications. *Sig Transduct Target Ther, 8*:201.
Zhong, X., et al. (2022). Warburg effect in colorectal cancer: the emerging roles in tumor microenvironment and therapeutic implications. *J Hematol Oncol, 15*:160.
Zhong, Z. A., et al. (2021). Regulation of Wnt receptor activity: implications for therapeutic development in colon cancer. *J Biolog Chem, 296*:100782.
Zhou, H., et al. (2016). A pilot phase II study of neoadjuvant triplet chemotherapy regimen in patients with locally advanced resectable colon cancer. *Chinese J Cancer Res, 28*(6), 598–605.
Zhou, Z., et al. (2024). The underlying mechanism of chimeric antigen receptor (CAR)-T cell therapy triggering secondary T-cell cancers: mystery of the Sphinx? *Cancer Letters, 597*:217083.
Zhu, Z., et al. (2022). FKBP4 regulates 5-fluorouracil sensitivity in colon cancer by controlling mitochondrial respiration. *Life Sci Alliance, 5*(11):e202201413.

Index

Pages in *italics* refer to figures and pages in **bold** refer to tables.

C

D

E

F

G

H

I

K

L

M

Q

R

S

T

For Product Safety Concerns and Information please contact our
EU representative GPSR@taylorandfrancis.com Taylor & Francis
Verlag GmbH, Kaufingerstraße 24, 80331 München, Germany